Conference Proceedings of the Society for Experimental Mechanics Series

Series Editor
Tom Proulx
Society for Experimental Mechanics, Inc.
Bethel, CT, USA

For further volumes:
http://www.springer.com/series/8922

Nancy Sottos • Robert Rowlands • Kathryn Dannemann
Editors

Experimental and Applied Mechanics, Volume 6

Proceedings of the 2014 Annual Conference on Experimental and Applied Mechanics

Springer

Editors
Nancy Sottos
Beckman Institute
University of Illinois
Urbana-Champaign
Urbana, IL, USA

Robert Rowlands
University of Wisconsin
Madison, WI, USA

Kathryn Dannemann
Southwest Research Institute
San Antonio, TX, USA

ISSN 2191-5644 ISSN 2191-5652 (electronic)
ISBN 978-3-319-06988-3 ISBN 978-3-319-06989-0 (eBook)
DOI 10.1007/978-3-319-06989-0
Springer Cham Heidelberg New York Dordrecht London

Library of Congress Control Number: 2011928691

Printed on acid-free paper

Springer is part of Springer Science+Business Media (www.springer.com)

Preface

Experimental and Applied Mechanics, Volume 6: Proceedings of the 2014 Annual Conference on Experimental and Applied Mechanics represents one of eight volumes of technical papers presented at the 2014 SEM Annual Conference & Exposition on Experimental and Applied Mechanics organized by the Society for Experimental Mechanics and held in Greenville, SC, June 2–5, 2014. The complete Proceedings also includes volumes on: *Dynamic Behavior of Materials*; *Challenges in Mechanics of Time-Dependent Materials*; *Advancement of Optical Methods in Experimental Mechanics*; *Mechanics of Biological Systems and Materials*; *MEMS and Nanotechnology*; *Experimental Mechanics of Composite, Hybrid, and Multifunctional Materials*; *Fracture, Fatigue, Failure and Damage Evolution*.

Each collection presents early findings from experimental and computational investigations of an important topic within the field of Experimental Mechanics. This volume includes papers on Residual Stress, Thermomechanics and Infrared Imaging, Hybrid Techniques and Inverse Problems.

Residual stresses are especially important in engineering systems and design. The insidious nature of residual stresses often causes them to be underrated or overlooked. However, they profoundly influence structural design and substantially affect strength, fatigue life, and dimensional stability. Since residual stresses are induced during most materials processing procedures, for example, welding/joining, casting, thermal conditioning, and forming, they must be included and addressed in engineering design and applications.

In recent years the applications of infrared imaging techniques to the mechanics of materials and structures have grown considerably. The expansion is marked by the increased spatial and temporal resolution of the infrared detectors, faster processing times, and much greater temperature resolution. The improved sensitivity and more reliable temperature calibrations of the devices have meant that more accurate data can be obtained than were previously available.

Advances in inverse identification techniques have been coupled with optical methods that provide surface deformation measurements and volumetric measurements of materials. In particular, inverse methodology was developed to more fully use the dense spatial data provided by optical methods to identify mechanical constitutive parameters of materials. Since its beginnings during the 1980s, creativity in inverse methods has led to applications for a wide range of materials, with different constitutive behavior, across heterogeneous material interfaces. Complex test fixtures have been implemented to produce the necessary strain fields for identification. Force reconstruction has been developed for high strain rate testing. As developments in optical methods improve for both very large and very small length scales, applications of inverse identification methods have expanded to include geological and atomistic events.

Urbana, IL, USA
Madison, WI, USA
San Antonio, TX, USA

Nancy Sottos
Robert Rowlands
Kathryn Dannemann

Contents

Chapter 1
Electrical Impedance Spectroscopy for Structural Health Monitoring

Geoffrey A. Slipher, Robert A. Haynes, and Jaret C. Riddick

Abstract Structural heath monitoring (SHM) can provide an estimate of the state of damage in a structure, and of the remaining useful life of that structure. The work presented here is an investigation of a proposed new SHM technique for composite structures composed of carbon-fiber-reinforced-polymers (CFRP). Electrical impedance spectroscopy (EIS) is employed to estimate the damage state of the composite. No modification to current CFRP processing methods is required, nor is the proposed technique invasive or destructive. EIS interfaces can be either permanently attached or temporarily connected. We hypothesize that EIS has the potential to be more sensitive and selective for damage detection by using a full complex-plane analysis, considering both impedance magnitude and phase angle. This is in contrast to electrical SHM approaches employing resistance measurement, the real component of impedance, which ignores phase angle and reactance information. In order to test our hypothesis we implemented three different experiments to evaluate the effectiveness of the EIS technique: (1) specimen load sensitivity; (2) specimen damage sensitivity; and (3) specimen fatigue sensitivity. Multiple electrical interrogation paths through the specimen are considered.

Keywords Electrical impedance spectroscopy • Structural health monitoring • Carbon fiber • Fatigue state • Damage state detection • Load state detection

1.1 Introduction

Carbon-fiber reinforced composite materials are rapidly gaining wide use for a variety of applications pertaining to aircraft, spacecraft and civil infrastructure. The recent trend toward condition-based maintenance of advanced structural platforms has given rise to sensor-based monitoring of critical structural components. Structural health monitoring refers to the use of sensor data as a means to provide an estimate of the state of damage in a structure, and of the remaining useful life of that structure. Pervasive use of SHM in composite structures in the future will enable new structural applications that exploit the presence of sensors to enable novel lightweight designs and reduced maintenance burdens.

Damage detection by structural self-sensing in carbon fiber composites has been investigated thoroughly in the literature. In particular concepts for the use of electrical resistance measurements to detect damage in fiber-reinforced composites have been investigated, showing that resistance changes irreversibly upon damage inflicted by flexure, tension, fatigue, and impact [1]. Resistance change can be associated with the mode of damage. For example, fiber breakage increases the longitudinal resistance, and delamination increases the through-thickness resistance. Embedding of carbon nanotubes (CNTs) into glass-epoxy composites to create percolating electrical networks for sensing damage has been shown to work based on measuring the change in resistance due to cracking [2–5].

Electrical resistance measurements have shown some promise in the detection of damage in fiber composites, particularly in those systems where embedded sensors or electrical networks were used. However, electrical impedance measurements have the potential to be more sensitive and selective for damage detection by using a full complex-plane analysis, considering both impedance magnitude and phase angle. Electrical resistance can be considered zero-phase angle impedance. Impedance-based methods have been employed for qualitative health monitoring by correlating variations in mechanical impedance to

G.A. Slipher (✉) • R.A. Haynes • J.C. Riddick
Vehicle Technology Directorate, U.S. Army Research Laboratory, Aberdeen Proving Ground, MD 21005, USA
e-mail: geoffrey.a.slipher.civ@mail.mil

N. Sottos et al. (eds.), *Experimental and Applied Mechanics, Volume 6: Proceedings of the 2014 Annual Conference on Experimental and Applied Mechanics*, Conference Proceedings of the Society for Experimental Mechanics Series,
DOI 10.1007/978-3-319-06989-0_1,

electrical impedance measurements of a piezoelectric patch [6]. Experimental measurements of carbon fiber composites with embedded piezoceramic patches have shown a close connection between mechanical properties and impedance [7]. Electrical conductivity measurements of embedded CNT thin films have been used to detect damage from low velocity impacts in glass fiber reinforced polymer composites [8]. The electrical impedance tomography response of the composite with the electrical conductive strain sensitive embedded thin film was measured by recording current-voltage measurements at the periphery of the composite. More recent work has shown the effectiveness of using piezoelectric patches to monitor the health of adhesive joints in composites subjected to various environmental conditions using electro-mechanical impedance [9].

Other recent approaches that rely on impedance and frequency related responses have been presented in the literature. An approach for probing embedded sensors wirelessly using inductive coupling depends on electrical impedance, acoustic response, and pulse-echo response of the system [10]. Another study shows the promise of using metamaterial electromagnetic lenses to detect delaminations in composite materials by extracting effective magnetic permeability and frequency measurements [11].

Carbon fiber composites have been shown to be excellent electromagnetic interference (EMI) shielding materials with low surface impedance and high reflectivity [12]. Several models have been presented for electromagnetic characterization of fiber composites [13]. Models for effective properties of fiber composites commonly used in aircraft EMI shielding often employ various levels of detail, and therefore capture various physical aspects in the estimation values such as the reflection and transmission coefficients.

The approach of embedding conductive networks or sensor/actuator patches has proven to be effective in exploiting the electromechanical impedance of composite material systems. The work presented here employs electrical impedance spectroscopy (EIS) to estimate the damage state of the carbon fiber reinforced composites without the modifications introduced by embedding conductive networks or sensors. EIS interfaces can be either permanently attached or temporarily connected. Multiple interrogation paths through the specimen are considered. If successful, the present method would offer the potential of an inspection method for SHM that depends wholly on the intrinsic properties of the material. This paper contains a description of the composite material considered, along with a description of the experimental procedure.

1.2 Motivation

We are seeking structural health monitoring techniques that are simple and readily implementable, involve minimal modification to existing CFRP component manufacturing techniques, can operate in real-time, have minimal cost, and have the potential to spatially resolve damage and/or damage precursor locations. Our hypothesis that we are testing, in part with the methods described in this paper, is that by moving beyond resistance measurement techniques to EIS we can make a significant step toward achieving these objectives, in particular minimal cost and impact to existing manufacturing techniques. Electrical interrogation techniques, such as resistance or impedance measurement, require only simple electrical connection to an electrically conductive specimen, such as carbon fiber reinforced components. For this reason they are attractive as minimally invasive techniques that require little or no modification to the way carbon fiber components are manufactured. As damage is introduced to a carbon fiber reinforced component, electrical properties change, and that change can be detected and correlated with the type and location of the damage. Fiber breakage is expected to manifest predominantly as variations in component resistivity as well as in higher frequency response characteristics due to variations in the effective antenna length paths within a component. Delamination is expected to manifest predominantly as variations in the capacitive reactivity of the specimen as the physical spacing between parallel layers varies. A method with the capability to simultaneously detect changes in resistivity, frequency response, and reactivity would thus be required to resolve as many damage modes as possible. Resistance measurement techniques provide only one dimension of information along the real axis of the complex plane to resolve damage in CFRP components. Impedance analysis techniques add two additional dimensions of information: reactance and phase, as shown in Fig. 1.1. We therefore propose the EIS method as a potential improvement in NDE methods for CFRP structural health monitoring in Army systems such as rotorcraft. This paper describes initial experimental efforts to validate these hypotheses.

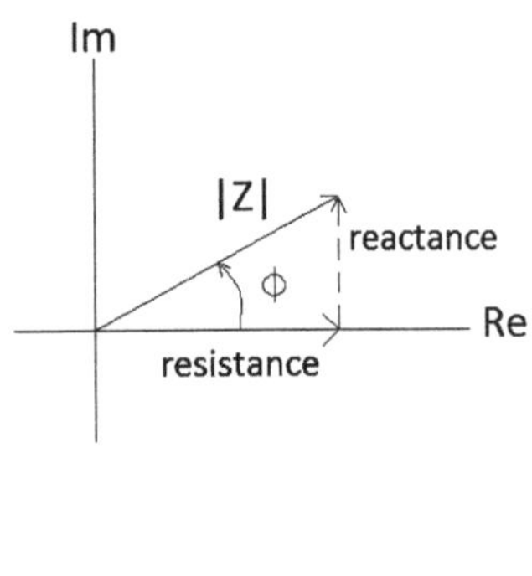

Fig. 1.1 Impedance complex plane

1.3 Experimental Procedure

Fifteen specimens were manufactured from a woven graphite/epoxy material system with a thickness of 0.125 in. and cut to dimensions of 1 in. by 12 in. Tabs of dimension 1.25 in. by 3 in. were attached to both sides and both ends of the specimen. The tabs extended 0.125 in. beyond the specimen on both sides and the end to reduce the likelihood that the wire leads would touch the testing machine; see Fig. 1.2. Five of the specimens were used to estimate the material's static strength as 68 kN, five specimens were used to estimate the material's fatigue life, and five specimens underwent cyclic loading with predefined pauses to record impedance measurements.

The cyclic loading was performed in an MTS 22-kip load frame with an MTS FlexTest 40 controller. An R-ratio of 0.1 was used with a maximum load of 55 kN and a frequency of 5 Hz. Silver epoxy was used to attach wire leads to the specimen in the configuration shown in Fig. 1.2. Five different electrical paths were interrogated through the specimen: A1-B1, A2-B2, A3-B2, A4-B3, and A5-B3. Electrical impedance was measured using an Agilent E5061B-LF network analyzer (NA) using a port 1–2 thru series method. Prior to initiating sample characterization, a calibration procedure was executed on the NA that pushed the calibration plane out to the specimen. The A1-A5 connections were used to inject the electrical signal through port 1, and the B1-3 connections were connected to port 2 for the NA return. For the fatigue tests, cyclic loading was applied as described above. Cyclic loading was stopped periodically and logarithmic frequency scans were performed with the NA from 1 kHz to 100 MHz for both statically loaded (55 kN) and unloaded states through each of the five electrical paths. Impedance magnitude and phase angle data were collected and saved. The electrical interrogation signal was a sine wave of varying frequency with power of 10 dBm (707 mV @ 50 Ω). During the first sample run, a manual testing protocol for running the impedance scans was used, and was identified as undesirable for two key reasons: (1) it consumed too much personnel time; (2) it introduced error in the data in the form of jump discontinuities in the cycle/failure domain as shown in Fig. 1.3. We believe the discontinuities arose from small variations in the calibration introduced by shutting down the equipment at night and recalibrating in the morning.

For both of the above stated reasons an around-the-clock automated testing routine was developed and implemented for the subsequent four samples described in this paper. The key component of the automated testing routine was the switching electronics that used magnetic reed relays triggered by a custom-programmed IC to switch between each of the five conductive paths through the specimen. Implementation of the automated routine completely eliminated the jump discontinuities in the cycle/failure domain. The automated experimental setup is shown in Fig. 1.4.

1.4 Results and Discussion

In order to test our hypotheses outlined above we implemented three different experiments to evaluate the effectiveness of the EIS technique: (1) specimen load sensitivity; (2) specimen damage sensitivity; and (3) specimen fatigue sensitivity. The specimen load sensitivity experiment involved incrementing the load applied to the specimen to identify if the EIS technique could detect changes in load levels. The damage sensitivity experiment involved scanning an undamaged specimen and subsequently introducing gross damage to the gage length and rescanning to determine if the EIS technique could detect the damage. The fatigue sensitivity experiment involved scanning the specimen at various numbers of loading cycles to identify whether or not the EIS technique can detect fatigue-related damage or damage precursors. The results of these three experiments are presented and discussed below.

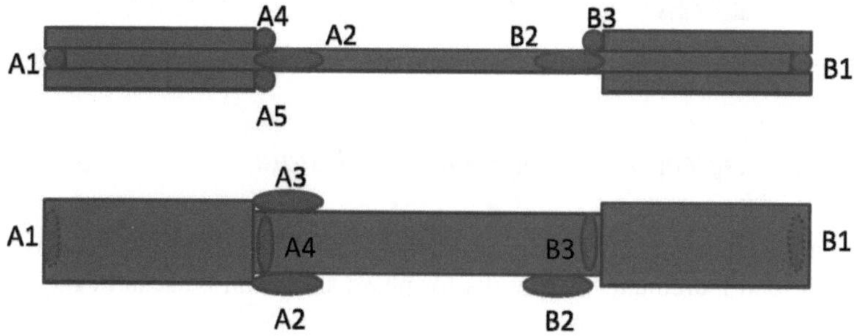

Fig. 1.2 Specimen configuration

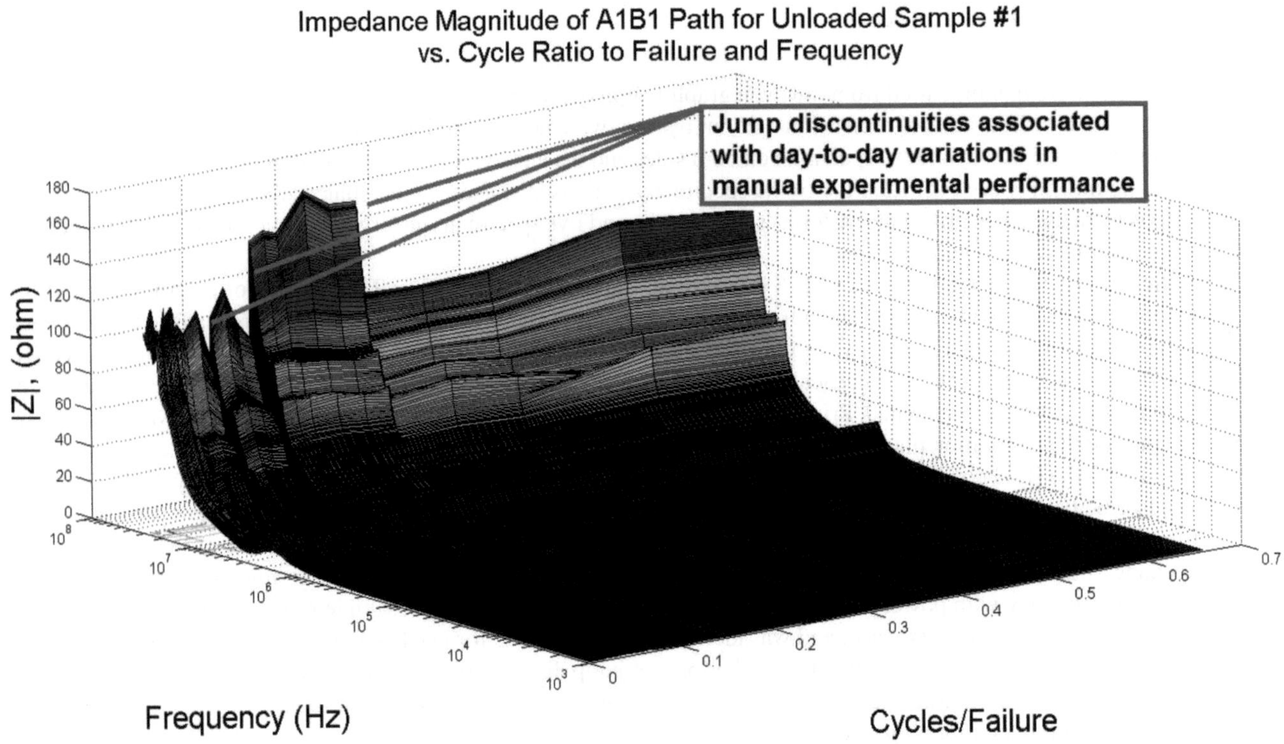

Fig. 1.3 Jump discontinuities in the data arising from manual data collection procedure, eliminated by implementing an automated procedure

Fig. 1.4 Automated experimental setup: (**a**) wide view; (**b**) close-in specimen view with switching electronics

1.4.1 Detection of Specimen Load State

Load on specimen 'S6' was incremented from 0 to 50 kN, and all five electrical paths were interrogated using the NA at each load increment. The results for phase angle can be seen in Fig. 1.5 and impedance magnitude in Fig. 1.6. Specimen load resulting in strain can be detected using both impedance magnitude and phase information from the EIS scans.

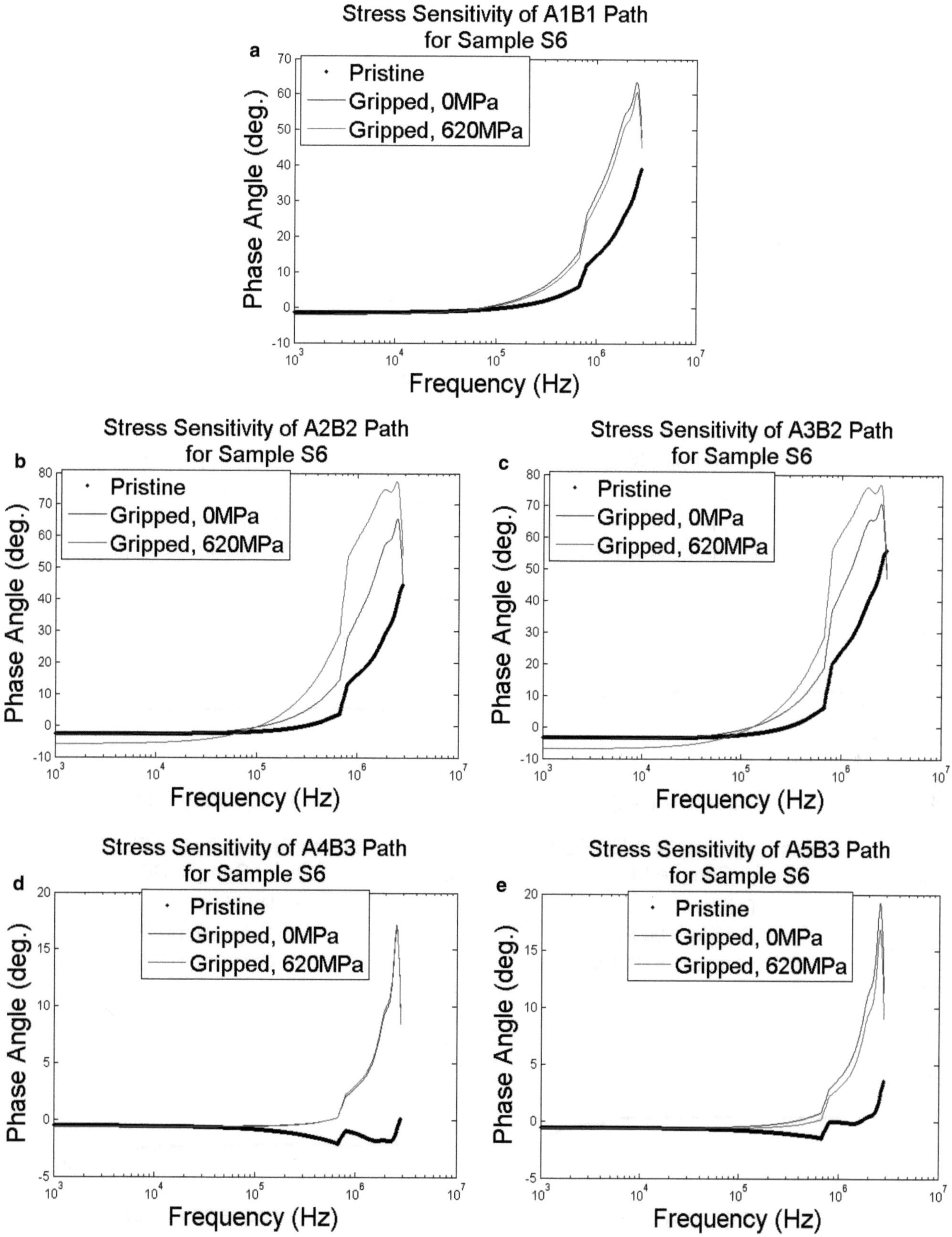

Fig. 1.5 Phase angle resolution of specimen S6 load state for different electrical paths (**a–e**)

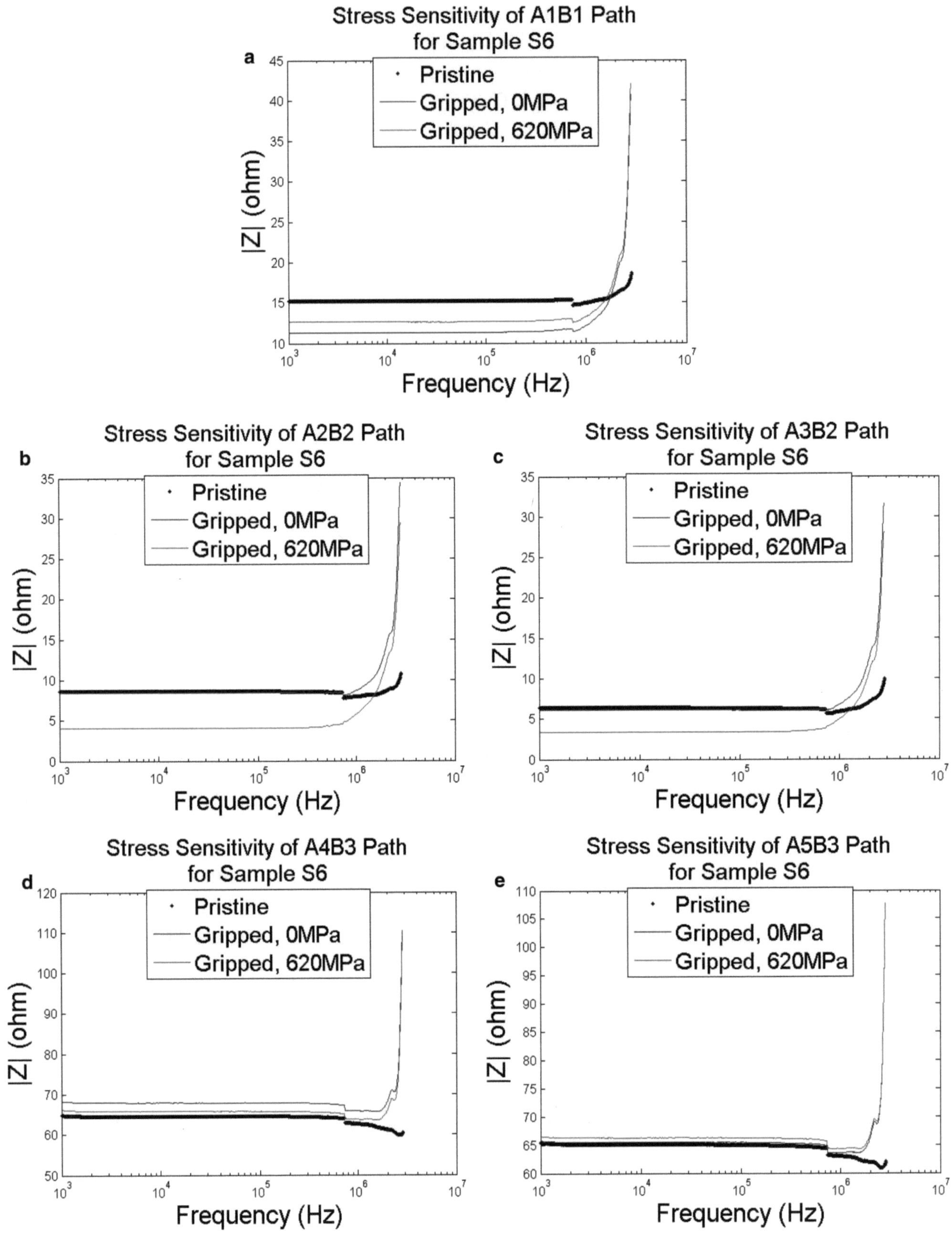

Fig. 1.6 Impedance magnitude resolution of specimen S6 load state for different electrical paths (**a–e**)

Fig. 1.7 Photograph of intentionally introduced damage to the specimen

Comparing the results from phase and impedance we conclude that phase is better at resolving sample load state at higher frequencies, whereas impedance magnitude is more sensitive at lower frequencies. For the specimens tested, the approximate cutoff frequency at which the phase becomes more effective at resolving specimen load state is about 500 kHz. Electrical path through the specimens also influences sensitivity to the load state, with paths A2B2 and A3B2 showing the highest level of sensitivity for both phase and impedance magnitude.

1.4.2 Detection of Specimen Damage State

Specimen ‘S6’ was scanned undamaged and unloaded. Damage was introduced to the specimen at the middle of the gage section without removing it from the grips using a spring-loaded center punch, as shown in Fig. 1.7. The specimen was then rescanned to determine if the EIS technique is capable of detecting the known presence of damage. The specimen was then subjected to 10,000 cycles at an r-ratio of 0.1 and a peak load of 50 kN to identify whether or not the EIS technique is able to detect changes in the specimen arising from a change in the damage due to the cyclic loading. All five electrical paths were interrogated using the NA at the undamaged, damaged, and damaged-cycled states. The results for phase angle can be seen in Fig. 1.8 and impedance magnitude in Fig. 1.9. The specimen damage state can be detected using both impedance magnitude and phase information from the EIS scans. The EIS technique is also capable of resolving changes in the damage state arising from the cyclic loading of the specimen. Comparing the results from phase and impedance we conclude that, as in the case with load state above, phase information is better at resolving damage state at higher frequencies, whereas impedance magnitude is more sensitive at lower frequencies. For the specimens tested, the approximate cutoff frequency at which the phase becomes more effective at resolving specimen load state is about 500 kHz, again the same result as for the load state. Electrical path through the specimens also influences sensitivity to the damage state, with paths A2B2, A3B2, and A4B3 showing the highest level of sensitivity for both phase and impedance magnitude.

1.4.3 Detection of Specimen Fatigue State

Five specimens were prepared for the fatigue testing (S1, S2, S3, S4, S5). The data for specimen S1 were inconclusive due to the aforementioned issues arising from the initial manual protocol. Samples S2 and S3 for the fatigue evaluation failed at very low numbers of cycles, and so are not included in the data presented in this paper. Therefore, the data for the fatigue sensitivity results presented in this paper are based on only two specimens, S4, and S5, and conclusions based on a sample size of two are thereby limited. Due to limitations in the way the initial experiments were performed we currently have no way to confidently discriminate between specimen fatigue damage and fatigue damage to the electrical lead bonds. Figure 1.10 shows an example of how obvious lead damage is in the phase and impedance data. Examination of the data was used to identify which leads failed, and when.

We have evidence to suggest that both types of damage are being detected. For sample S4, lead damage is apparent for the electrical path A2B2 in the impedance magnitude data shown in Fig. 1.10 at approximately 48 % cycle/failure. Similarly, lead damage was detected in S4 in electrical path A4B3 at approximately 25 % cycle/failure (see plot for S4 in Fig. 1.11). In both of these cases, the lead damage for paths A4B3 and A2B2 showed up in the scans along path A1B1 as shown in

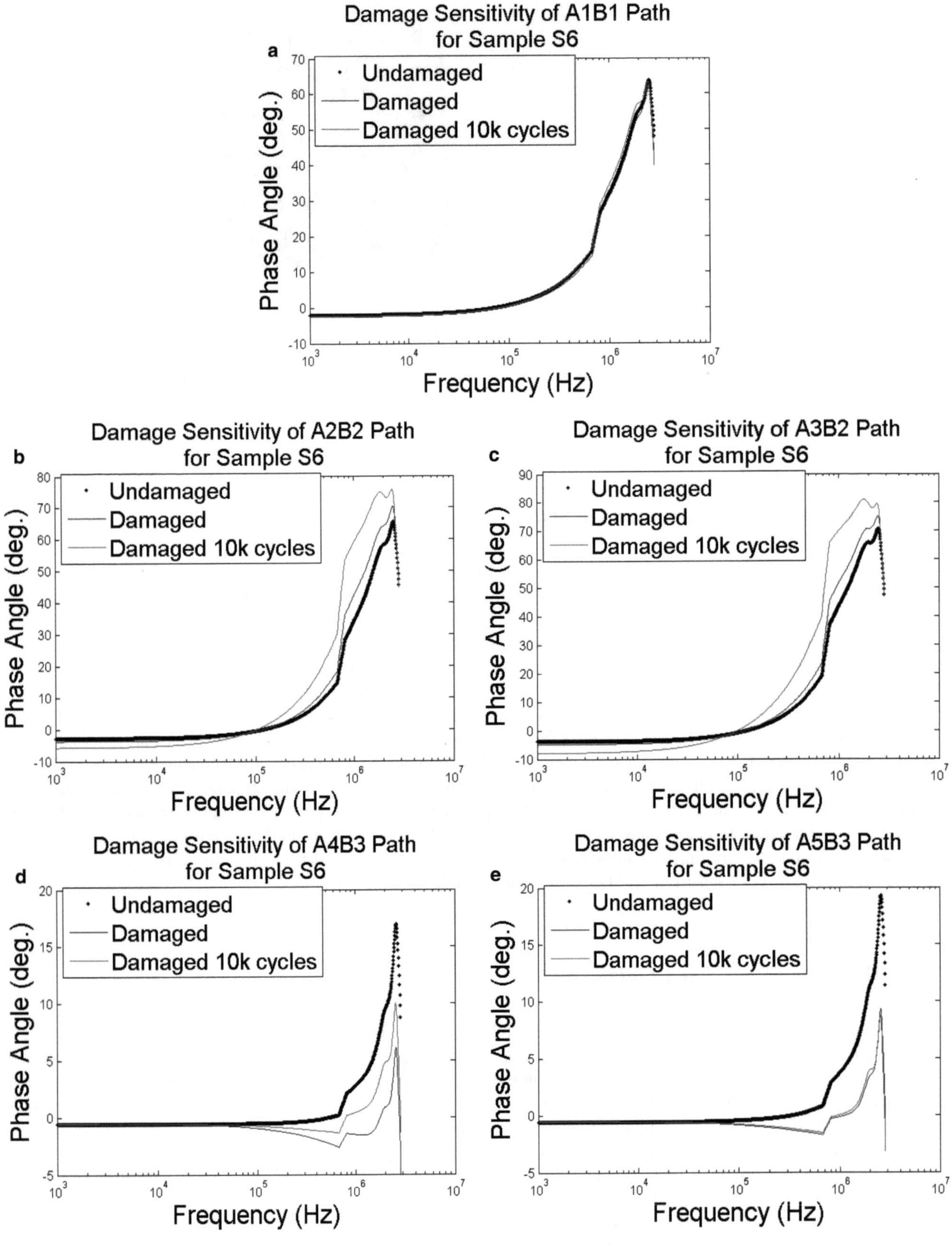

Fig. 1.8 Phase angle resolution of specimen S6 damage state for different electrical paths (**a–e**)

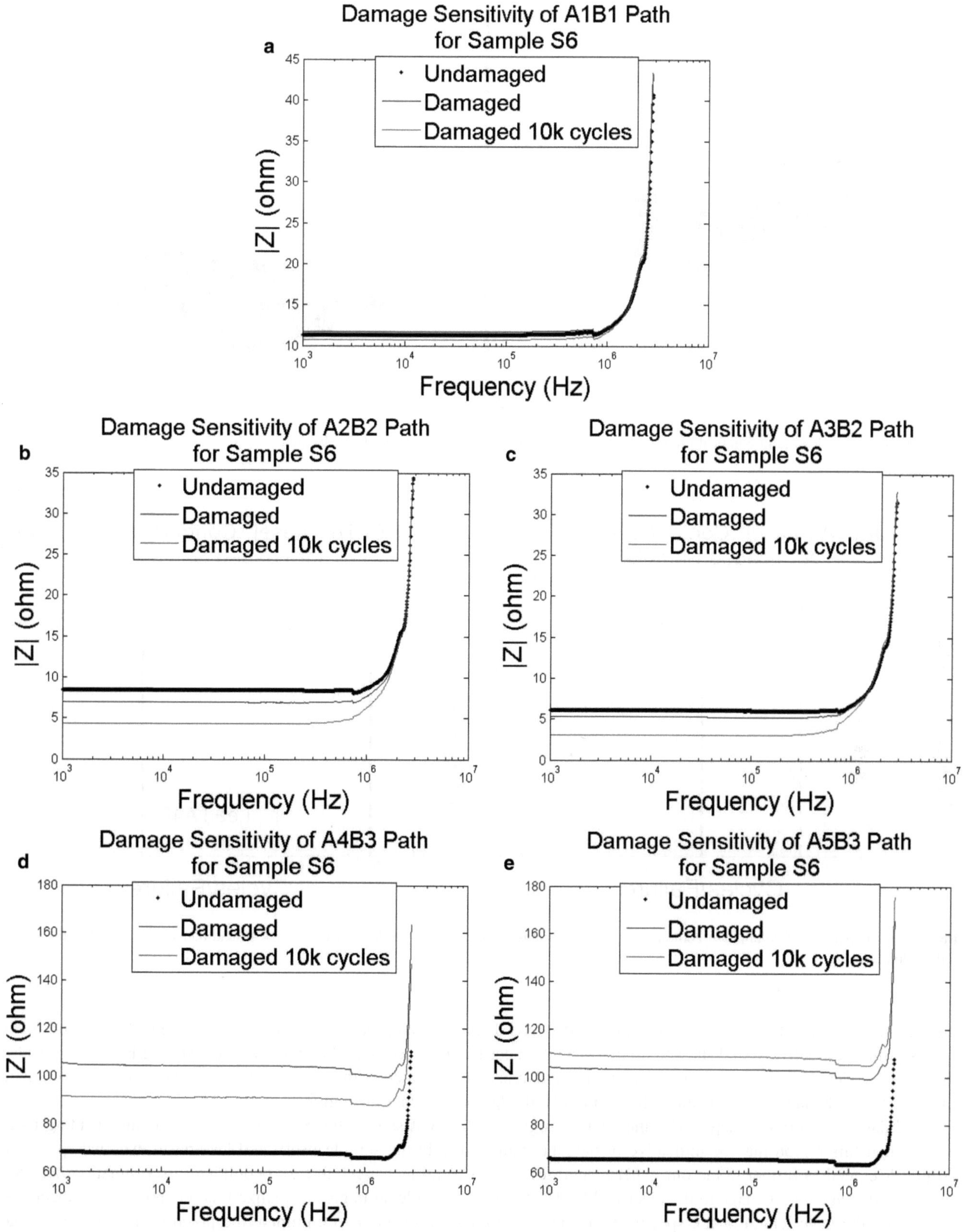

Fig. 1.9 Impedance magnitude resolution of specimen S6 damage state for different electrical paths (**a–e**)

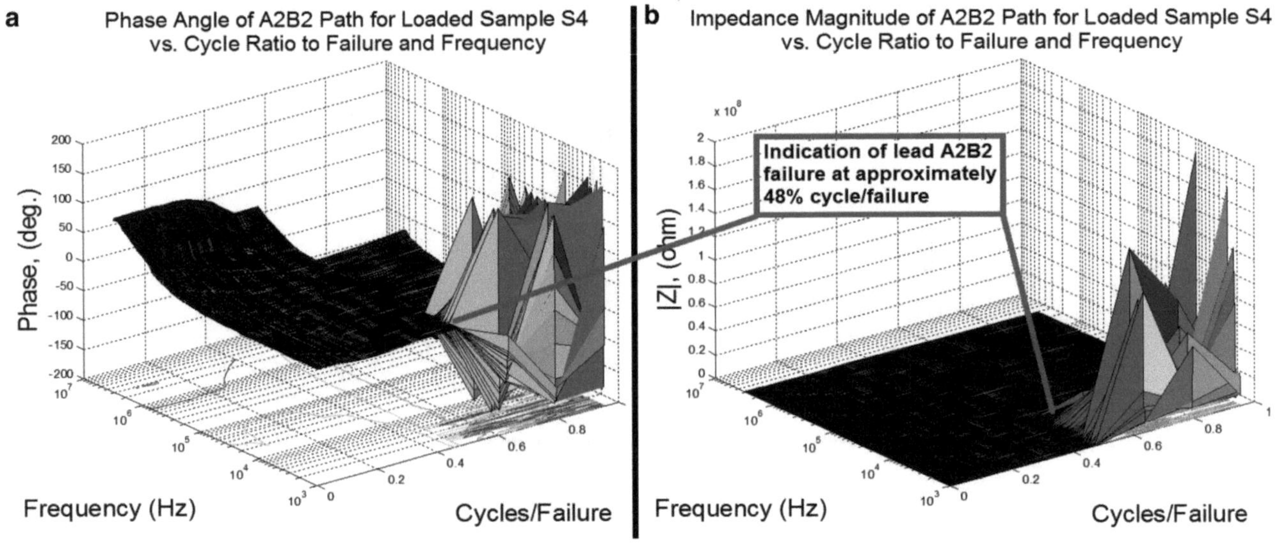

Fig. 1.10 Electrical lead failure indication in both phase angle (**a**) and impedance magnitude (**b**) data

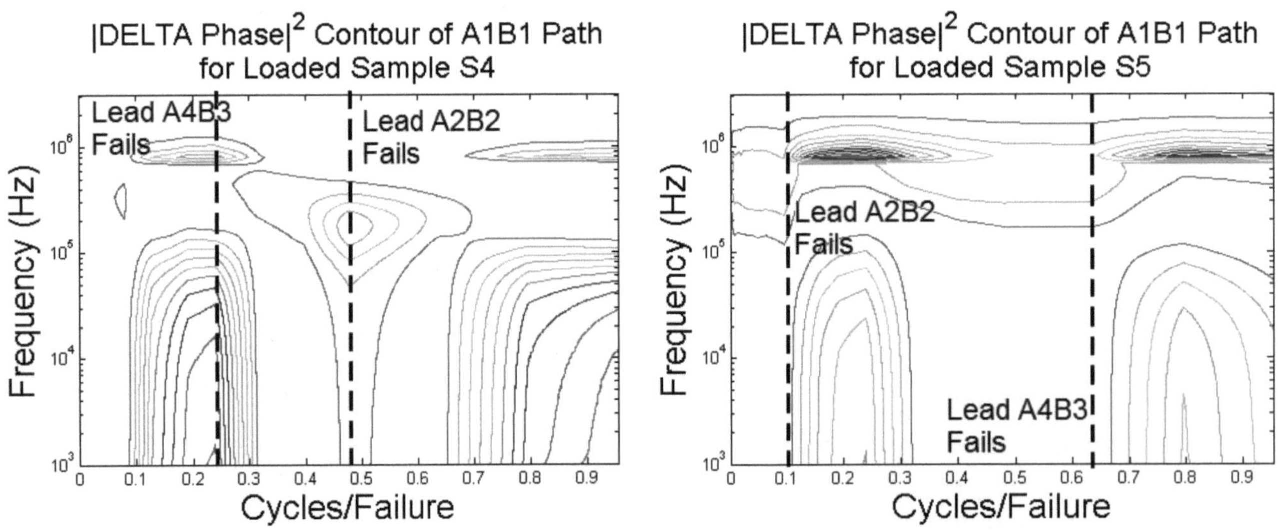

Fig. 1.11 Comparison of delta-phase contours for two specimens, S4 (*left*) and S5 (*right*), with electrical lead damage indicated in the cycle/failure domain

the left side of Fig. 1.11. Likewise, for sample 'S5,' lead damage in path A2B2 and A4B3 were detected at 10 and 62 % cycles/failure respectively, and showed up in the A1B1 electrical path scans as shown on the right side of Fig. 1.11. As also illustrated in Fig. 1.11, in the case of sample S4, the damage is correlated with the center of the contour regions, versus in sample S5, the damage is correlated with the onset (left edge) of the contour regions.

Lead damage is not correlated with the contour region at the right side of the S4 scan in Fig. 1.11, thus leading us to believe that this region may be indicative of damage manifesting in the sample itself, and leading to eventual specimen failure. We need to further refine our experimental techniques to allow for reliable and confident discrimination between specimen damage and damage to the electrical leads. We have multiple strategies to realize this: (1) remove the electrical leads from the gage length of the specimens to the extent possible in order to eliminate the potential for electrical lead damage; (2) analyze specimens with single pairs of electrical leads to eliminate the possibility of electrical paths into/out of the specimens via the passive non-scanning leads.

1.5 Conclusions

We have succeeded in an initial laboratory implementation of an EIS technique for damage detection in CFRPs. We have shown that EIS can be used to detect changes in both the loading and the gross damage states of a CFRP specimen. We found that impedance magnitude data has more resolution on specimen damage and loading state at lower frequencies, and that phase data has better resolution on specimen damage and loading state at higher frequencies. For the samples tested the cross-over point between when one would choose impedance magnitude versus phase angle to resolve specimen states occurred around 500 kHz. Electrical path through the specimen also needs to be carefully considered since we found that it also strongly influences sensitivity of the EIS technique to both damage and loading state. The principal conclusions based on our initial fatigue cycling results center mainly around refinements for our experimental methodology. At this point we cannot, with confidence, state that the EIS technique is able to resolve either damage precursors or internal damage to CFRP specimens caused by fatigue loading. This lack of confidence arises from the inability to confidently discriminate electrical lead damage from specimen damage, which, in turn, comes from flaws in the initial experimental methodology that we employed, especially electrical lead placement in the gage section of the evaluated specimens. However, our initial results clearly indicate directions of experimental refinement that we believe will allow us to discriminate between electrical contact and specimen damage in future experiments.

Acknowledgements The authors would like to acknowledge the expert and valuable assistance of Mr. Wosen Wolde with the design, production, assembly, and trouble shooting of the electronics switching board.

References

1. Chung DDL (2007) Damage detection using self-sensing concepts. Proc IME G J Aero Eng 221(4):509–520
2. Liu A, Wang KW, Bakis CE (2010) Damage detection of epoxy polymer via carbon nanotube fillers and external circuitry. In: 18th international conference on composites or nano engineering, Anchorage, Alaska
3. Gao L, Thostenson ET, Zhang Z, Chou T-W (2009) Coupled carbon nanotube network and acoustic emission monitoring for sensing of damage development in composites. Carbon 47(5):1381–1388
4. Thostenson ET, Chou T-W (2006) Carbon nanotube networks: sensing of distributed strain and damage for life prediction and self healing. Adv Mater 18(21):2837–2841
5. Chung DDL (2012) Carbon materials for structural self-sensing, electromagnetic shielding and thermal interfacing. Carbon 50(9):3342–3353
6. Ayres JW, Lalande F, Chaudhry Z, Rogers CA (1998) Qualitative impedance-based health monitoring of civil infrastructures. Smart Mater Struct 7(5):599
7. Pohl J, Herold S, Mook G, Michel F (2001) Damage detection in smart CFRP composites using impedance spectroscopy. Smart Mater Struct 10(4):834
8. Loyola BR, Briggs TM, Arronche L, Loh KJ, La Saponara V, O'Bryan G, Skinner JL (2013) Detection of spatially distributed damage in fiber-reinforced polymer composites. Struct Health Monitor 12(3):225–239
9. Na S, Tawie R, Lee HK (2011) Impedance-based non-destructive evaluation of the FRP adhesive joints in corrosive environment with re-usable technique. In: SPIE smart structures and materials + nondestructive evaluation and health monitoring. International Society for Optics and Photonics, pp 79811B–79811B
10. Zhong CH, Croxford AJ, Wilcox PD (2012) Inductively coupled transducer system for damage detection in composites. In: SPIE smart structures and materials + nondestructive evaluation and health monitoring. International Society for Optics and Photonics, pp 83480H–83480H
11. Grimberg R, Premel D, Lemistre MB, Balageas DL, Placko D (2001) Compared NDE of damages in graphite/epoxy composites by electromagnetic methods. In: 6th annual international symposium on NDE for health monitoring and diagnostics. International Society for Optics and Photonics, pp 65–72
12. Luo X, Chung DDL (1999) Electromagnetic interference shielding using continuous carbon-fiber carbon-matrix and polymer-matrix composites. Comp Part B Eng 30(3):227–231
13. Holloway CL, Sabrina Sarto M, Martin J (2005) Analyzing carbon-fiber composite materials with equivalent-layer models. IEEE Trans Electromagn Compat 47(4):833–844

Chapter 2
In situ Observation of NiTi Transformation Behaviour: A Micro–Macro Approach

Kasun S. Wickramasinghe, Rachel A. Tomlinson, and Jem A. Rongong

Abstract A novel experimental investigation is presented of thermally and stress induced transformation behaviour of a Polycrystalline NiTi Shape Memory Alloy (SMA) plate for flexural-type applications: In situ techniques are employed to allow simultaneous macroscopic and microstructural observation of the SMA in a 4-point flexural test. Forming part of a wider research towards realising a NiTi SMA Variable Stator Vane assembly for the gas turbine engine, the study explores variables critical to flexural-type morphing NiTi structures: (1) temperature; (2) strain; and (3) cyclic loading. It builds a relationship between the macro and micro response of the SMA under these key variables and lends critical implications for the future understanding and modelling of shape memory alloy behaviour for all morphing applications. This paper presents the methodological aspects of this study.

Keywords Shape memory • NiTi • In situ • Phase transformations • Micro–macro approach • Cyclic loading

2.1 Introduction

Gas turbine performance development has been governed traditionally by the "worst case" deterioration and operating condition [1]. This leads to severe compromises and large safety margins. Active control of the engine operation using smart materials could potentially improve engine efficiency. Shape Memory Alloys (SMAs), a class of smart materials, exhibit several desirable characteristics exploitable for this purpose. NiTi, based on an equiatomic compound of nickel and titanium is the most widely used SMA in commercial applications [2]. Besides the ability of tolerating relatively large amounts of shape memory strain, NiTi shows high stability in cyclic applications, possesses an elevated electrical resistivity, and is corrosion resistant [3].

An exciting possibility is the incorporation of SMA in plate form into blade structures of the Gas Turbine compressor. Airflow control is introduced into compressor designs through the use of Variable Inlet Guide Vanes and a number of stages incorporating Variable Stator Vanes. They operate by progressively closing as the compressor speed is reduced from the original design value to maintain an acceptable air angle value into the following rotor blades. Traditionally, such a system encompasses a complex structure employing control levers that are actuated through an electrical or bleed air system. Switching this system to a NiTi plate based actuation mechanism is intrinsically very attractive due to its high power density, solid-state actuation, high damping capacity, durability and fatigue resistance.

A concept actuator design using NiTi plates to form a solid-state actuator that replicates the behaviour of the VSV system is depicted in Fig. 2.1. In this concept, the activation of each NiTi plate translates to a deflection at the tip of the actuator. To achieve this, the system exploits a flexural type strain application/recovery using the Shape Memory Effect.

K.S. Wickramasinghe • R.A. Tomlinson (✉) • J.A. Rongong
Department of Mechanical Engineering, University of Sheffield, Mappin St., Sheffield S11 3JD, UK
e-mail: r.a.tomlinson@sheffield.ac.uk

N. Sottos et al. (eds.), *Experimental and Applied Mechanics, Volume 6: Proceedings of the 2014 Annual Conference on Experimental and Applied Mechanics*, Conference Proceedings of the Society for Experimental Mechanics Series,
DOI 10.1007/978-3-319-06989-0_2,

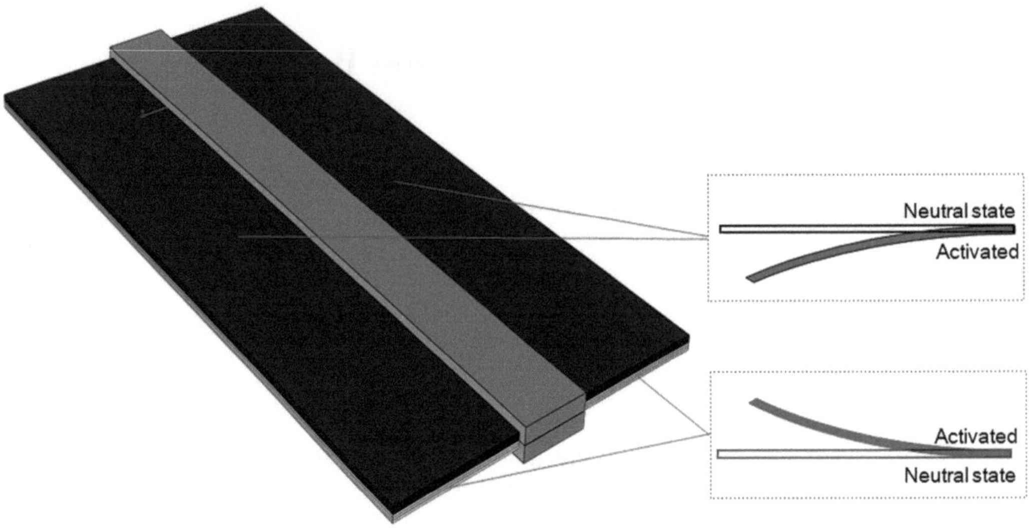

Fig. 2.1 Concept sketch

2.2 Previous Studies

A thorough understanding of the flexural performance of NiTi plate is required for the implementation of such a system. Previous studies have primarily evaluated material behaviour under tension employing mostly uni-axial loads. Of particular interest are studies that employ in-situ loading and microscopy [4–12]: they offer a rich understanding of SMA response that addresses both microstructural and macromechanical response simultaneously. In their micro–macro study, Brinson et al. [12] clarified several aspects of the transformation behaviour under loading at different strain rates and fatigue. This included observations of low-cycle localized plastic deformation, correlations between latent heat release and stress relaxation, and a refined definition of "full transformation" for polycrystalline materials.

However, only a few experimental studies have addressed SMA behaviour under compression or bending, despite the prevalence of these deformation modes in applications. Most studies have restrained from using traditional point-bending fixture tests; instead, they have opted for custom-built pure-bending apparatus. Studies on small diameter NiTi wire naturally restrict point-bending fixtures as a result of small radii of curvature and large displacements—conditions where undesirable axial loads can develop due to the support constraints. The studies conducted by Berg [13] and Bundara et al. [14] have employed moment-controlled experiments to assess the constitutive relationship between applied pure bending moments and the resulting curvatures. These studies, however, suffer a major limitation due to their lack of control over strain rate, leading to a significant impact on the stress/strain response during phase transformations [12]. More recent work by Rejzner et al. [15] and Reedlunn et al. [16] avoid this problem by using displacement-controlled, custom pure bending fixtures integrated into the load frame.

The study by Reedlunn et al. [16] is particularly interesting as they used larger diameter NiTi tube instead of wire, which negates the requirement of a small radius of curvature to incur strain. Additionally, the resulting smaller deformations and larger specimen dimensions allow the use of in-situ imaging techniques in the guise of Digital Image Correlation for strain measurement. Their experimental tests were able to discover a significant asymmetry between the compressive and tensile deformation modes. This resulted in a shift of the neutral axis towards the compressive side to equalise the distribution of tensile and compressive stresses in the cross section. One limitation that was noted, however, was the appearance of the *Brazier effect*; the tendency of thin walled tubes to ovalise in bending due to longitudinal tension and compression [17]. Ovalisation is particularly undesirable if it leads to variant re-orientation along this new deformation path at a granular level as it questions the homogeneity over compressive and tensile responses in their bend tests. This could also result in accelerated development of stress localisations under fatigue, which could lead to premature failure [12]. Another limitation encountered by Reedlunn et al. was through undertaking tests at room temperature without direct measurement of the specimen temperature. This procedure is not ideal because of the high temperature sensitivity of NiTi.

A common theme to most bending studies is the exclusive focus on SMA behaviour in its Superelastic phase. This can be attributed to the myriad of applications that already employ Superelastic SMAs in this deformation mode, making it a more lucrative research area. Moreover, most studies on bending have followed a macro level approach to exploring material behaviour. This has left the investigation of its microstructural behaviour largely unexplored, under this deformation mode.

2.3 Experimental Techniques

Inspired by the above stream of research, the present study explores NiTi behaviour in bending. What distinguishes this research is the analysis of both thermal and stress induced transformation behaviour of NiTi plate. Additionally, in-situ techniques allow simultaneous observation at both macroscopic and microstructural levels. The inclusion of these different aspects, however, presents the study with a new series of hurdles. The methodology employed to conduct this study is detailed below.

2.3.1 Testing Stage

Figure 2.2a illustrates the general experimental set up. The load frame—a tabletop MTS 858 servo-hydraulic test system—uses a cross-head assembly which includes a single moving upper grip, a stationary lower grip and a LVDT position sensor. The system is designed to function in a closed-loop configuration under computer control. The specimen sits within a temperature controlled environmental chamber—a Thermcraft LB-series Box Laboratory oven.

A custom 4-point bend fixture is built and integrated into the uniaxial load frame to perform the flexural tests, as depicted in Fig. 2.2b. The external and internal supports are inverted to limit the deflection of the optically investigated central section of the specimen. The fixture is designed with adequate spatial clearances in mind, allowing upper support deflection of up to 12.5 mm (equivalent to a 4 % strain at the outer fibre with our specimens).

2.3.2 Optical Microscopy

A digital camera is used to observe changes in the surface transformation characteristics at a macroscopic and fine-scale microstructural level as a function of loading parameters. All images are taken during static holds of constant temperature and strain. Figure 2.3 depicts the unique twin surface preparation employed to conduct this in-situ observation.

Digital Image Correlation (DIC) is employed to obtain local strain measurement without the need to rely solely on grip displacements. The specimen is prepared by applying a random speckle pattern to its surface. DIC starts with a reference image and followed by subsequent images recorded during the deformation process. A strain distribution map is created by calculating the correlation between the reference and deformed speckle patterns. In contrast, the microstructural changes are observed from the finely polished surface on the left hand side of the specimen. A polarized light interference filter is used

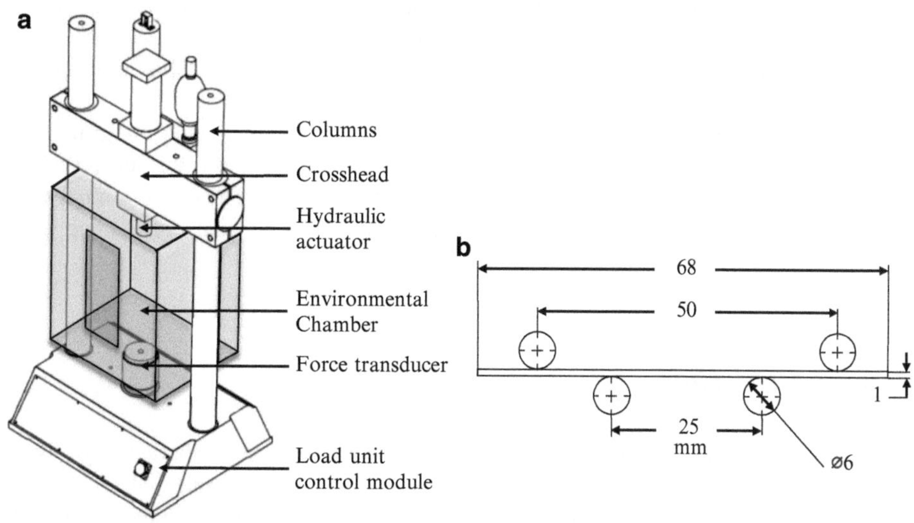

Fig. 2.2 (**a**) Testing stage; (**b**) custom 4-point bend fixture

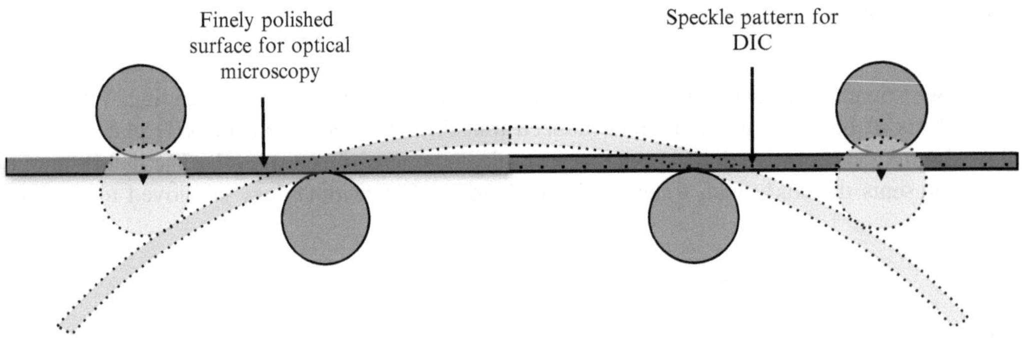

Fig. 2.3 Preparation of observational surface

for higher detection sensitivity of fine topological details. A maximum zoom which focuses on a surface area approximately 1 mm^2 is required to investigate observation of the material at a variant level. This allows the simultaneous viewing of both compressive and tensile deformation modes.

2.4 Specimen Preparation

Machining—Near-equiatomic 1 mm thick NiTi plate was obtained from Memry Metalle GmbH with manufacturer specified Austenite finish (A_f) temperature of 65 °C. The plate is machined to dimensions of 66 mm × 8 mm using a wire cut electrical discharge machining process. This minimizes heat distortion and cold work induced on the specimen. As the tests optically investigate the specimen surface and relate it to the material behaviour as a whole, the homogeneity over their behaviours is important.

Annealing—The specimens are heat treated in an oven at 450 °C for 1 h followed by water quench. This annealing condition ensures a large grain size—the amorphous bands start to crystallize without completely annealing the structural defects in the material, as crystallite nucleation is suppressed [18]. This step is essential for the material to recover from any previous cold work or damage it has sustained and to stabilize transformation characteristics [18].

Thermal cycling—Additionally, the specimens are thermally cycled through martensite and austenite phases using a water bath. Miyazaki et al. [6] indicate that M_f and M_s temperatures decrease with increasing number of thermal cycles. The decrease is rapid at first but more gradual with increasing cycles as values stabilize. Their TEM observations reveal the introduction of dislocations from the first thermal cycle onwards, which act as obstacles to dislocation movement and stabilizes behaviour through further thermal cycles. However, their thermal cycling had also shown to activate R-Phase transformation in NiTi SMAs that had a previously suppressed R-Phase. In our study, the specimens are subjected to 30 thermal cycles despite possible R-Phase activation, as it is paramount to obtain stable and repeatable transformation characteristics throughout.

Micropreparation—The thermally prepared specimens are then subjected to several steps of grinding and polishing using the Bueler Automet 250, to allow observation of variant microstructure at the specimen surface. As the surface was initially rough, all sides of the specimen are grinded using P800 and P1200 abrasive discs. The observational side of the specimen is then sequentially polished using 3 μm MetaDi Supreme Diamond abrasive for coarse polishing and then finally using 0.02–0.06 μm MasterMet Silica for fine polishing.

2.4.1 Specimen Characterisation

Differential Scanning Calorimetry (DSC) is used to identify phase transformation temperatures of the NiTi plates. This is a thermoanalytical method commonly used in NiTi SMA studies that determines absolute phase transformation temperatures. It measures the difference between the amount of heat required to increase the temperature of a sample and reference, as a function of temperature. Our DSC tests adopt a temperature profile of 15 °C to +100 °C with a constant scanning rate of 10 K/min. This is considered the most efficient temperature rate to measure the intrinsic transition quantities

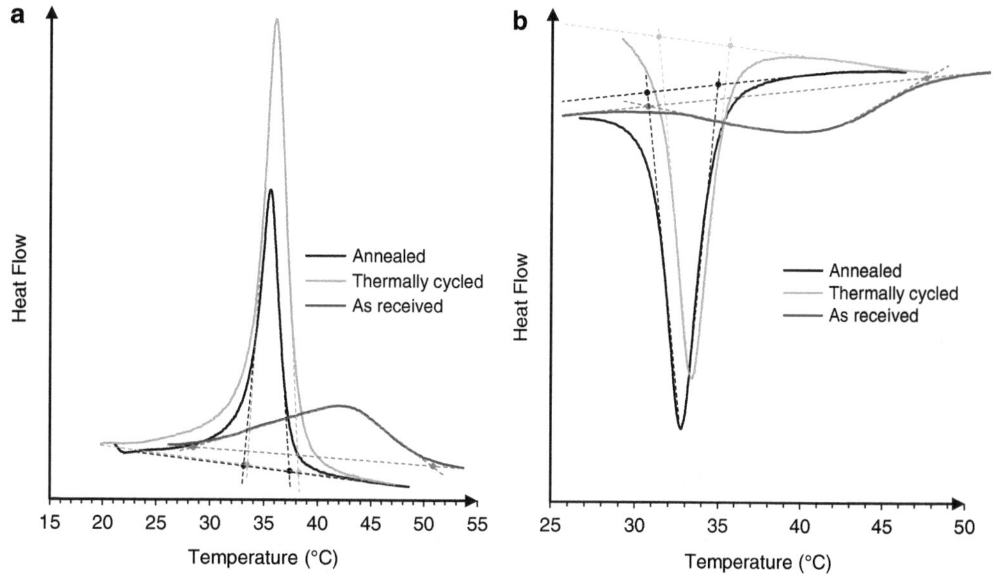

Fig. 2.4 (**a**) DSC Thermogram for heating (M → A); (**b**) DSC Thermogram for cooling (A → M)

Table 2.1 Phase transformation temperatures

Specimen	M_s (°C)	M_f (°C)	A_s (°C)	A_f (°C)
As received	47.8	30.7	28.5	50.9
Annealed	34.9	30.6	33.1	37.4
Thermally cycled	35.6	31.3	33.5	38.2

and also a common value within published literature. The characterisation of the transformation temperatures is essential in establishing temperature ranges for the experimental study.

DSC scans of the NiTi plate specimens in the following conditions are presented: as received (grey), annealed (red), and thermally cycled (blue). Figure 2.4a, b depict thermograms of each of these states under heating (M → A) and cooling (A → M), respectively. The transformation temperatures for each of the specimen states are recorded in Table 2.1.

2.5 Test Plan

The experimental test study can be subcategorized into two sections: (1) constant temperature tests with load increments, and (2) constant load tests with temperature increments. The tests employ a temperature range between A_f −25 °C → A_f + 25 °C and a peak flexural strain of 2 %. Using these load parameters, the study performs a comprehensive evaluation of the material behaviour within the operating envelope of the planned NiTi SMA actuator.

The extremities of these parameters are established due to the requirement of the maximum strain falling within the de-twinning phase at the lowest temperature, A_f −25 °C. However, it must be noted that due to the non-linear response of the material, localised strains could be considerably more, especially during the de-twinning phase where large heterogeneities can exist between grain boundaries. An adequate safety margin is thus kept from the end of the strain plateau observed from previous data from tensile tests on this material: ~4 % at room temperature. This strain magnitude range ensures that thermo-mechanical cycling effects are minimised and life span of each specimen is improved. Furthermore, the lower strains reflect the behaviour of a high cycle element such as the variable stator vane where uniformity of performance must be established throughout the life cycle.

Three NiTi plate specimens are used in this study. The thermal and fatigue history of each of the specimens is recorded and carefully balanced to enhance comparability between results. In each test cycle, the loading is temporarily paused at planned intervals and the specimen is held at constant displacement, while photographs are taken of the observational surface.

Constant temperature tests with load increments attempt to replicate the deflection behaviour of the NiTi plate actuator concept over a range of static temperatures. Table 2.2 highlights the parameters used in this study. The tests are sequenced so

Table 2.2 Constant temperature tests

Strain	Strain rate	Temperature	Optical imaging	Total tests
2 %	$1 \times 10^{-4}\,s^{-1} \rightarrow 1 \times 10^{-2}\,s^{-1}$ (3 steps)	A_f −25 °C → A_f + 25 °C (12 steps + 12 steps reverse)	Static strain holds 0.25 % → 2 % → 0 % (16 steps)	72

Table 2.3 Constant load tests

Load	Temperature	Temperature ramp rate	Optical imaging	Total tests
0.1 % → 2 % (at A_f −25 °C/ 20 steps)	A_f −25 °C→ A_f + 25 °C→ A_f −25 °C	$1 \times 10^{-1}\,°Cs^{-1}$	Static temperature holds (11 steps/every 5 °C)	60

that strain is always recovered after a cycle that is loaded from the martensitic phase: the subsequent cycle runs above A_f and activates the SMA. Loading sequences for each of the initial 36 tests are repeated in reverse order to account for the temperature history sensitivity of the material. The strain is regulated by a load input to allow strain relaxation in non-isothermal type loading.

Constant load tests with temperature increments attempt to replicate the actuation behaviour of the NiTi plate actuator concept over a range of static loads. Strain recovery characteristics of the NiTi plate specimens at different load magnitudes are analysed to accomplish this. Table 2.3 highlights the parameters used in this study, which allows the evaluation of macroscopically observable material characteristics and microstructural changes over the stated temperature range. The strain is regulated by a load input to allow strain recovery. The specimen is heated past its Austenite finish (A_f) temperature prior to each subsequent test to ensure that strains are not carried over between tests. This is executed at zero strain, using an identical temperature ramp rate.

2.6 Predictions and Limitations

In this section, we evaluate the possible outcomes of the experimental tests. While the broader aim of the study is to evaluate material behaviour under the possible operating envelope of the NiTi plate actuator, this study also assesses several NiTi SMA behavioural aspects that may have academically significant implications, as discussed below.

Asymmetry between tension and compression—The bend tests allow the simultaneous evaluation of the material behaviour under both tensile and compressive deformation modes. While classical Euler–Bernoulli bending kinematics assumes a linear strain profile across the cross section, the heterogeneous nature of the bending strain fields in NiTi result in a much more complex strain profile. In their bend study, Reedlunn et al. [16] report a significant shift of the neutral axis under bending. However, the appearance of the Brazier effect, due to their choice in specimen geometry, could have affected this to a certain extent. In contrast, the use of plate specimens in our study negates such effects. Moreover, this study potentially allows a richer understanding of the two deformation modes as we use broader testing conditions and follow a micro–macro approach.

Relationships between macroscopic behaviour and microstructural observations—This study combines macroscopic and microstructural analysis via DIC and optical microscopy. Most previous studies have either focused on a micro or macro level approach: micromechanical studies are often theoretical and not directly applicable to end applications whereas macro level studies do not contain enough depth to explain the observed phenomena. The employment of both approaches allows relationships to be constructed, explaining material behaviour in greater detail.

A complete map of transformation behaviour—Phase transformations, temperatures and applied stresses are closely intertwined where one affects the other. As this study varies both temperature and load in its tests, a complete phase diagram can be developed, showing the effect of applied flexural deflection on the phase transformation temperatures for the tested specimens.

Effects of loading rates—The constant temperature tests repeat every loading sequence at three loading rates: $1 \times 10^{-4}\,s^{-1}$, $1 \times 10^{-3}\,s^{-1}$ and $1 \times 10^{-2}\,s^{-1}$. Uniaxial studies by Brinson et al. [12] report on slower strain rates following an isothermal behaviour where martensite initiates and grows from a single band, while at higher strain rates, the formation of multiple bands is present. To our knowledge, however, the implications on loading rate for bend tests using full-field, in-situ methods such as DIC are largely unexplored.

Localised phase transformations—Under tension, NiTi is known to exhibit strain localization and propagation during portions of stress induced transformations [19]. Under flexural loading, however, the development of strain localisations can be non-linear due to the presence of compressive and tensile deformation modes in varying magnitudes through the specimen thickness. In their study of Superelastic NiTi tube, Reedlunn et al. highlight this effect, reporting on strain localisations in tension and on the absence of strain localisations in compression [16]. Our study will further evaluate this behaviour, linking the evolution of macroscopic transformation bands to the microstructural changes as a function of both applied load and temperature.

Due the testing techniques employed, our study contains several limitations. These limitations and the extent to which they can affect the outcomes are discussed below.

Bend fixture—The use of 4-point bend fixtures can be somewhat problematic in NiTi studies due to the creation of undesired axial loads at the support constraints. These stress concentrations, if large enough, can lead to the premature transformation to martensite near the constraints, and result in inaccurate displacement measurements. While these issues exist in our study, they are outweighed as a result of the easy integration of the 4-point bend fixture in the loading frame and the subsequent flexibility offered in tests. Furthermore, we try to negate drawbacks associated with point-bending fixtures by employing several measures: (1) a flat contact surface between specimen and constraints to reduce stress concentrations, (2) relatively small peak deformations in tests, (3) 4-point (as opposed to a 3-point) bend fixture as they have a larger region of deflection, and (4) contact surfaces polished to reduce shear.

In-situ material characterisation—This study uses DIC and optical microscopy to capture phase transformation behaviour at a kinematic level. Another characteristic of phase transformation behaviour that can be captured is its thermal signature. This can be accomplished by using an Infrared camera, which can detect the instantaneous local heating of the domains due to the released latent heat during phase transformation. The synchronization of both kinematic and thermal measurements would lead to a highly attractive complete thermo-mechanical characterisation of phase transformation behaviour. However, the integration of such a technique alongside DIC and optical microscopy can be logistically very difficult (antagonist surface texture requirements, imaging devices having different pixel and acquisition rates [20]). A novel technique developed by Maynadier et al. [20]—Infra-Red Image Correlation—synchronizes kinematic and thermal measurements and thus negates many of these drawbacks. This will be explored in future studies.

Fatigue—Previous studies have reported on a continuous change in the functional properties of the NiTi such as the transformation temperatures, transformation hysteresis and strain response under thermo-mechanical cycling [18]. Our experimental work comprises 132 tests, which effectively subject the 3 test specimens to thermo-mechanical cycling throughout the study; hence, the stability of the material response is a concern. Several measures have been taken to dramatically reduce such effects, including careful thermal preparation of the specimens, use of low peak deformation magnitudes and careful workload management between the specimens in the experimental tests. The option to completely ignore thermo-mechanical cycling effects—employing a vast number of specimens—was not explored as it does not truly reflect the required characteristics of the NiTi plate actuator, a high cycle element with uniformity of performance required throughout its life cycle.

Temperature range—All the tests are conducted within a temperature range of A_f −25 °C → A_f + 25 °C. This range is restricted due to the lack of active cooling in the environmental chamber, which prevents tests being undertaken below ambient temperature. The temperature range should be sufficient to capture forward and reverse transformation at most load levels. Nevertheless, the integration of active cooling in future studies is a vital step, as it would allow recreation of the extreme temperatures that would be experienced by the NiTi plate actuator.

2.7 Closing Remarks

A review of previous studies has demonstrated a gap in the knowledge concerning the thermo-mechanical behaviour of NiTi, especially in the bending deformation mode. In this paper, we have presented the methodological aspects of a novel experimental investigation that addresses these issues. Our attempt is to evaluate thermal and stress induced transformation behaviour of NiTi plate for flexural-type applications. We have employed in-situ techniques, which allow simultaneous observation of both macroscopic and microstructural material behaviour in 4-point bend flexure tests. A special emphasis is placed on material preparation not only to attain the desired surface characteristics for in-situ observation, but also to stabilize the thermo-mechanical response of the material for accurate tests. The experimental study is designed to assess several key characteristics of NiTi SMA behaviour. The tests were broadly split into two categories: (1) constant temperature tests to evaluate deflection characteristics, and (2) constant load tests to evaluate strain recovery and relaxation characteristics. The speculated outcomes and their implications were hypothesised. In particular, the following aspects may have academically significant implications: (1) Asymmetry between tension and compression, (2) Relationships between macroscopic behaviour and microstructural observations, (3) A complete map of transformation behaviour, (4) Effects of loading rates, and (5) Localised phase transformations. We were also able to identify several limitations in our study. They include limitations related to the experimental techniques used and to the range of the test study, and limitations universally inherent to NiTi testing. Future work will attempt at addressing these limitations.

References

1. Horn W, Ardey S, Grauer F, Schmidt K, Staudacher S (2008) Opportunities and challenges for more intelligent gas turbine engines. Deutscher Luft- und Raumfahrtkongress
2. Wu M, Schetky L (2000) Industrial application for shape memory alloys. In: Proceedings of the international conference on shape memory and superelastic technologies, Pacific Grove, CA, pp 171–182
3. Lagoudas D (2008) Shape memory alloys: modelling and engineering applications. Springer, Texas
4. Miyazaki S, Kimura S, Otsuka K, Suzuki Y (1983) Shape memory effect and pseudoelasticity in a Ti–Ni single crystal. Scr Metall 18:1057–1062
5. Miyazaki S, Imai T, Igo Y, Otsuka K (1986) Effect of cyclic deformation on the pseudoelasticity characteristics of Ti-Ni alloys. Metall Trans A 17(1):115–120
6. Abeyaratne R, Chu C, James R (1994). Kinetics and hysteresis in martensitic single crystals. In: Proceeding of symposium on the mechanics of phase transformations and shape memory alloys
7. Vivet A, Lexcellent C (1998) Observations and analysis of martensitic phase transformation on CuZnAl single crystals. J Phys IV France 9:411–418
8. Fang D, Lu W, Hwang K (1999) Pseudoelastic behavior of a CuAlNi single crystal under uniaxial loading. Metall Mater Trans A 30 (8):1933–1943
9. Jost N (1999) Thermal fatigue of Fe–Ni–Co–Ti shape-memory-alloys. Mater Sci Eng A 275:649–653
10. Liu Y, Xie Z, Van Humbeeck J, Delaey L (1999) Effect of texture orientation on the martensite deformation of NiTi shape memory alloy sheet. Acta Mater 47(2):645–660
11. Zheng Y, Huang B, Zhang J, Zhao L (1999) High resolution electron microscopy studies on the interface structure of deformed stress induced martensite variants in a Ti–Ni–Nb shape memory alloy. Mater Sci Eng A 275:271–274
12. Brinson L, Schmidt I, Lammering R (2004) Stress-induced transformation behavior of a polycrystalline NiTi shape memory alloy: micro and macromechanical investigations via in situ optical microscopy. J Mech Phys Solids 52(7):1549–1571
13. Berg B (1995) Bending of superelastic wires, Part I: experimental aspects. J Appl Mech 62:459
14. Bundara B, Tokuda M, Kuselj B, Ule B, Tuma J (2000) Superelastic tension and bending characteristics of shape memory alloys. Metals Mater Int 6:293–299
15. Rejzner J, Lexcellent C, Raniecki B (2002) Pseudoelastic behaviour of shape memory alloy beams under pure bending: experiments and modelling. Int J Mech Sci 44(4):665–686
16. Reedlunn B, Churchill C, Nelson E, Shaw J, Daly H (2014) Tension, compression, and bending of superelastic shape memory alloy tubes. J Mech Phys Solids 63:506–537
17. Brazier L (1927) On the flexure of thin cylindrical shells and other "thin" sections. Proc R Soc Lond A 116(773):104
18. Saikrishna C, Ramaiah K, Prabhu S, Bhaumik S (2009) On stability of NiTi wire during thermo-mechanical cycling. Bull Mater Sci 32(3):343–352
19. Bechle J, Kyriakides S (2014) Localization in NiTi tubes under bending. Int J Solid Struct 51:967–980
20. Maynadier A, Poncelet M, Lavernhe-Taillard K, Roux S (2013). One-shot thermal and kinematic field measurements: infra-red image correlation. In: Conference proceedings of the society for experimental mechanics series 2013, pp 243–250

Chapter 3
Bio-Inspired Design of a Multi-scale Pass Band Frequency Sensor Using Local Resonance Phenomena

Riaz Ahmed and Sourav Banerjee

Abstract With growing interest in nanotechnology, the manufacturing industries for Micro-Electro-Mechanical Systems (MEMS) and Nano-Electro-Mechanical-Systems (NEMS) are constantly thriving towards extraordinary precision in the machining and etching tools. It is a common practice, during manufacturing, a set of instructions are provided to a manufacturing tool, by actuating them at certain frequencies to perform their respective tasks. Every different task (e.g. cutting nano-channels, drilling micro-holes, nano-welding etc.) has unique instruction with unique frequency input. In such cases, other than the desired frequencies, remaining possible frequencies in the system needs to be filtered or stopped. It is extremely challenging to avoid system noises electronically and select or actuate specific frequencies. Hence, in a noisy environment (e.g. fluctuation of temperature, external vibration etc.) it is extremely difficult to provide a unique frequency to a tool to perform a task precisely without having an uncertainty. In this study, we intend to propose a mechanical model to precisely sense, pass and actuate desired frequencies and filter unwanted input frequencies, which in turn will result a mechanical pass band sensor. Traditionally researches are interested in stopping undesired frequencies to pass desired frequencies through local resonance phenomena. However, if only certain frequencies are required, it is extremely difficult to filter all unnecessary frequencies by creating frequency band gaps. Hence, in this effort, bio-inspired logistic, adopting local resonator physics is employed by extracting the benefit of unique frequency sensing, mechanically. Human cochlea senses only sonic (20 Hz to 20 kHz) frequencies by filtering all other frequencies in the environment. Basilar membrane is the principal component of the human cochlea with logarithmically decreasing stiffness from its basal to apical end. Basilar membrane composed of series of thin micro beams attached to each other, where each beam holds unique bending rigidity and hence, capable of resonating at a particular sonic frequency. Similarly, in this study, to replicate the functionality of a basilar membrane, a discrete mass-in-mass (DMM) metamaterial model is proposed while using a complete different physics of local resonance. It is hypothesized that, systematic arrangement of such DMM cells can select of band of frequencies, predictively.

Keywords Acoustic metamaterial • Frequency selectivity • Basilar membrane • Pass band sensor • Local resonance

3.1 Introduction and Literature Review

In this study, the principal aim was to demonstrate the feasibility of designing a mechanical pass band sensor that can predictively introduce the capability of sensing, passing and actuating acoustic waves with desired specific frequencies or a band of frequencies, mechanically. Two possible techniques can be considered to achieve the proposed objective. Either an introduction of a model that can create numerous frequency band gaps by stopping all the undesired system frequencies, called mechanical stop band filter. Alternatively, introduction of a model that can sense only the desired frequencies by satisfying the primary objective of this work, called mechanical pass band filter.

R. Ahmed • S. Banerjee (✉)
Department of Mechanical Engineering, University of South Carolina, Columbia, SC 29208, USA
e-mail: ahmed@email.sc.edu; banerjes@cec.sc.edu

N. Sottos et al. (eds.), *Experimental and Applied Mechanics, Volume 6: Proceedings of the 2014 Annual Conference on Experimental and Applied Mechanics*, Conference Proceedings of the Society for Experimental Mechanics Series,
DOI 10.1007/978-3-319-06989-0_3,

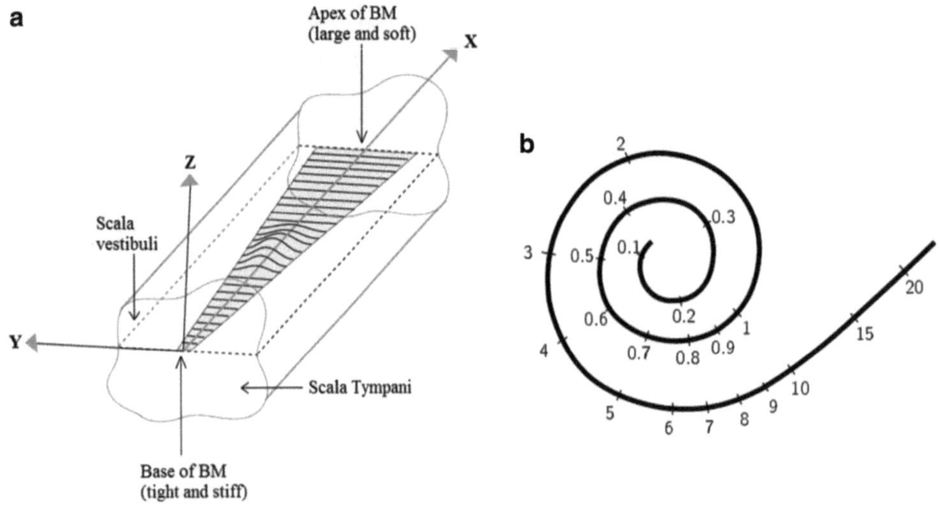

Fig. 3.1 (**a**) Simplified representation of unrolled basilar membrane with tentative fiber orientation (**b**) Approximate frequency (in kHz) selecting zone on the basilar membrane

Traditionally, researchers are interested in creating frequency band gaps exploiting either Bragg scattering or the local resonance phenomena to stop acoustic waves. It is well-established that in any metamaterial, frequency band gaps are the results of either local resonance or Bragg scattering [1, 2]. Resonating the idea, numerous researchers have improved the concept of wave filtration capabilities by designing new and innovative materials [3–11]. However, still no model exists that are capable of filtering all desired frequencies form a system. Generally the stop band frequency ranges are considerably lower (couple of kHz) from the previously proposed models and nearly impossible to fit for our requirement. With our previous study we were only able to open a wider band gap, which proves that it is difficult to pass a very narrow band of frequency using such Bragg scattering or local resonance models [11]. Hence, the novel alternative is to think differently, where principal objective is to propose a model that can mechanically sense/pass the user-defined frequencies only. After rigorous study, it was observed that in nature there are devices that are capable of performing similar task. The auditory organ, human cochlea, allows human to hear only a certain range of frequencies (20 Hz to 20 kHz) and filters all other frequencies, mechanically. Hence, mimicking the concept, cochlear mechanics has been utilized, where series of hair cells are sensitive to the selected frequencies travelling through the basilar membrane depending on their spatial position inside the organ of corti [12]. However, we fused this concept with the local resonance physics to present our new model. Before discussing our model it would worthwhile to review the concept of cochlea.

There are few important components in the cochlea which performs mechanical action to filter infrasonic and ultrasonic waves. Basilar membrane plays the most important role in a human cochlea by responding to sonic frequencies through changing its material properties from basal to apical end. The basilar membrane is not an isotropic stretched membrane; it consists of long, thin, beamlike fibers running across the width [12]. The basilar membrane is stiff and narrow (about 100 μm) near the base, and flexible and wide (about 500 μm) near the apex, with a smooth transition along its length (See Fig. 3.1a). Each fiber possesses distinct stiffness and hence senses a unique sonic frequency. The position of maximum displacement of the basilar membrane varies approximately logarithmically with input frequency, above ~1 kHz. Frequencies lower than 1 kHz are more compressed along the length of the cochlea, as shown in Fig. 3.1b.

In 1961, Hungarian biophysicist Georg Von Bekesy was awarded novel prize for his experimental investigation on human cochlea [13]. Referencing Von Bekesy's finding, in last couple of decades several attempts were made to develop mathematical model for the human cochlea [14–17]. Although most of the cochlear research were limited to mathematical modeling, only recently a very few experimental efforts were reported to exactly replicate the human cochlea [18–23] with the similar design configurations. Such models could be good argument for cochlear implantation study, however, may not be adequate for industrial applications where specific frequency sensing are essential from external vibration. As indicated, we fused the concept of cochlea with the local resonance physics, in this study, a mass-in-mass metamaterial system is envisioned where each unit cell of the proposed metamaterial is idealized to a specific fiber of basilar membrane and capable of sensing a unique frequency. Here we demonstrate the feasibility of sensing a specific frequency from the local resonator using a comparatively macro size structural dimensions, however, it is hypothesized that similar physics can be applied to lower scales for micro scale device applications.

3.2 Numerical Approach

Local resonance phenomenon is emphasized in this work to sense signal from a metamaterial. For computational convenience, unit metamaterial cell is studied using finite element method. Proposed locally resonant metamaterial, consists of a heavy lead core embedded into a soft matrix (Rubber). Lead-Rubber combination is then placed in a considerably stiff structure (Aluminum) to complete the metamaterial (See Fig. 3.2a) system. A 0.98″ diameter hole is considered in aluminum structure to accommodate the lead-rubber components. The unit cell is a 0.55″ block with 1.47″ in length and width. It is assumed that, diameter of the lead core (0.49″) is slightly less than the thickness of the block to enclosed it inside. Material properties of the components of the unit cell are listed in Table 3.1.

A frequency domain analysis is performed for a frequency range of 20–5,000 Hz with unit harmonic displacement on two opposite sides (Ref. Fig. 3.2b) of the block. A local resonance phenomenon is observed at around ~370 Hz on lead and rubber component, whereas aluminum structure seems vibrationless. In such cases, it appeared that the energy is trapped inside the lead-rubber component. Since lead ball is the principal resonating component at that frequency, it can create high compressive stress on rubber element. Hence, placing a sensor into rubber component can be useful to collect/sense mechanical energy into electric energy, which resembles the hair cells in organ of corti [12].

3.3 Experimental Approach

3.3.1 *Fabrication*

Since the work targeted for industrial applications, it is essential to validate the simulation results with experimental response. Hence, to validate the model, first exactly same metamaterial unit cell is fabricated. Machined aluminum 6061 is considered as the boundary structure with a cylindrical hole to place rubber and lead components. Since lead ball's diameter is slightly lower than the thickness of aluminum block and it is required to place the ball at the middle (concerning all three dimensions) of the aluminum hole, a cylindrical support is designed and fabricated through 3D printing technology (See Fig. 3.3). The cylindrical support consist of thee parts (Insider, Base and Handle) where diameter of the Insider portion is exactly same as the diameter of the aluminum hole. A small arc indentation is considered at the middle of the insider, such that it can hold the lead ball at the middle. The insider goes inside of the aluminum hole and its dimensions are set to support the lead core at the middle of the structure with high precision. Diameter of the Base is slightly higher than the width of the aluminum block, so that it can carry the whole structure.

The fabrication process is divided into two steps. First, the lead core is placed inside the middle of the aluminum block with the help of cylindrical support. A liquid rubber (OOMOO 300, contains two parts, mixed slowly to avoid bubbles and

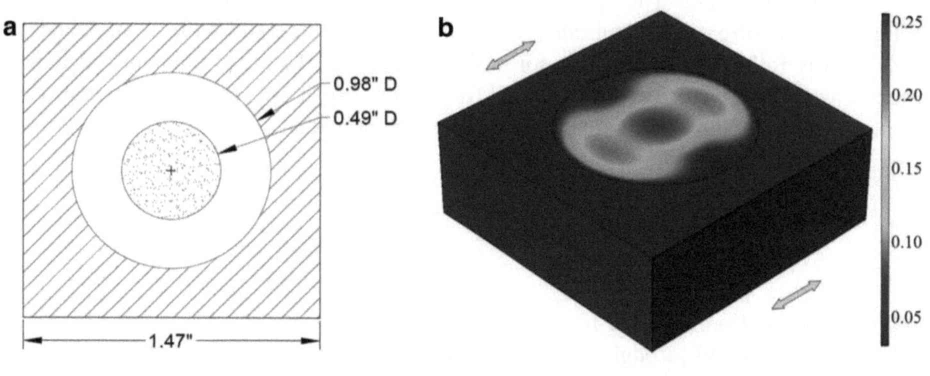

Fig. 3.2 Unit cell metamaterial (**a**) geometric configuration (**b**) displacement pattern at 370 Hz (in inch)

Table 3.1 Properties of the components enclosed in unit cell

	Lead	Rubber	Aluminum
Young's modulus (Pa)	13.5e9	0.9942e6	68.9e9
Density (kg/m^3)	11,340	1,600	2,700
Poisson's ratio	0.435	0.47	0.33

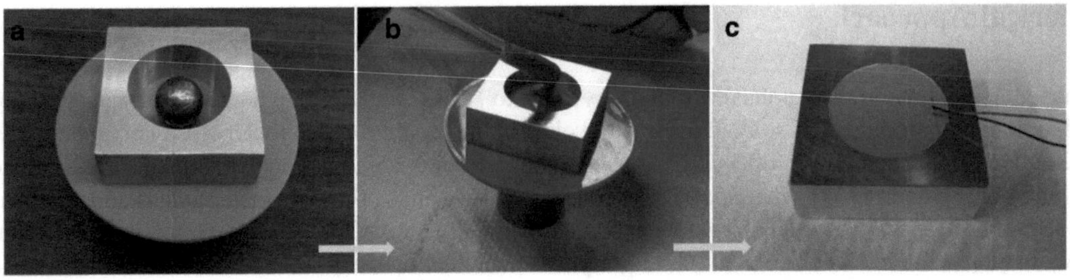

Fig. 3.3 Fabrication steps of unit cell metamaterial. (**a**) Fabrication setup (**b**) Liquid rubber placement (**c**) Fabricated final form with embedded piezoelectric sensors to capture the vibration in the rubber only

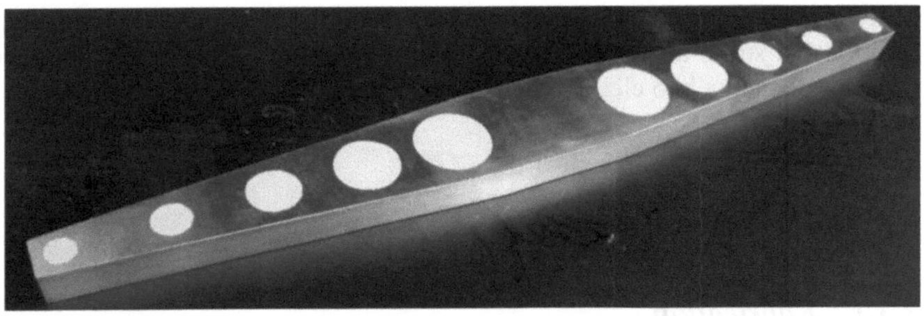

Fig. 3.4 Fabricated multi-cell metamaterial with linearly varying geometric parameters for each cell

homogeneous strength) is then used to fill the hole in aluminum block. Since it is necessary to sense/transfer signal from rubber component, a piezoelectric disk (with soldered wire) is fully submerged into the liquid rubber in such a way that it remains untouched with both lead ball and aluminum structure.

Usual rubber curing time is 6 h. However, it is required to start the second step of fabrication at around 3–4 h after the first step. During the initial steps, cylindrical support was used to hold the lead ball at the middle. Hence, an empty space was open at the bottom of the structure after removing the cylindrical support. In the second step, the new empty space was filled with rubber following the same procedure described above. Since it is required to have a good bonding between the rubber, the second step was started before the full curing time in the first step. Please note that during the first step a piezoelectric sensor is placed vertically and allowed to be embedded once set.

Following the outcome of the unit cell, a multi-cell metamaterial model is fabricated to sense multiple frequencies from one structure at the same time. The new model is a combination of five (5) unit cells with linearly varying design parameters (e.g. lead ball diameter, hole diameter etc.). It is hypothesized that each unit cell may idealize a specific fiber in the basilar membrane and can take part in sensing a unique frequency. However, constant thickness is considered for the entire structure. Considering the shape of basilar membrane (which is narrow at base and wide at apex end), width of the main structure is decreased linearly following the function for basilar membrane. To avoid moment imbalance during harmonic excitation, main structure was mirrored to complete the total frame (Ref. Fig. 3.4). Same fabrication procedure is followed to make the multi-cell metamaterial and the piezoelectric sensors were embedded.

3.3.2 Testing

A Vibration Exciter (type 4809, from B & K Instruments) is used in testing the harmonic response of the fabricated metamaterial. The exciter is controlled through a Sine-Random Generator (type 1024) and Power Amplifier (type 2706) from Bruel and Kjaer. An aluminum support is manufactured to hold the metamaterial system on the vibration exciter. Schematic diagram of test setup with a unit cell is shown in Fig. 3.5a. Experimental setup with the multicellular structure is shown in Fig. 3.5b. The test job is clamped in two opposite sides and a simple harmonic displacement excitation (identical with simulation input) was applied using the Sine-Random Generator (SRG). A frequency sweep operation is not available on used SRG, however, it is only capable of generating a single frequency excitation at one time and excitation frequency can be tuned with the SRG knob. Tuning range of SRG is limited to 20 Hz to 20 kHz. An Oscilloscope (TDS 2004C, Tektronix) was connected to the piezoelectric wafer to sense the vibrational response from the rubber component.

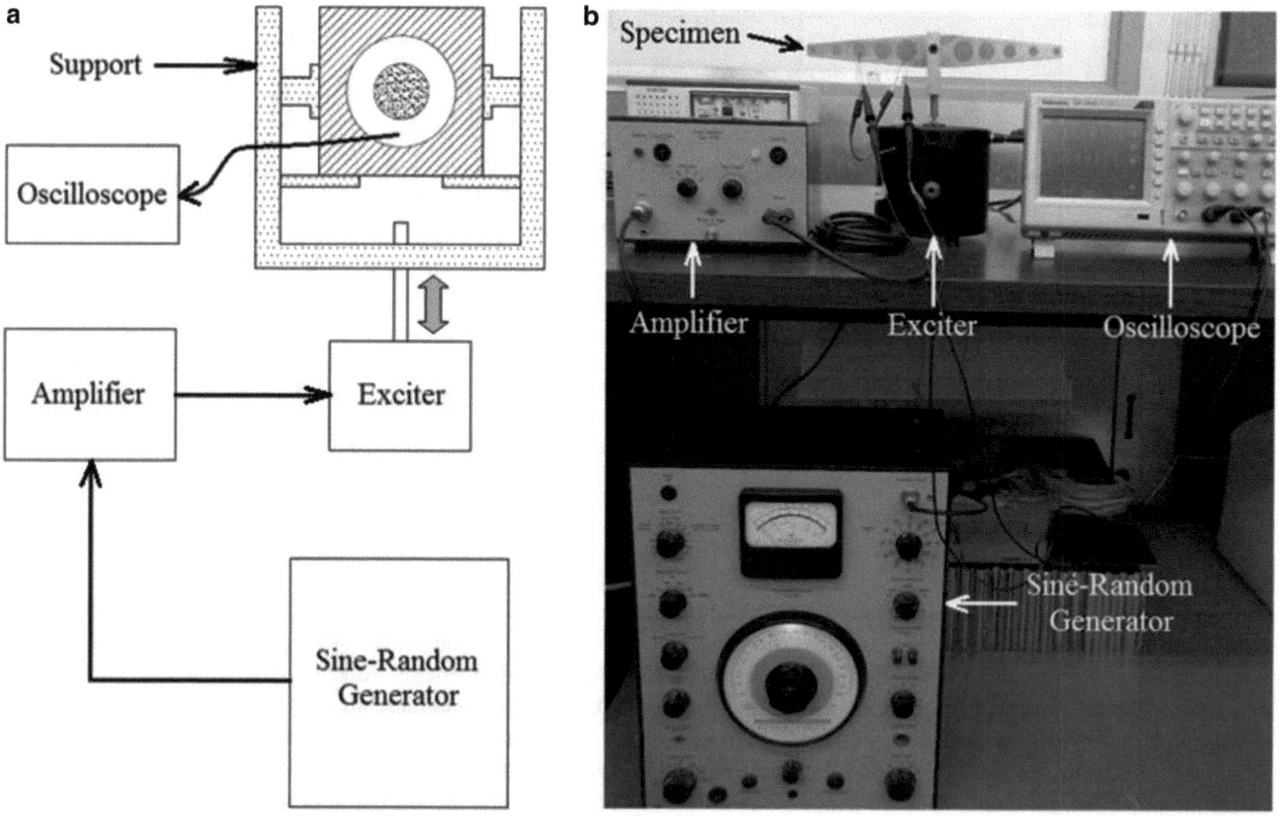

Fig. 3.5 (**a**) Schematic diagram of test setup for unit cell (**b**) Real test setup for multi-cell specimen

3.4 Results and Discussion

In experimental study, a frequency range of 20 Hz to 20 kHz is considered to analyze the response amplitude with frequency. From numerical analysis it was noticed that at around ~370 Hz excitation, local resonance takes place at rubber-lead combination in the unit cell, which seems the best state to sense the signals from rubber through a piezoelectric sensor. Experimental study shows that, for unit cell, sensed voltage amplitude (from rubber) is significantly high at ~345 Hz compared to any other sonic frequencies. Voltage amplitude drops very quickly right before/after ~345 Hz, which signifies the occurrence of resonance phenomenon in rubber material. Numerical and experimental outcome matches quite well for a single unit cell model, keeping the uncertainties in the material properties that are used in the model in mind. However it is still not confirmed whether different frequencies can be sensed using different design parameters and how a band of frequency can be sensed. Hence, the multi-cell metamaterial is proposed. The proposed model composed of ten cells with linearly varying design parameters. To demonstrate the feasibility, response from only two cells are tested and reported in this article. Extensive investigations using all the ten cells are underway.

Although the study was performed within the entire sonic frequency band, only two cell shows maximum voltage output at only two unique frequencies. Responses from both the cells on those frequencies are reported in Fig. 3.6. In multi-cell metamaterial, largest cell dimension is almost identical (except the cell boundary) with the unit cell that was studies earlier. From the largest cell, maximum voltage output was recorded at ~360 Hz, which is very close to the sensed frequency level (~345 Hz) through the single unit cell approach. However, maximum voltage output is recorded at ~1,380 Hz from the smaller cell. It is also noticed that, at each frequency, response from the respective cells are predominant compared to other cells. Hence, it is confirmed that, any specific frequency can be sensed from only one cell of specific geometry. Further, we hypothesize that with further study and optimization of our predictive model the process could help design a mechanical pass band sensor that can only sense desired frequency bands on demand.

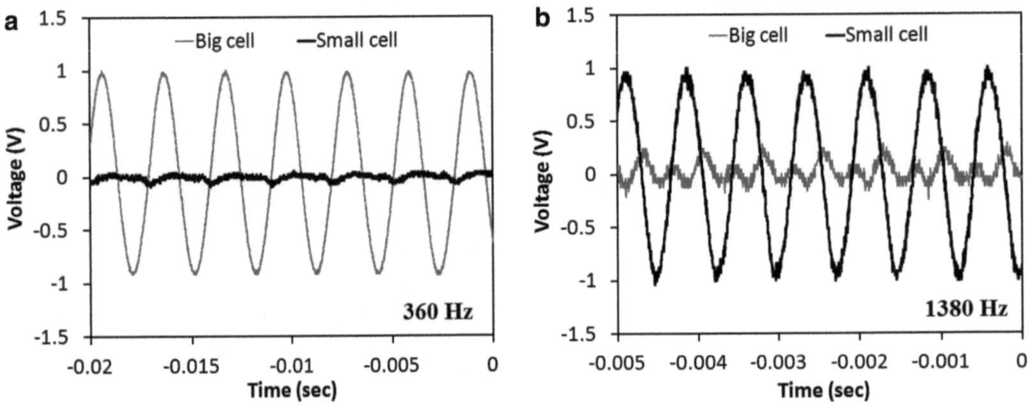

Fig. 3.6 Normalized voltage response due to harmonic displacement excitation at (**a**) 360 Hz (**b**) 1,380 Hz

3.5 Conclusion

Principal objective of this study is to propose a mechanical model that can select, sense and actuate specific frequencies or a range of frequencies, predictively. Two possible ways can be adopted to achieve this objective. One approach is to create stop band for all the possible frequencies in the system expect the desired frequencies. Alternatively, modeling a system that can sense only the preferred frequencies will be more beneficial. From early studies, no model was found which is close enough to create large stop bands to filter undesired frequencies from a system. Hence, in this study, the alternative technique is adopted. Since this method of sensing specific frequencies wasn't studied before, it was critical to develop a model to fulfill the objective. In nature, basilar membrane in human cochlea performs the similar action, where it senses only sonic frequencies and filters all other frequencies. Basilar membrane is composed of a series of thin fibers. Each fiber contains unique stiffness and hence able to sense specific frequencies. A discrete mass-in-mass metamaterial is considered in this study to replicate the functionality of basilar membrane fibers but using different physics. Each unit of the metamaterial represents a single fiber in the basilar membrane and should be able to sense a unique frequency. Initially an unit cell metamaterial was studied both numerically and experimentally. Both method shows almost same frequency sensing capability from the studied model. Although, the unit metamaterial shows the sensing capability of a specific frequency, but it wasn't confirmed whether the sensed frequency is geometry dependent or not. Hence, a multi-cell metamaterial is proposed with five (5) unit cells of linearly varying geometry. Main frame of the structure was also tapered following the shape of basilar membrane. For handling convenience, two cells were analyzed. Note that, inner geometry (core and rubber) of the largest cell of the multi-cell model is identical with the unit cell studied earlier and almost same frequency was sensed from both the cells. However, a higher frequency was sensed from the smaller cells of the multi-cell model. Hence, it can be concluded that, frequency selectivity of a cell is independent of its outer structure geometry, however strongly dependent on inner structure geometry. In other word, the multi-cell metamaterial can be represented as a one-dimensional spring-mass system, where no structural geometric parameter is involved. It is also found that, any frequency can be sensed only from one cell of specific model parameters. In this study, it can be summarized that, by using the proposed technique; desired frequencies can be sensed with appropriate inner structure geometry. It is also possible to sense a band of frequencies by using a multi-cell model. This study can be a useful note for many engineering applications where selecting specific frequencies is critical.

Acknowledgement This project was founded by the Office of the Vice-President for Research, University of South Carolina. Authors are thankful to Prof. Juan Caicedo, Department of Civil and Environmental Engineering, University of South Carolina, for providing support for experimental studies. Authors are also thankful to Prof. Anthony P. Reynolds, Department of Mechanical Engineering, University of South Carolina, for providing necessary aluminum material for making the device in to the desired shape.

References

1. Caballero D, Sánchez-Dehesa J, Rubio C, Mártinez-Sala R, Sánchez-Pérez JV, Meseguer F, Llinares J (1999) Large two-dimensional sonic band gaps. Phys Rev E 60:R6316–R6319
2. Phani AS, Woodhouse J, Fleck NA (2006) Wave propagation in two-dimensional periodic lattices. J Acoust Soc Am 119:1995–2005
3. Huang HH, Sun CT, Huang GL (2009) On the negative effective mass density in acoustic metamaterials. Int J Eng Sci 47:610–617

4. Huang GL, Sun CT (2010) Band gaps in a multiresonator acoustic metamaterial. J Vib Acoust 132:031003
5. Krynkin A, Umnova O, Chong AYB, Taherzadeh S, Attenborough K (2010) Predictions and measurements of sound transmission through a periodic array of elastic shells in air. J Acoust Soc Am 128:3496–3506
6. Oudich M, Li Y, Assouar BM, Hou Z (2010) A sonic band gap based on the locally resonant phononic plates with stubs. New J Phys 12:083049
7. Hsu J-C (2011) Local resonances-induced low-frequency band gaps in two-dimensional phononic crystal slabs with periodic stepped resonators. J Phys D Appl Phys 44:55401–55409
8. Fan L, Zhang S-Y, Zhang H (2011) Transmission characteristics of double negativity acoustic metamaterials studied with fluid impedance theory. J Acoust Soc Am 129:2483
9. Norris A (2011) Periodic metal structures for acoustic wave control. J Acoust Soc Am 130:2359
10. Chesnais C, Boutin C, Hans S (2012) Effects of the local resonance on the wave propagation in periodic frame structures: generalized Newtonian mechanics. J Acoust Soc Am 132:2873–2886
11. Riaz A, Banerjee S (2013) Wave propagation in metamaterial using multiscale resonators by creating local anisotropy. Int J Mod Eng 13:9
12. Watts L (1992) Cochlear mechanics: analysis and analog VLSI. Ph.D. thesis, California Institute of Technology
13. Bekesy GV (1989) Experiments in hearing. McGraw-Hill, New York
14. Neely ST (1977) Mathematical models of the mechanics of the Cochlea. California Institute of Technology
15. Holmes M (1980) Low frequency asymptotics for a hydroelastic model of the cochlea. SIAM J Appl Math 38:445–456
16. Allen JB, Sondhi MM (1979) Cochlear macromechanics: time domain solutions. J Acoust Soc Am 66:123–132
17. Kolston PJ (1989) Towards a better understanding of cochlear mechanics: a new cochlear model. University of Canterbury
18. Chen F, Cohen HI, Bifano TG, Castle J, Fortin J, Kapusta C, Mountain DC, Zosuls A, Hubbard AE (2006) A hydromechanical biomimetic cochlea: experiments and models. J Acoust Soc Am 119:394–405
19. Tanaka K, Abe M, Ando S (1998) A novel mechanical cochlea “Fishbone” with dual sensor/actuator characteristics. IEEE/ASME Trans Mechatron 3:98–105
20. Wittbrodt MJ, Steele CR, Puria S (2006) Developing a physical model of the human cochlea using micro-fabrication methods. Audiol Neurotol 11:104–112
21. White RD, Grosh K (2005) Microengineered hydromechanical cochlear model. Proc Natl Acad Sci USA 102:1296–1301
22. Shintaku H, Kobayashi T, Zusho K, Kotera H, Kawano S (2013) Wide-range frequency selectivity in an acoustic sensor fabricated using a microbeam array with non-uniform thickness. J Micromech Microeng 23:115014
23. Tanujaya H, Shintaku H, Kitagawa D, Adianto S, Kawano S (2013) Experimental and analytical study approach of Artificial basilar membrane prototype (ABMP). J Eng Technol Sci 45:61–72

Chapter 4
Dynamic Equations for an Isotropic Spherical Shell Using Power Series Method and Surface Differential Operators

Reza Okhovat and Anders Boström

Abstract Dynamic equations for an isotropic spherical shell are derived by using a series expansion technique. The displacement field is split into a scalar (radial) part and a vector (tangential) part. Surface differential operators are introduced to decrease the length of the shell equations. The starting point is a power series expansion of the displacement components in the thickness coordinate relative to the mid-surface of the shell. By using the expansions of the displacement components, the three-dimensional elastodynamic equations yield a set of recursion relations among the expansion functions that can be used to eliminate all but the four of lowest order and to express higher order expansion functions in terms of these of lowest orders. Applying the boundary conditions on the surfaces of the spherical shell and eliminating all but the four lowest order expansion functions give the shell equations as a power series in the shell thickness. After lengthy manipulations, the final four shell equations are obtained in a more compact form which can be represented explicitly in terms of the surface differential operators. The method is believed to be asymptotically correct to any order. The eigenfrequencies are compared to exact three-dimensional theory and membrane theory.

Keywords Spherical shell • Shell equations • Surface differential operators • Dynamic • Eigenfrequency • Power series

4.1 Introduction

Shells are commonly used in many branches of engineering and the demands for shell structures are increasing. Therefore they have been investigated for a number of different types of shells and new methods are developed for designing and using shells in structures. Shells appear in many aspects like designing pressure vessels, fuselages of airplanes, boat and ship hulls, roof structures, bodies of cars, trains, and aeroplanes.

A shell can be considered as a curved plate having small thickness compared to the other geometrical dimensions as well as to the wavelengths of importance. The most important superiority of shells in comparison to plates is that shell structures can provide high strength and low weight because of their membrane stiffness.

Particularly, spherical shells appear in many structures and have many applications. Some dynamic shell theories have thus been developed for this case. All these theories seem to depend on more or less ad hoc kinematical assumptions and/or other approximations.

The literature on shells is large. A comprehensive introduction to shell theory is given by Tuner [1] and shells with general shapes are studied by Heyman [2]. Basic equations for spherical and cylindrical shells are discussed by Niordson [3]. For the present purposes the most relevant references seem to be those of Shah et al. [4, 5] and Niordson [6]. Shah et al. [4] seem to be the first to give the exact three-dimensional solution for the eigenfrequencies of a spherical shell (of arbitrary thickness), drawing on earlier work by Morse and Feshbach [7]. In particular, governing equations for spherical and cylindrical shells are given by Soedel [8], and thin cylindrical shells are studied by Leissa [9].

R. Okhovat • A. Boström
Department of Applied Mechanics, Chalmers University of Technology, SE-41296 Gothenburg, Sweden
e-mail: reza.okhovat@chalmers.se

N. Sottos et al. (eds.), *Experimental and Applied Mechanics, Volume 6: Proceedings of the 2014 Annual Conference on Experimental and Applied Mechanics*, Conference Proceedings of the Society for Experimental Mechanics Series,
DOI 10.1007/978-3-319-06989-0_4,

In this paper the dynamic equations for a spherical shell are derived by using a power series method developed during the last decade for bars, plates, beams and rods [10]. It has, in particular, been developed for a number of different structures, such as anisotropic, layered, and piezoelectric plates, see e.g. [11–13], anisotropic rods [14, 15], piezoelectric layers [16] and porous plates [17]. Dynamic equations for isotropic and anisotropic cylindrical shells are also derived using the power series method [18, 19]. Dynamic equations for an isotropic spherical shell are derived in [20], but the final six equations are very lengthy, even for the second order approximation each equation would cover several pages, so in that paper just the terms that are independent of shell thickness are given. The method is very systematic and very useful for developing the equations to any order. Also the resulting structural equations are believed to be asymptotically correct to any order [11].

The present paper can be described as follows. In the next section the problem is stated and the three-dimensional equations of elasticity are given. The displacement field is split into scalar and vector parts and surface differential operators are introduced. Next, the expansion of the displacement components in series in the thickness coordinate is performed, leading to the recursion relations for the expansion functions. By using the recursion relations, all the higher order expansion functions can be expressed in terms of some of the lowest ones. This is the key ingredient in the present approach. Applying the boundary conditions at the inner and outer surfaces of the spherical shell and using the recursion relations to eliminate all but the four lowest-order expansion functions give the four shell equations. These can in principle be given to any order and are believed to be asymptotically correct to any order. The lengthy analytical calculations are performed using the commercial program Mathematica.

Numerical results for eigenfrequencies are computed by using the power series approximation for different shell thicknesses and for different orders of approximation. Then results are compared with the results given by the exact solution.

In general, computations of the eigenfrequencies from the shell equations are in good correspondence with the results from the exact solution.

4.2 Problem Formulation

Consider a spherical shell with mean radius R and thickness $2h$. The material is assumed to be isotropic and linearly elastic with Lamé constants λ and μ and density ρ. Introduce spherical coordinates r, θ, and φ, where r is the radial coordinate, θ the polar coordinate, and φ the azimuthal coordinate. The main goal is to derive a set of dynamic shell equations for this case, i.e. a set of differential equations that depend on the two angular spherical coordinates and time, but where the radial dependence has disappeared.

The starting point is the three-dimensional dynamic equations of elasticity for the displacement field $\mathbf{u}$

$$(\lambda + 2\mu)\nabla(\nabla \cdot \mathbf{u}) - \mu\nabla \times (\nabla \times \mathbf{u}) = \rho\frac{\partial^2 \mathbf{u}}{\partial^2 t}. \tag{4.1}$$

This equation is written in a way that clearly shows the decoupling into compression and shear waves. In spherical coordinates this equation is written for the radial component w, the polar component u, and the azimuthal component v. Surface differential operators are introduced which decrease the length of the shell equations considerably. First the displacement field is split into a scalar (radial) part and a vector (tangential) part, i.e. w and $\mathbf{S}$, respectively, as

$$\mathbf{u} = w\,\mathbf{e}_r + \mathbf{S}, \tag{4.2}$$

$$\mathbf{S} = u\,\mathbf{e}_\theta + v\,\mathbf{e}_\varphi. \tag{4.3}$$

Now surface operators are defined by considering the equations of motion. There are five differential operators where two of them are related to the scalar part, i.e. w, and three of them are related to the vector part, i.e. $\mathbf{S}$. The two surface operators that apply to the scalar part are

$$\nabla_a w = \mathbf{e}_\theta \partial_\theta w + \frac{\mathbf{e}_\varphi}{\sin\theta}\partial_\varphi w, \tag{4.4}$$

$$\nabla_a^2 w = \frac{1}{\sin\theta}\partial_\theta(\sin\theta\partial_\theta w) + \frac{1}{\sin^2\theta}\partial_\varphi^2 w = \nabla_a \cdot (\nabla_a w). \tag{4.5}$$

Two surface operators that apply to the vector part are

$$\nabla_a \cdot \mathbf{S} = \frac{1}{\sin\theta}\partial_\theta(\sin\theta u) + \frac{1}{\sin\theta}\partial_\varphi v, \tag{4.6}$$

$$\begin{aligned} \nabla_a^2 \mathbf{S} = {} & \frac{\mathbf{e}_\theta}{\sin\theta}\left(\partial_\theta(\sin\theta\partial_\theta u) + \frac{1}{\sin\theta}\partial_\varphi^2 u - \frac{u}{\sin\theta} - 2\cot\theta\partial_\varphi v\right) \\ & + \frac{\mathbf{e}_\varphi}{\sin\theta}\left(\partial_\theta(\sin\theta\partial_\theta v) + \frac{1}{\sin\theta}\partial_\varphi^2 v - \frac{v}{\sin\theta} + 2\cot\theta\partial_\varphi u\right). \end{aligned} \tag{4.7}$$

From the definition of the ∇_a operator and by considering $\nabla_a \cdot \mathbf{S}$ there follows

$$\begin{aligned} \nabla_a(\nabla_a \cdot \mathbf{S}) = {} & \mathbf{e}_\theta\left(\partial_\theta^2 u + \cot\theta\partial_\theta u - \frac{u}{\sin^2\theta} + \frac{1}{\sin\theta}\partial_\theta\partial_\varphi v - \frac{\cos\theta}{\sin^2\theta}\partial_\varphi v\right) \\ & + \frac{\mathbf{e}_\varphi}{\sin\theta}\left(\partial_\theta\partial_\varphi u + \cot\theta\partial_\varphi u + \frac{1}{\sin\theta}\partial_\varphi^2 v\right). \end{aligned} \tag{4.8}$$

Here ∂_r, ∂_θ, and ∂_φ denote partial derivatives with respect to r, θ, and φ, respectively.

By applying the surface differential operators to the three-dimensional dynamic equations of elasticity and splitting the displacement field into scalar and vector parts, the equations of motion can be written

$$(\lambda+2\mu)\left(r^2\partial_r^2 w + 2r\partial_r w - 2w\right) + \mu\nabla_a^2 w + (\lambda+\mu)r\partial_r\nabla_a \cdot \mathbf{S} - (\lambda+3\mu)\nabla_a \cdot \mathbf{S} = r^2\rho\partial_t^2 w, \tag{4.9}$$

$$(\lambda+\mu)\left(\nabla_a(\nabla_a \cdot \mathbf{S}) + (r\partial_r + 2)\nabla_a w\right) + \mu\left(\nabla_a^2\mathbf{S} + (r^2\partial_r^2 + 2r\partial_r)\mathbf{S} + 2\nabla_a w\right) = r^2\rho\partial_t^2\mathbf{S}. \tag{4.10}$$

These equations are to be supplemented with boundary conditions on the inner and outer surfaces of the shell. These are for simplicity taken as vanishing traction on both surfaces, although, e.g., an applied pressure would also be possible.

4.3 The Shell Equations

To derive the shell equations the substitution $r = R + \xi$ is made where the variable ξ is introduced as $-h \le \xi \le h$. It should be noted that h is considered to be small enough, i.e. to be relatively small compared to the radius R and to relevant wavelengths. The displacement components are then expanded in power series in the thickness coordinate ξ as

$$w(r,\theta,\varphi,t) = \sum_{k=0} w_k(\theta,\varphi,t)\xi^k, \tag{4.11}$$

$$\mathbf{S}(r,\theta,\varphi,t) = \sum_{k=0} \mathbf{S}_k(\theta,\varphi,t)\xi^k. \tag{4.12}$$

These sums are in principle infinite but only a few terms are actually used. No truncation is performed at this stage in contrast to most other methods and the truncation scheme is discussed later on. The expansions involve both even and odd powers of ξ as symmetric and antisymmetric motions with respect to ξ do not decouple. The displacement expansions are inserted into the equations of motion and by identifying equal powers of ξ and solving for the highest order expansion functions recursion relations are derived as

$$w_{k+2} = \frac{1}{R^2(k+1)(k+2)(\lambda+2\mu)}\Big(\rho(R^2\partial_t^2 w_k + 2R\partial_t^2 w_{k-1} + \partial_t^2 w_{k-2}) - (\lambda+\mu)(2R(k+1)^2 w_{k+1} + (k+2)(k-1)w_k) - \mu\nabla_a^2 w_k - R(k+1)(\lambda+\mu)\nabla_a\cdot\mathbf{S}_{k+1} - ((k-1)\lambda + (k-3)\mu)\nabla_a\cdot\mathbf{S}_k\Big), \quad (4.13)$$

$$\mathbf{S}_{k+2} = \frac{1}{R^2\mu(k+1)(k+2)}\Big(\rho(R^2\partial_t^2\mathbf{S}_k + 2R\partial_t^2\mathbf{S}_{k-1} + \partial_t^2\mathbf{S}_{k-2}) - \mu(2R(k+1)^2\mathbf{S}_{k+1} + k(k+1)\mathbf{S}_k + \nabla_a^2\mathbf{S}_{k+1}) - (\lambda+\mu)\nabla_a(\nabla_a\cdot\mathbf{S}_k) - (\lambda+\mu)(k+1)R\nabla_a w_{k+1} - ((k+2)\lambda + (k+4)\mu)\nabla_a w_k\Big), \quad (4.14)$$

where $k = 0, 1, \ldots$. The advantage with these recursion relations is that all the higher order expansion functions can be expressed in terms of the four lowest order ones w_0, w_1, $\mathbf{S}_0$, and $\mathbf{S}_1$ by using these equations recursively.

It is noticed that the procedure so far does not depend on any assumption about the thickness of the shell. The recursion relations can be used also to get a good representation of the displacements in the shell for other purposes. To obtain the shell equations the boundary conditions on the shell are now applied. The relevant stress components in spherical coordinates are as follow.

The radial stress, i.e. σ_{rr} is written

$$\sigma_{rr} = (\lambda+2\mu)\partial_r w + \lambda\left(\frac{2w}{r} + \frac{\nabla_a\cdot\mathbf{S}_k}{r}\right), \quad (4.15)$$

and the shear stress, i.e. $\mathbf{t}$ is written

$$\mathbf{t} = \mu\left(\partial_r\mathbf{S} - \frac{\mathbf{S}}{r} + \nabla_a w\right). \quad (4.16)$$

The expansions of the displacement components are inserted into the stress components σ_{rr} and $\mathbf{t}$ which yield

$$\sigma_{rr} = \frac{1}{R+\xi}\sum_{k=0}((\lambda+2\mu)((k+1)Rw_{k+1} + kw_k) + \lambda(2w_k + \nabla_a\cdot\mathbf{S}_k)\ \xi^k, \quad (4.17)$$

$$\mathbf{t} = \frac{\mu}{R+\xi}\sum_{k=0}\Big((k+1)R\mathbf{S}_{k+1} + (k-1)\mathbf{S}_k + \nabla_a w_k\Big)\xi^k. \quad (4.18)$$

The boundary conditions at $\xi = \pm h$ are then applied, usually this means that both two stress components are zero. It is convenient to take the sum and difference between the equations at the two boundaries.

By combining the stresses according to $\sum\sigma = \sigma(h,\theta,t) + \sigma(-h,\theta,t) = 0$ and $\Delta\sigma = \sigma(h,\theta,t) - \sigma(-h,\theta,t) = 0$ and using the expansions of σ_{rr} and $\mathbf{t}$ and the recursion relations, these four boundary conditions can be written in terms of the four displacements w_0, w_1, $\mathbf{S}_0$, and $\mathbf{S}_1$ which deliver four partial differential equations. These are given as an expansion in h, which can, in principle, be given to any order.

Finally, by lengthy manipulations which have been carried out using the commercial program Mathematica, the four shell equations including terms up to h^2 become

$$\begin{aligned}
&2(\lambda+2\mu)w_0-\mu\nabla_a^2w_0+R^2\rho\partial_t^2w_0+R(\lambda-2\mu)w_1+(\lambda+3\mu)\nabla_a\cdot\mathbf{S}_0-R\mu\nabla_a\cdot\mathbf{S}_1\\
&\quad+\frac{h^2}{6R^2}\bigg(32\mu(w_0-Rw_1)-9R\lambda\nabla_a^2w_1+\frac{1}{\lambda+2\mu}\Big(-2(9\lambda^2+34\lambda\mu+40\mu^2)\nabla_a^2w_0\\
&\quad-\lambda\mu\nabla_a^4w_0+2R^2\rho(\lambda+6\mu)\partial_t^2w_0+R^2\rho(\lambda-\mu)\partial_t^2\nabla_a^2w_0+R^3\rho(\lambda-2\mu)\partial_t^2w_1\\
&\quad+16\mu(\lambda+3\mu)\nabla_a\cdot\mathbf{S}_0-(9\lambda^2+33\lambda\mu+32\mu^2)\nabla_a^2\nabla_a\cdot\mathbf{S}_0+3R^2\rho(3\lambda+5\mu)\partial_t^2\nabla_a\cdot\mathbf{S}_0\\
&\quad-4R\mu(9\lambda+14\mu)\nabla_a\cdot\mathbf{S}_1+R\mu(3\lambda+4\mu)\nabla_a^2\nabla_a\cdot\mathbf{S}_1-R^3\rho(2\lambda+3\mu)\partial_t^2\nabla_a\cdot\mathbf{S}_1\Big)\bigg)=0,
\end{aligned}\tag{4.19}$$

$$\begin{aligned}
&2\lambda w_0+R(\lambda+2\mu)w_1+\lambda\nabla_a\cdot\mathbf{S}_0\\
&\quad+\frac{h^2}{2R^2}\bigg(8\mu(Rw_1-w_0)+R\lambda\nabla_a^2w_1+R^3\rho\partial_t^2w_1+(\lambda+2\mu)\nabla_a^2\nabla_a\cdot\mathbf{S}_0\\
&\quad R^2\rho\partial_t^2\nabla_a\cdot\mathbf{S}_0+\frac{1}{\lambda+2\mu}\Big(2(\lambda^2+4\lambda\mu+6\mu^2)\nabla_a^2w_0+2R\rho\lambda\partial_t^2w_0-4\mu(\lambda+3\mu)\nabla_a\cdot\mathbf{S}_0\\
&\quad+4R\mu(2\lambda+3\mu)\nabla_a\cdot\mathbf{S}_1\Big)\bigg)=0,
\end{aligned}\tag{4.20}$$

$$\begin{aligned}
&\mu\nabla_a^2\mathbf{S}_0+(\lambda+\mu)\nabla_a\nabla_a\cdot\mathbf{S}_0-R^2\rho\partial_t^2\mathbf{S}_0+2R\mu\mathbf{S}_1+2(\lambda+2\mu)\nabla_aw_0+R\lambda\nabla_aw_1\\
&\quad+\frac{h^2}{6R^2}\bigg(10\mu\nabla_a^2\mathbf{S}_0-\mu\nabla_a^4\mathbf{S}_0-3(\lambda+\mu)\nabla_a^2\nabla_a\nabla_a\cdot\mathbf{S}_0-4R^2\rho\partial_t^2\mathbf{S}_0+2R^2\rho\partial_t^2\nabla_a^2\mathbf{S}_0\\
&\quad+12R\mu\mathbf{S}_1-6R\mu\nabla_a^2\mathbf{S}_1+2R^3\rho\partial_t^2\mathbf{S}_1+4(3\lambda+11\mu)\nabla_aw_0+2R(3\lambda-7\mu)\nabla_aw_1\\
&\quad+\frac{1}{\lambda+2\mu}\Big(2(3\lambda^2+10\lambda\mu+13\mu^2)\nabla_a\nabla_a\cdot\mathbf{S}_0-2R\mu(6\lambda+7\mu)\nabla_a\nabla_a\cdot\mathbf{S}_1\\
&\quad-2(3\lambda^2+8\lambda\mu+9\mu^2)\nabla_a\nabla_a^2w_0-R(3\lambda^2+8\lambda\mu+4\mu^2)\nabla_a\nabla_a^2w_1\Big)\\
&\quad+\frac{1}{\mu(\lambda+2\mu)}\Big(R^2\rho(\lambda+\mu)(\lambda+4\mu)\partial_t^2\nabla_a\nabla_a\cdot\mathbf{S}_0+2R^2\rho(\lambda^2+3\lambda\mu+5\mu^2)\partial_t^2\nabla_aw_0\\
&\quad+R^3\rho(\lambda^2+4\lambda\mu+2\mu^2)\partial_t^2\nabla_aw_1\Big)-\frac{R^4\rho^2}{\mu}\partial_t^4\mathbf{S}_0\bigg)=0,
\end{aligned}\tag{4.21}$$

$$\begin{aligned}
&-\mathbf{S}_0+R\mathbf{S}_1+\nabla_aw_0\\
&\quad+\frac{h^2}{2R^2}\bigg(3\nabla_a^2\mathbf{S}_0+4R\mathbf{S}_1-R\nabla_a^2\mathbf{S}_1+\frac{1}{\lambda+2\mu}\Big(\lambda\nabla_a\nabla_a^2w_0-2R(\lambda+\mu)\nabla_a\nabla_a\cdot\mathbf{S}_1\Big)\\
&\quad+\frac{1}{\mu}\Big(R^2\rho\partial_t^2(R\mathbf{S}_1-\mathbf{S}_0)+4(\lambda+3\mu)\nabla_aw_0+R(\lambda-\mu)\nabla_aw_1\Big)\\
&\quad+\frac{1}{\mu(\lambda+2\mu)}\Big(2(\lambda^2+3\lambda\mu+3\mu^2)\nabla_a\nabla_a\cdot\mathbf{S}_0-R^2\rho\lambda\partial_t^2\nabla_aw_0\Big)\bigg)=0.
\end{aligned}\tag{4.22}$$

If only the lowest, h-independent, terms are considered, from the second and fourth equation it is of course straightforward to eliminate w_1 and $\mathbf{S}_1$ and this leads to the classical membrane equations. Including also the quadratic terms in h, a type of bending theory is obtained, and it is noted this theory does not include any shear correction factor and that it is believed to be asymptotically correct. It should be noted that with increasing orders, the shell equations become extremely complex and if surface operators are not used, just the h^2 terms of the shell equations will cover several pages. But by using the introduced surface operators, this complexity is tackled and the shell equations can be represented in an explicit and more compact form.

4.4 Results

To validate the shell equations, results from the approximation method are compared with results from the exact solution. First the spherically symmetric vibrations is considered which means that only the radial component w is nonzero. As a second case, the torsional modes are investigated for which only v is nonzero and independent of the azimuthal coordinate (called modes of the first class by Shah et al. [4]). Results from different cases are compared with results from the exact solution (Shah et al. [4]).

For the first case which is the spherically symmetric vibrations the shell equations simplify to

$$\begin{aligned} &2(\lambda+2\mu)w_0+(\lambda-2\mu)Rw_1+\rho\partial_t^2 w_0+\frac{h^2}{6R^2(\lambda+2\mu)}\Big(32\mu(\lambda+2\mu)(w_0-Rw_1) \\ &+2\rho R^2(\lambda+6\mu)\partial_t^2 w_0+\rho R^3(\lambda-2\mu)\partial_t^2 w_1+\rho^2R^2\partial_t^4 w_0\Big)=0, \end{aligned} \tag{4.23}$$

$$\begin{aligned} &2\lambda w_0+(\lambda+2\mu)Rw_1+\frac{h^2}{2R^2(\lambda+2\mu)}\Big(8\mu(\lambda+2\mu)(Rw_1-w_0) \\ &+\rho R^2\lambda\partial_t^2 w_0+\rho R^3(\lambda+2\mu)\partial_t^2 w_1\Big)=0. \end{aligned} \tag{4.24}$$

With a harmonic time dependence these equations give the lowest eigenfrequency expanded to second order in the shell thickness

$$(k_sR)^2=\frac{4(1+\nu)}{1-\nu}+\frac{4h^2(1+\nu)(-1+9\nu)}{3R^2(1-\nu)^2}, \tag{4.25}$$

where ν is the Poisson's ratio and the wave number $k_s=\omega\sqrt{\rho/\mu}$. This result agrees exactly with that of Niordson [6].

Table 4.1 presents numerical values of the results for the spherically symmetric case. The three first eigenfrequencies are computed for different ratio of shell thickness to the shell mean radius, i.e. $h/R=0.01$, $h/R=0.1$, $h/R=0.25$. The eigenfrequency is normalized so that $\Omega=\omega h/(\pi c_s)$, similarly to Shah et al. [4].

First, the eigenfrequencies are calculated from the exact solution. Then, the results from the second order approximation according to Eqs. (4.1) and (4.2) are given. It is seen that the shell equations up to and including h^2 terms give values for the first eigenfrequencies exactly the same as those from the exact solution for low ratios of the shell thickness. For higher ratios,

Table 4.1 The three first eigenfrequencies for radial modes for exact and power series theory for different powers of h and for $h/R=0.01$, $h/R=0.1$ and $h/R=0.25$

h/R	Mode	Exact	h^2	h^4	h^6	h^8	h^{10}
$h/R=0.01$	Ω_0	0.00868	0.00868	0.00868	0.00868	0.00868	0.00868
	Ω_1	0.93546	0.84222	0.94836	0.93493	0.93547	0.93546
	Ω_2	1.87085	1.45871	1.83202	1.83336	1.87508	1.87059
$h/R=0.1$	Ω_0	0.08710	0.08710	0.08710	0.08710	0.08710	0.08710
	Ω_1	0.93981	0.84697	0.95269	0.93928	0.93982	0.93981
	Ω_2	1.87302	1.46143	1.83415	1.83555	1.87725	1.87275
$h/R=0.25$	Ω_0	0.22209	0.22170	0.22206	0.22209	0.22209	0.22209
	Ω_1	0.96507	0.87220	0.97890	0.96447	0.96509	0.96507
	Ω_2	1.88538	1.47495	1.84353	1.84732	1.88975	1.88510

Table 4.2 The three first eigenfrequencies for torsional modes for exact and power series theory for different powers of h and for $h/R = 0.01$, $h/R = 0.1$ and $h/R = 0.25$

h/R	Mode	Exact	h^2	h^4	h^6	h^8	h^{10}
$h/R = 0.01$	Ω_0	0.00637	0.00637	0.00637	0.00637	0.00637	0.00637
	Ω_1	0.50010	0.45027	0.50699	0.49982	0.50011	0.50010
	Ω_2	1.00005	0.77976	0.97929	0.98001	1.00230	0.99991
$h/R = 0.1$	Ω_0	0.06314	0.06314	0.06314	0.06314	0.06314	0.06314
	Ω_1	0.51018	0.46137	0.51700	0.50990	0.51019	0.51018
	Ω_2	1.00511	0.78615	0.98432	0.98513	1.00737	1.00497
$h/R = 0.25$	Ω_0	0.15109	0.15108	0.15111	0.15109	0.15109	0.15109
	Ω_1	0.56485	0.51883	0.57285	0.56446	0.56487	0.56485
	Ω_2	1.03353	0.81886	1.00708	1.01260	1.03604	1.03337

for example for $h/R = 0.25$, the second order approximation still gives results for the first eigenfrequency very close to the values from the exact theory. For the second radial eigenfrequency, if terms with second power of h are used, it will not give acceptable results and they are not as exact as those for the first eigenfrequency Ω_0. This is even worse for the third eigenfrequency Ω_2. Second order theory approximates the results from the exact theory with an error about − 22 %, which is not good enough.

By considering the values for the three first eigenfrequencies for different ratios of h/R in both the approximate theory and the exact theory, it is seen that the first eigenfrequency increases considerably by increasing the ratio of h/R. It can even be approximated as a linear relation between the first radial eigenfrequency and the ratio h/R, which can be roughly estimated as $\Omega_{0,i} = 0.87h/R$, i = 0. 01, 0. 1, 0. 25. On the other hand, the second and third eigenfrequencies do not change much by increasing the h/R ratio.

Numerical results for the eigenfrequencies are given in the rest part of the table by using higher powers of h in the approximation theory, up to and including the h^{10} terms. The second and the third eigenfrequencies are not so dependant on the shell thickness ratio, but these two eigenfrequencies tend to change by increasing the power of h, specially by going to h^4 from h^2. Finally, by comparing the values for all the eigenfrequencies and by going to higher orders of h, the numerical computations obtained from the approximation theory converge very well to the values from the exact solution.

The second case to investigate is the torsional modes for which only v is nonzero and independent of the azimuthal coordinate φ (called modes of the first class by Shah et al. [4]). The numerical values for the eigenfrequencies for the torsional case for the approximate method for different power of h and for the exact solution are presented in Table 4.2.

It is seen in Table 4.2 that torsional eigenfrequencies are in general lower than their radial counterparts. So, as expected, torsional vibration rather than radial vibration are easier to excite at low frequencies. Considering the numerical values in Table 4.2, it is clear that the first eigenfrequencies change roughly linearly with respect to the shell thickness, and can be estimated like $\Omega_{0,i} = 0.63h/R$, i = 0. 01, 0. 1, 0. 25. It is also shown that the first eigenfrequencies do not alter at all by going to higher orders of h and they are exactly the same as the values for the exact solution.

For the two next eigenfrequencies, it is seen that the second order approximation does not give acceptable results for the second torsional modes and it can not approximate the third eigenfrequencies as good as the first one. The error in the later case is about − 22 % which is not good enough. By going to higher power of h, for example h^4, the results are in a very good agreement with the exact theory.

It is concluded from the above that there is very good agreement between the results for the first eigenfrequency from the exact solution and the power series approximation for both radial and torsional modes. But for the second and third eigenfrequencies, the approximate method presents frequencies too low in comparison with the exact solution. By using higher order terms in the differential equations the results become much more accurate.

4.5 Conclusions

In the present work, dynamic equations for a spherical shell are derived. The power series method is used to obtain the shell equations as a power series in the shell thickness. The four partial differential equations can in principle be truncated to any order and are believed to be asymptotically correct. Surface differential operators are used to decrease the length of these shell equations.

First the spherically symmetric eigenfrequencies are computed and results are given for different truncations ranging from membrane (h-independent) to h^{10} terms. These eigenfrequencies are then compared with results from exact theory. It is seen that the power series method gives satisfactory results which are in good agreement with results from exact theory and by including higher powers of thickness, the power series theory converges to the results from exact theory.

Then the torsional eigenfrequencies are computed and results are given for the same truncations. It is shown again that the power series theory approximates the results from exact theory very well.

The shell equations obtained from the power series approximation can be derived also for more complicated cases like general anisotropy, piezoelectricity, and layered shells.

Acknowledgements The present project is funded by the Swedish Research Council and this is gratefully acknowledged.

References

1. Tuner CE (1965) Plate and shell theory. Longmans, London
2. Heyman J (1977) Equilibrium of shell structures. Oxford University Press, Oxford
3. Niordson FI (1985) Shell theory. Elsevier Science, New York
4. Shah AH, Ramkrishnan CV, Datta SK (1969) Three-dimensional and shell-theory analysis of elastic waves in a hollow sphere. part I: analytical foundation. J Appl Mech 36:431–439
5. Shah AH, Ramkrishnan CV, Datta SK (1969) Three-dimensional and shell-theory analysis of elastic waves in a hollow sphere. part II: numerical results. J Appl Mech 36:440–444
6. Niordson FI (2001) An asymptotic theory for spherical shells. Int J Solids Struct 38:8375–8388
7. Morse PM, Feshbach H (1953) Methods of theoretical physics. McGraw-Hill, New York
8. Soedel W (1993) Vibrations of shells and plates. Marcel Dekker, New York
9. Leissa AW (1993) Vibration of shells. Acoustical Society of America, New York
10. Boström A (2000) On wave equations for elastic rods. Z Angew Math Mech 80:245–251
11. Boström A, Johansson J, Olsson P (2001) On the rational derivation of a hierarchy of dynamic equations for a homogeneous, isotropic, elastic plate. Int J Solids Struct 38:2487–2501
12. Mauritsson K, Folkow PD, Boström A (2011) Dynamic equations for a fully anisotropic elastic plate. J Sound Vib 330:2640–2654
13. Mauritsson K, Boström A, Folkow PD (2008) Modelling of thin piezoelectric layers on plates. Wave Motion 45:616–628
14. Martin PA (2005) On flexural waves in cylindrically anisotropic elastic rods. Int J Solids Struct 42:2161–2179
15. Martin PA (2004) Waves in woods: axisymmetric waves in slender solids of revolution. Wave Motion 40:387–398
16. Johanson G, Niklasson AJ (2003) Approximation dynamic boundary conditions for a thin piezoelectric layer. Int J Solids Struct 40:3477–3492
17. Folkow PD, Johanson M (2009) Dynamic equations for fluid-loaded porous plates using approximate boundary conditions. J Acoust Soc Am 125:2954–2966
18. Okhovat R, Boström A (2013) Dynamic equations for an anisotropic cylindrical shell using power series method. In: Topics in modal analysis, vol 7. The society for experimental mechanics. Springer, New York
19. Hägglund AM, Folkow PD (2008) Dynamic cylindrical shell equations by power series expansions. Int J Solids Struct 45:4509–4522
20. Okhovat R, Boström A (2012) Dynamic equations for an isotropic spherical shell using power series method. In: Eleventh international conference on computational structural technology. Civil-Comp Press

Chapter 5
Hydrogen Embrittlement and "Cold Fusion" Effects in Palladium During Electrolysis Experiments

A. Carpinteri, O. Borla, A. Goi, S. Guastella, A. Manuello, and D. Veneziano

Abstract Recent experiments provided evidence of piezonuclear reactions occurring in condensed matter during fracture of solids, cavitation of liquids, and electrolysis. These experiments were characterized by significant neutron and alpha particle emissions, together with appreciable variations in the chemical composition. A mechanical reason for the so-called Cold Nuclear Fusion was recently proposed by the authors. The hydrogen embrittlement due to H atoms produced by the electrolysis plays an essential role for the observed microcracking in the electrode host metals (Pd, Ni, Fe, etc.). Consequently, our hypothesis is that piezonuclear fission reactions may occur in correspondence to the microcrack formation. In order to confirm the first results obtained by Co-Cr and Ni-Fe electrodes, electrolytic tests have been conducted using 100 % Pd at the cathode. As a result, relevant compositional changes and traces of elements previously absent have been observed on the Pd and Ni-Fe electrodes after the experiments and significant neutron emissions were observed during the test.

Keywords Hydrogen embrittlement • Cold fusion • Electrolysis • Piezonuclear reactions • Neutron emission

5.1 Introduction

Several evidences of anomalous nuclear reactions occurring in condensed matter were observed by different authors [1–34]. These experiments are characterized by extra heat generation, neutron emission, and alpha particle detection. Some of these studies, using electrolytic devices, reported also significant evidences of compositional variation after the microcraking of the electrodes [35–40].

In most of these experiments, the generated heat was calculated to be several times the input energy, and the neutron emission rate, during electrolysis, was measured to be about three times the natural background level [6]. In 1998, Mizuno presented the results of the measurements conducted by means of neutron detectors and compositional analysis techniques related to different electrolytic experiments. According to Preparata, "despite the great amount of experimental results observed by a large number of scientists, a unified interpretation and theory of these phenomena has not been accepted and their comprehension still remains unsolved" [6–9, 26, 27]. On the other hand, as shown by most of the articles devoted to Cold Nuclear Fusion, one of the principal features is the appearance of microcracks on electrode surfaces after the tests [26, 27]. Such evidence might be directly correlated to hydrogen embrittlement of the material composing the metal electrodes (Pd, Ni, Fe, Ti, etc.). This phenomenon, well-known in Metallurgy and Fracture Mechanics, characterizes metals during forming or finishing operations [41]. In the present study, the host metal matrix (Pd and Ni) is subjected to mechanical

A. Carpinteri (✉) • O. Borla • A. Manuello • D. Veneziano
Department of Structural, Geotechnical and Building Engineering, Politecnico di Torino, Corso Duca degli Abruzzi 24, 10129 Torino, Italy
e-mail: alberto.carpinteri@polito.it

A. Goi
Private Consultant

S. Guastella
Department of Applied Science and Technology, Geotechnical and Building Engineering, Politecnico di Torino, Corso Duca degli Abruzzi 24, 10129 Torino, Italy

N. Sottos et al. (eds.), *Experimental and Applied Mechanics, Volume 6: Proceedings of the 2014 Annual Conference on Experimental and Applied Mechanics*, Conference Proceedings of the Society for Experimental Mechanics Series, DOI 10.1007/978-3-319-06989-0_5, © The Society for Experimental Mechanics, Inc. 2015 damaging and fracturing due to hydrogen atoms penetrating into the lattice structure and forcing it, during the gas loading. Hydrogen effects are largely studied especially in metal alloys where the hydrogen absorption is particularly high. The hydrogen atoms generate an internal stress that lowers the fracture resistance of the metal, so that brittle crack growth can occur with a hydrogen partial pressure below 1 atm [41, 42]. Some experimental evidence shows that neutron emissions may be strictly correlated to fracture of non-radioactive or inert materials. From this point of view, anomalous nuclear emissions and heat generation had been verified during fracture in fissile materials [2–4] and in deuterated solids [5, 8, 30]. The experiments recently carried out by Carpinteri et al. [36–39] represent the first evidence of neutron emissions from piezonuclear fissions occurring during the failure of inert, stable, and non-radioactive solids under compression, as well as from non-radioactive liquids under ultrasound cavitation [37, 38]. In the present paper, we analyze neutron and alpha particle emissions during tests conducted on an electrolytic cell, where the electrolysis is obtained using Ni-Fe and Pd electrodes in aqueous potassium carbonate solution. Voltage, current intensity, solution conductivity, temperature, alpha and neutron emissions were monitored. The compositions of the electrodes were analyzed both before and after the tests. Strong evidences suggest that the so-called Cold Nuclear Fusion, interpreted under the light of hydrogen embrittlement, may be explained by piezonuclear fission reactions occurring in the host metal, rather than by nuclear fusions of hydrogen isotopes adsorbed in the lattice. These results give an important confirmation about the hypothesis proposed by the authors and reported in previous papers on electrolysis with Ni-Fe and Cr-Co electrodes [43, 44].

5.2 Experimental Set Up

Over the last 10 years, specific experiments have been conducted on an electrolytic reactor (owners: Mr. A. Goi et al.). The aim was to investigate whether the anomalous heat generation may be correlated to a new type of nuclear reactions during electrolysis phenomena. The reactor was built in order to be appropriately filled with a salt solution of water and Potassium Carbonate (K_2CO_3). The electrolytic phenomenon was obtained using two metal electrodes immersed in the aqueous solution. The solution container, named also reaction chamber in the following, is a cylinder-shaped element of 100 mm diameter, 150 mm high, and 5 mm thick. For the reaction chamber Inox AISI 316 L steel was used. The two metallic electrodes were connected to a source of direct current: a Ni-Fe-based electrode as the positive pole (anode), and a Pd-based electrode as the negative pole (cathode).

The reaction chamber base consists of a ceramic plate preventing the direct contact between liquid solution and Teflon lid. Two threaded holes host the electrodes, which are screwed to the bottom of the chamber successively filled with the solution. A valve at the top of the cell allows the vapor to escape from the reactor and condense in an external collector. Externally, two circular Inox steel flanges, fastened by means of four threaded ties, hold the Teflon layers. The inferior steel flange of the reactor is connected to four supports isolated from the ground by means of rubber-based material. As mentioned before, a direct current passes through the anode and the cathode electrodes, provided by a power circuit connected to the power grid through an electric socket. The components of the circuit are an isolating transformer, an electronic variable transformer (Variac), and a diode bridge linked in series.

Electric current and voltage probes were positioned in different parts of the circuit. The voltage measurements were performed by a differential voltage probe of 100 MHz with a maximum rated voltage of 1,400 V. The current was measured by a Fluke I 310S probe with a maximum rated current of 30 A. Current intensity and voltage measurements were also taken by means of a multimeter positioned at the input line. From the turning on to the switching off of the electrolytic cell, current and voltage were found to vary in a range from 3 to 5 A and from 20 to 120 V, respectively.

For an accurate neutron emission evaluation, a He^3 proportional counter was employed (Xeram, France) with pre-amplification, amplification, and discrimination electronics directly connected to the detector tube. The detector is supplied by a high voltage power (about 1.3 kV) via NIM (Nuclear Instrument Module). The logic output producing the TTL (transistor–transistor logic) pulses is connected to a NIM counter. The logic output of the detector is enabled for analog signals exceeding 300 mV. This discrimination threshold is a consequence of the sensitivity of the He^3 detector to the gamma rays ensuing neutron emission in ordinary nuclear processes. This value has been determined by measuring the analog signal of the detector by means of a Co-60 gamma source. The detector is also calibrated at the factory for the measurement of thermal neutrons; its sensitivity is 65 cps/$n_{thermal}$ (± 10 % declared by the factory), i.e., the flux of thermal neutrons is one thermal neutron/s cm^2, corresponding to a count rate of 65 cps. Finally, before and after the experiments, Energy Dispersive X-ray Spectroscopy has been performed in order to recognise possible direct evidence of piezonuclear reactions that can take place during the electrolysis. The elemental analyses were performed by a ZEISS Auriga field emission scanning electron microscope (FESEM) equipped with an Oxford INCA energy-dispersive X-ray detector (EDX) with a resolution of 124 eV @ MnKa. The energy used for the analyses was 18 KeV.

5.3 Neutron Emission Measurements

Neutron emission measurements performed during the experimental activity are represented in Fig. 5.1. The measurements performed by the He^3 detector were conducted for a total time of about 24 h. The background level was measured for different time intervals before and after switching on the reaction chamber. These measurements reported an average neutron background of about $(3.23 \pm 1.49) \times 10^{-2}$ cps. Furthermore, when the reaction chamber is active, it is possible to observe that, after a time interval of about 7.5 h (460 min), neutron emissions of about 3 times the background level may be detected. After 9 h (545 min) from the beginning of the measurements, it is possible to observe a neutron emission level of about one order of magnitude greater than the background level. Similar results were observed after 20 h (1,200 min) when neutron emissions up to 7 times the background were measured.

5.4 Compositional Analysis of the Pd Electrode

In the present section, the chemical compositions before and after the experiment will be taken into account, as well as the concentrations measured for each element identified on the surfaces of the two electrodes (see Tables 5.1 and 5.2).

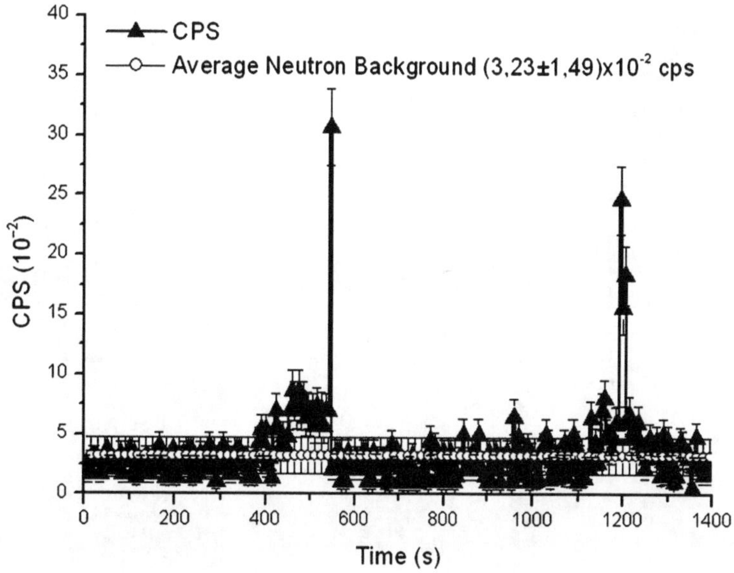

Fig. 5.1 Neutron emission measurements. Emissions between 3 times and 10 times the background level have been observed during the experiment

Table 5.1 Element concentrations before and after the electrolysis (Pd electrode)

Element	Average concentration: Before the experiment (%)	Average concentration: After 20 h (%)	Average variation (Negative for decrease) (%)
O	0.0	18.6	+18.5
Na	0.0	0.2	+0.2
Mg	0.0	1.0	+1.0
Al	0.0	0.4	+0.4
Si	0.0	1.1	+1.1
K	0.0	1.5	+1.5
Ca	0.0	0.2	+0.2
Cr	0.0	0.5	+0.5
Fe	0.0	2.0	+2.0
Co	0.0	0.2	+0.2
Cu	0.0	3.0	+3.0
Br	0.0	0.0	+0.0
Pd	100.0	71.3	−28.6

Table 5.2 Element concentrations before and after the electrolysis (Ni-Fe electrode)

	Average concentration		Average variation
Element	Before the experiment (%)	After 20 h (%)	(Negative for decrease) (%)
C	0.0	0.8	+0.8
O	2.0	21.5	+19.5
Al	0.0	1.8	+1.8
Si	0.3	1.1	+0.8
K	0.0	2.7	+2.7
Ti	3.4	3.0	−0.4
Cr	0.2	0.1	−0.2
Mn	0.0	0.1	+0.1
Fe	2.4	0.4	−2.0
Ni	91.6	68.5	−23.1

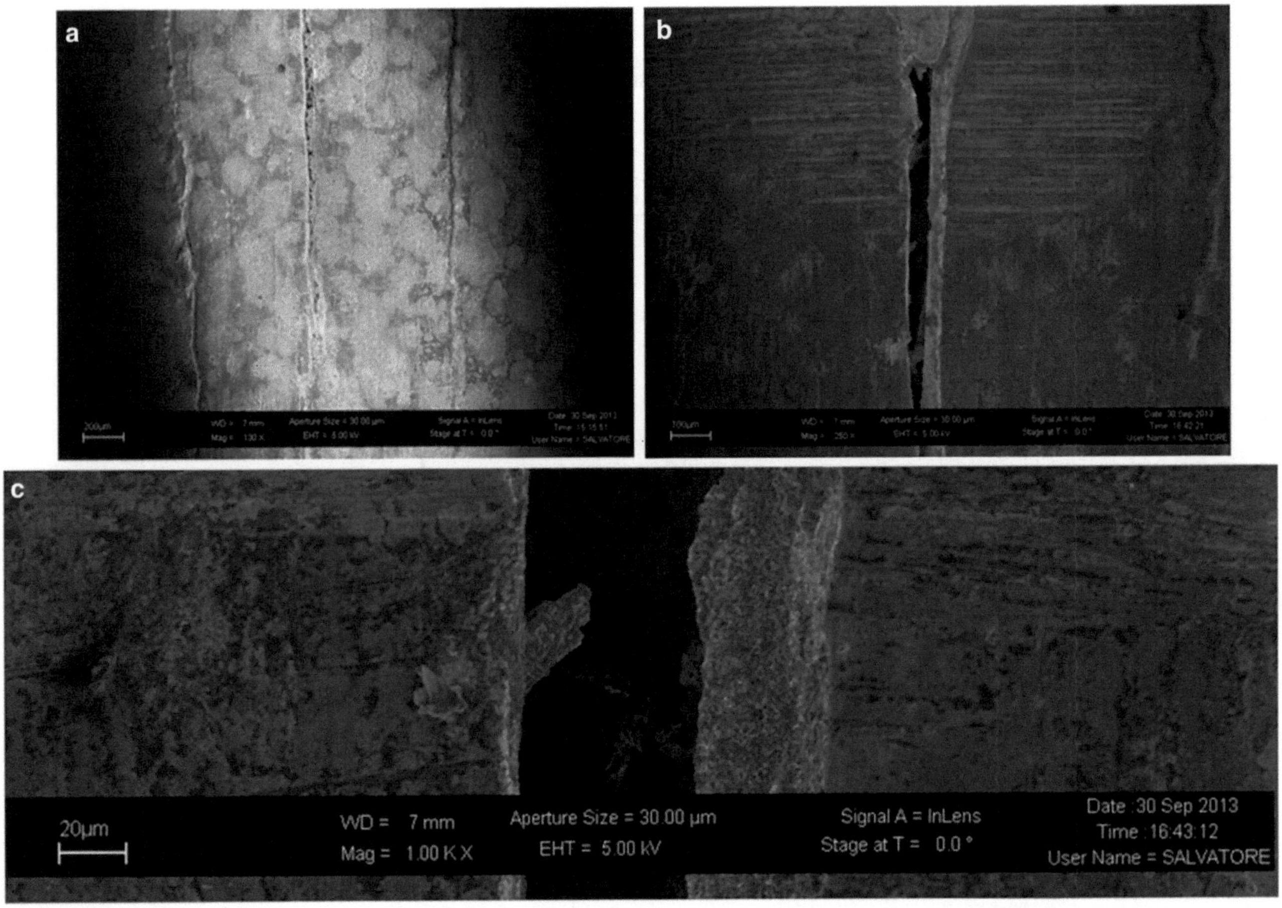

Fig. 5.2 Images (**a, b, c**) of the Palladium electrode surface: the fracture presented a width of about 40 μm and was observable at naked eyes

In particular, under the light of what can be deduced from the neutron emissions measurements and according to the hydrogen embrittlement hypothesis suggested by Carpinteri et al. [43, 44], the presence of microcracks and macrocracks on the electrode surface (Fig. 5.2a, b, c) is accounted in the mechanical interpretation of the phenomena. These evidences are particularly strong in the case of the Pd electrode, where a macroscopic fracture took place during the test. The fracture presented a width of about 40 μm observable at naked eyes (see Fig. 5.2c).

Considering the average decrement of Pd (−28.6 %), reported in Table 5.1, a first and fundamental fission reaction can be assumed:

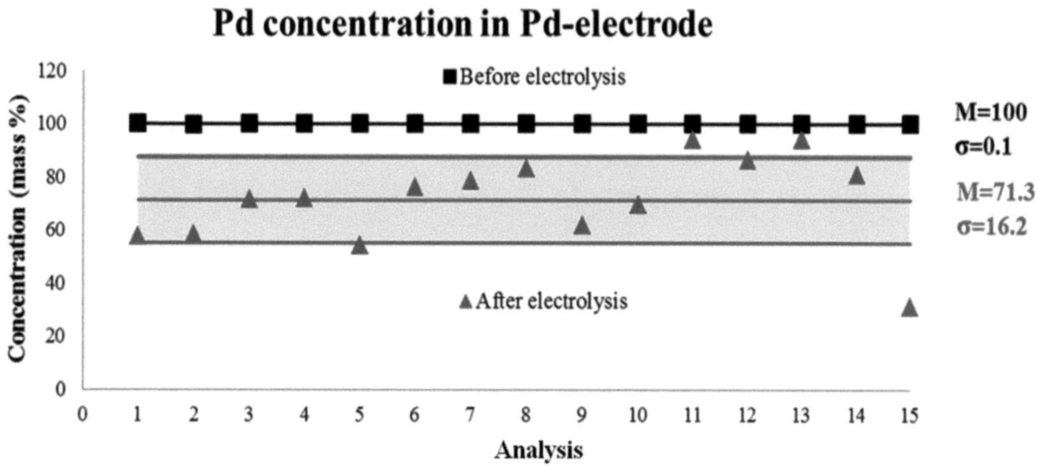

Fig. 5.3 Pd concentrations measured on 15 different points of the electrode surface, before (*square*) and after (*triangle*) the electrolysis; the average concentration M and the corresponding stand. dev. σ are reported

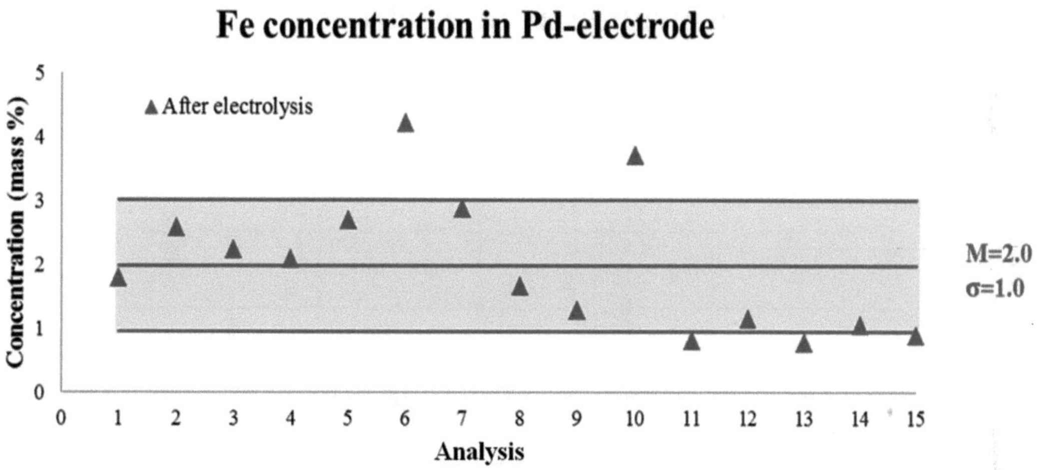

Fig. 5.4 Fe concentrations measured on the electrode surface after the electrolysis; Iron has not been detected before the experiment

$$Pd_{46}^{106} \rightarrow Ca_{20}^{40} + Fe_{26}^{56} + 10 \text{ neutrons} \quad (5.1)$$

According to reaction (5.1), the Pd decrement is counterbalanced by Ca and Fe increments by the following quantities: 10.8 and 15.1 %, for Ca and Fe, respectively. These variations may be accompanied by a neutron emissions corresponding to the remaining 2.7 % of the mass concentration (see also Figs. 5.3, 5.4, 5.5, and 5.6). The whole iron increment, according reaction (5.1), could be entirely considered as the starting element for the production of other elements. Hence, a second hypothesis can be proposed involving Fe as starting element and O as the product, together with alpha and neutron emissions:

$$Fe_{26}^{56} \rightarrow 3O_{8}^{16} + He_{2}^{4} + 4 \text{ neutrons} \quad (5.2)$$

According to reaction (5.2), the iron depletion produces 12.9 % of oxygen with alpha particles (He) and neutron emissions. The total measured increment in oxygen after the experimental test is equal to 18.5 % (see Table 5.1). This quantity seems to be only partially explained by reaction (5.2). The remaining 5.6 % of O concentration could be explained considering other reactions involving Ca (product in reaction 5.1) as the starting element:

$$Ca_{20}^{40} \rightarrow O_{8}^{16} + Mg_{12}^{24} \quad (5.3)$$

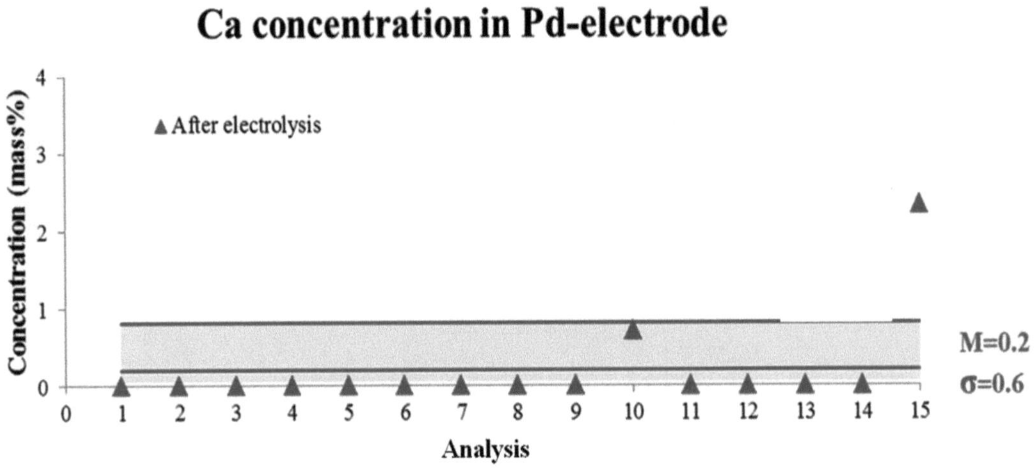

Fig. 5.5 Traces of Ca concentrations have been detected after the electrolysis

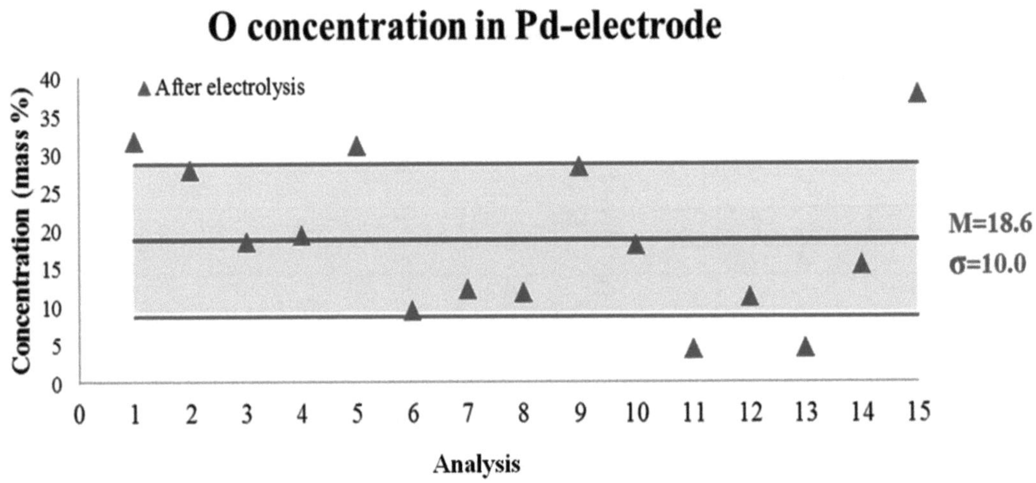

Fig. 5.6 The presence of the oxygen is remarkable after the experiments

$$Ca_{20}^{40} \rightarrow 2O_8^{16} + 4H_1^1 + 4 \text{ neutrons} \tag{5.4}$$

From reaction (5.3), we can consider a decrease in Ca concentration of 1.6 % and the formation of 1.0 % of Mg and 0.6 % of O. On the other hand, from reaction (5.4), we obtain a decrease in Ca equal to 5.9 %. This decrease gives an increase of 4.7 % in O and 0.6 in H, with neutron emissions equal to 0.6 %.

Considering the O increases coming from reactions (5.2), (5.3), and (5.4), totally equal to 18.2 %, and the experimental evidence reporting a total measured O concentration of 18.5 %, we may assume that O seems to be almost perfectly justified by the proposed reactions. At the same time, the Mg increment observed after the experiment can be explained by the results of reaction (5.3), see also Table 5.1. According to reactions (5.3) and (5.4), the following balances may be considered: Ca (−1.6 %) = Mg (+1.0 %) + O (+0.6 %); Ca (−5.9 %) = O (+4.7 %) + H (+0.6 %) + neutrons (+0.6 %). Taking into account the same reactions, and considering the Ca increase coming from reaction (5.1) (10.8 %), a concentration of 3.3 % of Ca remains to be counterbalanced. To this purpose, it is possible to take into account additional reactions involving Ca as starting element and Si, K and C as the results:

$$Ca_{20}^{40} \rightarrow K_{19}^{39} + H_1^1, \tag{5.5}$$

$$Ca_{20}^{40} \rightarrow C_6^{12} + Si_{14}^{28}. \tag{5.6}$$

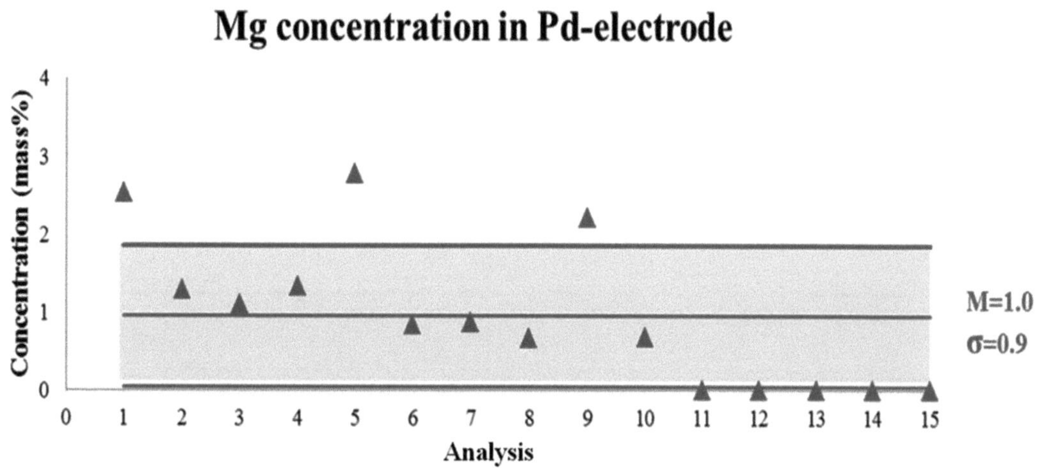

Fig. 5.7 The magnesium presence is evident after the experiment, while there was no trace of it before

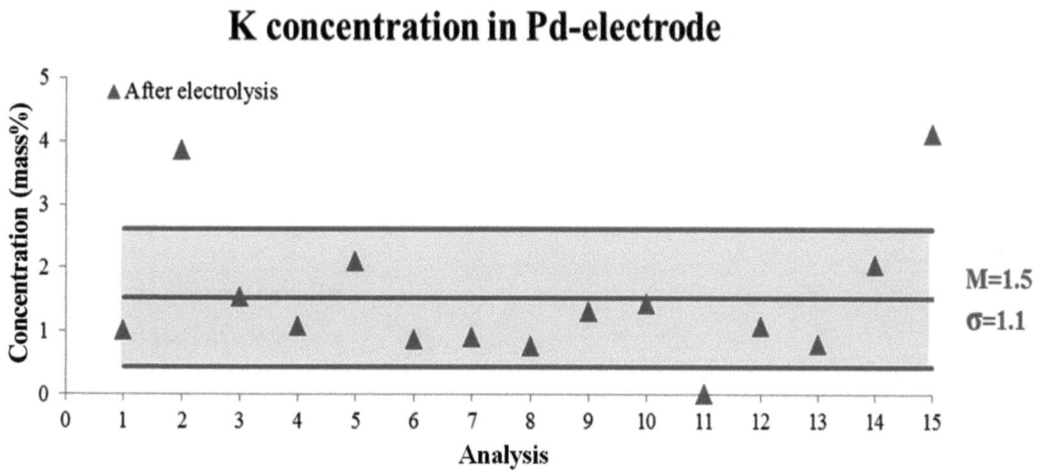

Fig. 5.8 Potassium evidences have been detected only after the experiment

From these reactions, the following balances may be considered: Ca (−1.6 %) = Si (+1.1 %) + C (+0.5 %); Ca (−1.5 %) = K (+1.5 %). Finally, considering the formation of 0.2 % of Ca after the experiment, the total Ca increase assumed by reaction (5.1) may be perfectly matched. The total Ca increment is equal to 10.8 %, the total Ca decrement by reactions (5.3), (5.4), (5.5), and (5.6) is 10.6 %, which returns a 0.2 % Ca concentration as measured. With regards to iron, its experimental compositional increase, measured on the Pd electrode, could be partially interpreted as an electrochemical effect due to the Fe decreasing in the Ni electrode. Finally, as for the final average variations in Mg (1.0 %), Si (1.1 %), and K (1.5 %), the hypotheses are perfectly consistent with the measured values (see Table 5.1 and Figs. 5.7, 5.8, and 5.9).

5.5 Nickel Electrode Composition Analysis

Let us consider the Nickel Electrode. Table 5.2 summarizes the concentration variations after the electrolysis. Nickel diminishes of 23.1 %, whereas the most apparent positive variation is that of oxygen (+19.5 %). It is worth noticing that the average concentration decrease in Fe (−2.0 %) is comparable to the average increase in Aluminum (+1.8 %). The Figs. 5.10, 5.11, 5.12, and 5.13 show the series of the concentrations measured before and after the experiment. In particular, when an element average concentration is zero, the corresponding series is absent: for instance, on the Ni electrode surfaces the Aluminum has not been detected before the electrolysis (Fig. 5.13).

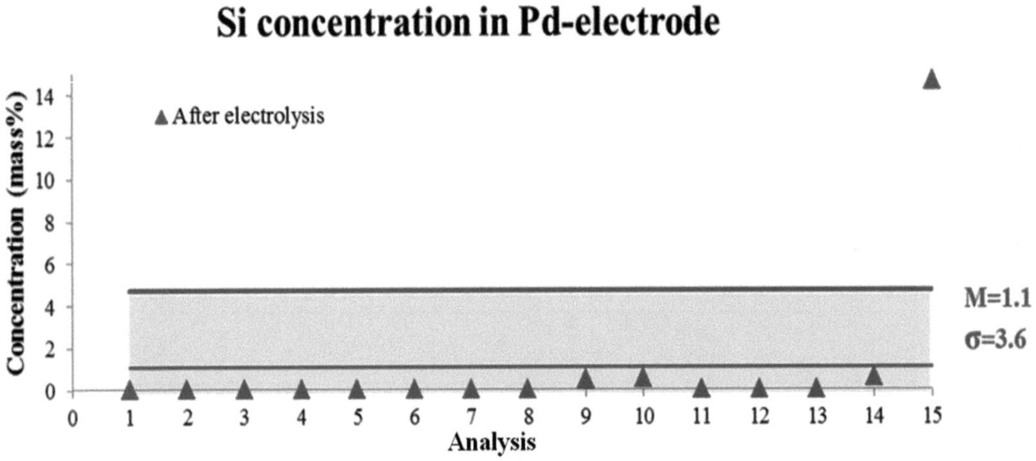

Fig. 5.9 EDS measures show evident traces of silicon only after electrolysis

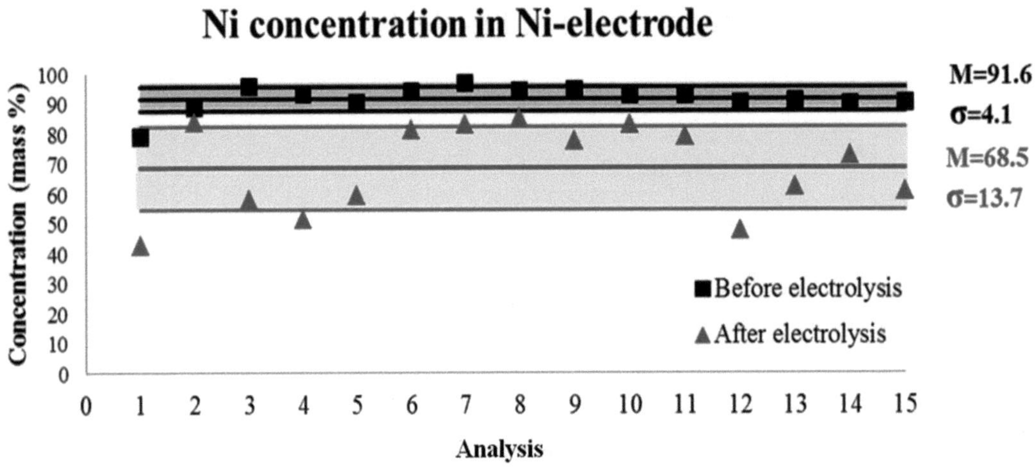

Fig. 5.10 Nickel electrode: the series of the Ni concentrations measured before and after 20 h of electrolysis are showed. An average variation between the two series of about 23.1 % may be considered

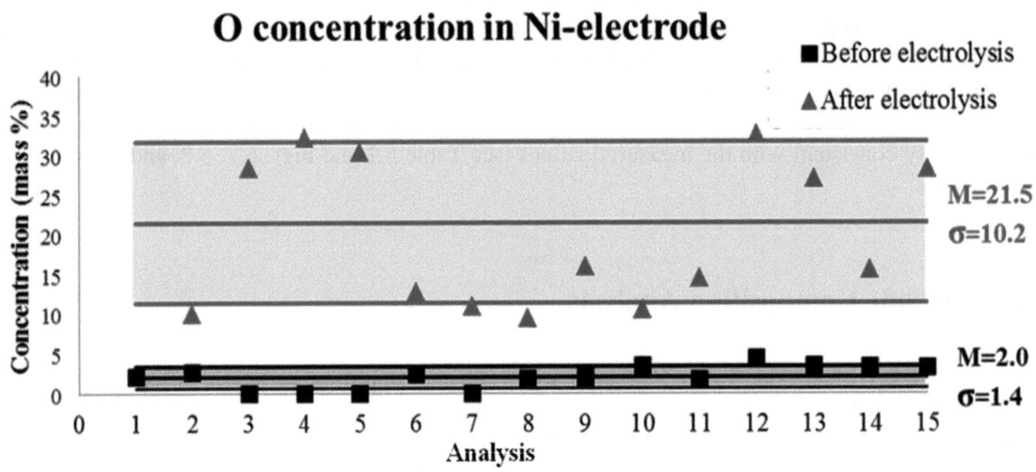

Fig. 5.11 O concentration before and after the experiment. The average values of O concentration change from a mass percentage of 2.0 % at the beginning of the experiment to 21.5 %

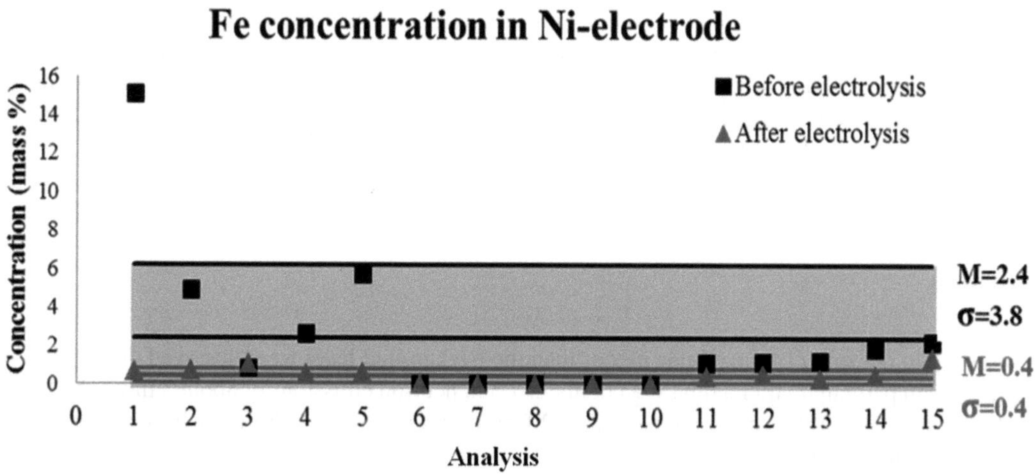

Fig. 5.12 Fe concentration before and after the experiment. The average values of Fe concentration change from a mass percentage of 0.4 % at the beginning of the experiment to 2.4 %

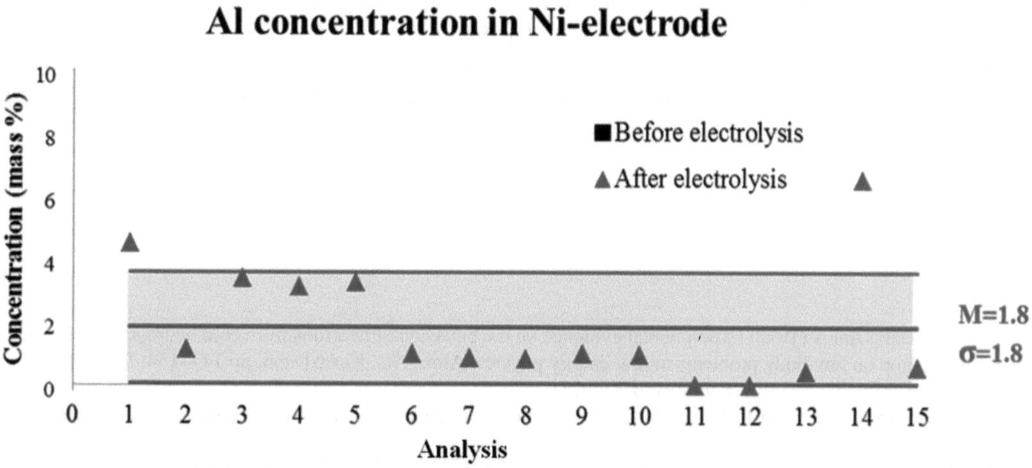

Fig. 5.13 Al concentration before and after the experiment. The average values of Al concentration (1.8 %) appear at the end of the test

On the basis of the piezonuclear reaction conjecture, we could assume the oxygen average variation as a nuclear effect caused by the following hypothesis:

$$\mathrm{Ni}_{28}^{59} \rightarrow 3\mathrm{O}_8^{16} + 2\mathrm{He}_2^4 + 3 \text{ neutrons}. \tag{5.7}$$

A second hypothesis could be considered for the Al average variation, which could be consistent with the following:

$$\mathrm{Fe}_{26}^{56} \rightarrow 2\mathrm{Al}_{13}^{27} + 2 \text{ neutrons}. \tag{5.8}$$

A third hypothesis could be made considering the Silicon average variation:

$$\mathrm{Ni}_{28}^{59} \rightarrow 2\mathrm{Si}_{14}^{28} + 3 \text{ neutrons} \tag{5.9}$$

Reactions (5.7), (5.8), and (5.9) imply emissions of neutrons, which will provide a great support to the hypotheses based on piezonuclear reactions. The main idea underlying the hypothesis is that an average decrease of 23.1 % in Nickel underwent a reaction producing at least 18.8 % of oxygen together with alpha and neutron emissions. Secondly, an average depletion of 2 % in Fe would have produced about 1.8 % of Aluminum accompanied by neutron emissions. Thirdly, another average decrease in Ni of 1 % could have gone into Silicon (+0.9 %) and more neutrons. Considering the standard deviations corresponding to each average concentration measured before and after the experiment (Figs. 5.10, 5.11, 5.12, and 5.13), the three considerations outlined above could be summarized according to the following balances: (1) Ni (−23.1 %) ≈ O (+18.8 %) + Si (+0.9 %) + alpha + neutrons, and (2) Fe (−2.0 %) ≈ Al (+1.8 %) + neutrons. Except for the Potassium increment (+2.7 %), which could be explained considering the deposition of the salt contained in the solution. The variations of the other elements' are lower than 1 %.

5.6 Conclusions

Neutron emissions up to one order of magnitude higher than the background level were observed during the operating time of an electrolytic cell. In particular, after a time span of about 7.5 h, neutron emissions of about 3 times the background level were measured. After 9 h and at the end of the test (after 20 h), it was possible to observe neutron emissions of about one order of magnitude greater than the background level.

By the EDX analysis performed on the two electrodes, significant compositional variations were recorded. In general, the decrements in Pd at the first electrode seem to be almost perfectly counterbalanced by the increments in lighter elements like oxygen. As far as the Ni-Fe electrode is concerned, the Ni decrement is almost perfectly counterbalanced by the O and Si increments according to reactions (5.7) and (5.9). At the same time, the Fe decrement may be considered to explain the final concentration of Al according to reaction (5.8). Chemical variations and energy emissions may be accounted for direct and indirect evidence of mechano-nuclear fission reactions correlated to microcrack formation and propagation due to hydrogen embrittlement. The so-called Cold Nuclear Fusion, interpreted under the light of hydrogen embrittlement, may be explained by piezonuclear fission reactions occurring in the host metal, rather than by nuclear fusion reactions of hydrogen isotopes forced into the metal lattice.

References

1. Borghi DC, Giori DC, Dall'Olio A (1992) Experimental evidence on the emission of neutrons from cold hydrogen plasma. In: Proceedings of the international workshop on few-body problems in low-energy physics, Alma-Ata, Kazakhstan, pp 147–154; Unpublished Communication (1957); Comunicacao n. 25 do CENUFPE, Recife Brazil (1971)
2. Diebner K (1962) Fusionsprozesse mit Hilfe konvergenter Stosswellen – einige aeltere und neuere Versuche und Ueberlegungen. Kerntechnik 3:89–93
3. Kaliski S (1978) Bi-conical system of concentric explosive compression of D-T. J Tech Phys 19:283–289
4. Winterberg F (1984) Autocatalytic fusion–fission implosions. Atomenergie-Kerntechnik 44:146
5. Derjaguin BV et al (1989) Titanium fracture yields neutrons? Nature 34:492
6. Fleischmann M, Pons S, Hawkins M (1989) Electrochemically induced nuclear fusion of deuterium. J Electroanal Chem 261:301
7. Bockris JO'M, Lin GH, Kainthla RC, Packham NJC, Velev O (1990) Does tritium form at electrodes by nuclear reactions? In: The first annual conference on cold fusion. National Cold Fusion Institute, University of Utah Research Park, Salt Lake City
8. Preparata G (1991) Some theories of cold fusion: a review. Fusion Technol 20:82
9. Preparata G (1991) A new look at solid-state fractures, particle emissions and "cold" nuclear fusion. Il Nuovo Cimento 104A:1259–1263
10. Mills RL, Kneizys P (1991) Excess heat production by the electrolysis of an aqueous potassium carbonate electrolyte and the implications for cold fusion. Fusion Technol 20:65
11. Notoya R, Enyo M (1992) Excess heat production during electrolysis of H_2O on Ni, Au, Ag and Sn electrodes in alkaline media. In: Proceedings of the third international conference on cold fusion, Nagoya, Japan. Universal Academy Press, Inc., Tokyo, Japan
12. Miles MH, Hollins RA, Bush BF, Lagowski JJ, Miles RE (1993) Correlation of excess power and helium production during D_2O and H_2O electrolysis using palladium cathodes. J Electroanal Chem vol 346:99–117
13. Bush RT, Eagleton RD (1993) Calorimetric studies for several light water electrolytic cells with nickel fibrex cathodes and electrolytes with alkali salts of potassium, rubidium, and cesium. In: Fourth international conference on cold fusion. Lahaina, Maui. Electric Power Research Institute, Palo Alto, CA
14. Fleischmann M, Pons S, Preparata G (1994) Possible theories of cold fusion. Nuovo Cimento Soc Ital Fis A 107:143
15. Szpak S, Mosier-Boss PA, Smith JJ (1994) Deuterium uptake during Pd-D codeposition. J Electroanal Chem 379:121
16. Sundaresan R, Bockris JOM (1994) Anomalous reactions during arcing between carbon rods in water. Fusion Technol 26:261
17. Arata Y, Zhang Y (1995) Achievement of solid-state plasma fusion ("cold-fusion"). Proc Jpn Acad 71(Ser. B):304–309
18. Ohmori T, Mizuno T, Enyo M (1996) Isotopic distributions of heavy metal elements produced during the light water electrolysis on gold electrodes. J New Energy 1(3):90
19. Monti RA (1996) Low energy nuclear reactions: experimental evidence for the alpha extended model of the atom. J New Energy 1(3):131
20. Monti RA (1998) Nuclear transmutation processes of lead, silver, thorium, uranium. In: The seventh international conference on cold fusion. ENECO Inc., Vancouver, Canada; Salt Lake City, UT

21. Ohmori T, Mizuno T (1998) Strong excess energy evolution, new element production, and electromagnetic wave and/or neutron emission in light water electrolysis with a tungsten cathode. Infinite Energy 20:14–17
22. Mizuno T (1998) Nuclear transmutation: the reality of cold fusion. Infinite Energy Press, Concord
23. Little SR, Puthoff HE, Little ME (1998) Search for excess heat from a Pt electrode discharge in K_2CO_3-H_2O and K_2CO_3-D_2O electrolytes
24. Ohmori T, Mizuno T (1999) Nuclear transmutation reaction caused by light water electrolysis on tungsten cathode under incandescent conditions. Infinite Energy 5(27):34
25. Ransford HE (1999) Non-stellar nucleosynthesis: transition metal production by DC plasma-discharge electrolysis using carbon electrodes in a non-metallic cell. Infinite Energy 4(23):16
26. Storms E (2000) Excess power production from platinum cathodes using the Pons-Fleischmann effect. In: 8th international conference on cold fusion. Lerici (La Spezia). Italian Physical Society, Bologna, Italy, pp 55–61
27. Storms E (2007) Science of low energy nuclear reaction: a comprehensive compilation of evidence and explanations about cold fusion. World Scientific, Singapore
28. Mizuno T et al (2000) Production of heat during plasma electrolysis. Jpn J Appl Phys A 39:6055
29. Warner J, Dash J, Frantz S (2002) Electrolysis of D2O with titanium cathodes: enhancement of excess heat and further evidence of possible transmutation. In: The ninth international conference on cold fusion. Tsinghua University, Beijing, China, p 404
30. Fujii MF et al (2002) Neutron emission from fracture of piezoelectric materials in deuterium atmosphere. Jpn J Appl Phys 41:2115–2119
31. Mosier-Boss PA et al (2007) Use of CR-39 in Pd/D co-deposition experiments. Eur Phys J Appl Phys 40:293–303
32. Swartz M (2008) Three physical regions of anomalous activity in deuterated palladium. Infinite Energy 14:19–31
33. Mosier-Boss PA et al (2010) Comparison of Pd/D co-deposition and DT neutron generated triple tracks observed in CR-39 detectors. Eur Phys J Appl Phys 51(2):20901–20911
34. Kanarev M, Mizuno T (2002) Cold fusion by plasma electrolysis of water. New Energy Technol 1:5–10
35. Cardone F, Mignani R (2004) Energy and geometry. World Scientific, Singapore, Chap. 10
36. Carpinteri A, Cardone F, Lacidogna G (2009) Piezonuclear neutrons from brittle fracture: early results of mechanical compression tests. Strain 45:332–339. (2009) Atti dell'Accademia delle Scienze di Torino 33:27–42
37. Cardone F, Carpinteri A, Lacidogna G (2009) Piezonuclear neutrons from fracturing of inert solids. Phys Lett A 373:4158–4163
38. Carpinteri A, Cardone F, Lacidogna G (2010) Energy emissions from failure phenomena: mechanical, electromagnetic, nuclear. Exp Mech 50:1235–1243
39. Carpinteri A, Lacidogna G, Manuello A, Borla O (2012) Piezonuclear fission reactions: evidences from microchemical analysis, neutron emission, and geological transformation. Rock Mech Rock Eng 45:445–459
40. Carpinteri A, Lacidogna G, Manuello A, Borla O (2013) Piezonuclear fission reactions from earthquakes and brittle rocks failure: evidence of neutron emission and nonradioactive product elements. Exp Mech 53:345–365
41. Milne I, Ritchie RO, Karihaloo B (2003) Comprehensive structural integrity: fracture of materials from nano to macro, vol 6. Elsevier, Amsterdam, pp 31–33, Chapter 6.02
42. Liebowitz H (1971) Fracture an advanced treatise. Academic, New York
43. Carpinteri A, Borla O, Goi A, Manuello A, Veneziano D (2013) Mechanical conjectures explaining cold nuclear fusion. In: Proceedings of the conference and exposition on experimental and applied mechanics (SEM), Lombard, Illinois, CD-ROM, Paper N. 481
44. Veneziano D, Borla O, Goi A, Manuello A, Carpinteri A (2013) Mechanical conjectures based on hydrogen embrittlement explaining cold nuclear fusion. In: Proceedings of the 21° congresso nazionale di meccanica teorica ed applicata (AIMETA), Torino, Italy, CD-ROM

Chapter 6
Torque Arm Actuated Bi-Stable Buckled Energy Harvester Characterization

D.A. Porter and T.A. Berfield

Abstract A bi-stable energy harvester utilizing PVDF strips driven via two torque arms with end masses and pseudo pinned in the middle is evaluated. A sinusoidal acceleration is applied to the base of the device with varying frequencies and magnitudes while the compression of the center beam is achieved by applying a small displacement to the center beam. Frequency sweeps will be done forwards as well as backwards to evaluate hysteresis performance. Peak voltages, natural frequencies, snap-through acceleration values, static actuation displacement values, and material properties for unknowns are derived experimentally.

While many parametric values such as beam length, compliance arm length, and proof mass can be varied, the focus of this study is on the effects of the compliance arm width on bi-stability switching and energy harvesting potential. For vibration-based energy harvesting, performance parameters such as power generated, power density, frequency broadening, frequency shifting, and optimal load impedance will be quantified. Results show that wider compliance arms decrease buckling amplitude, but increase the bi-stability switching regime and the overall power production. Current data also indicates that an optimal compression load exists for a given acceleration value.

Keywords Bi-stable structures • Buckling • Energy harvesting • Piezoelectric materials • Vibrations

6.1 Introduction

Vibration-based energy harvesting devices represent an intriguing class of power supplies that may serve as potential replacements for batteries in specific operating environments. In particular, vibration driven energy scavengers offer an alternative power source solution for inaccessible or remote wireless sensors, such as for structural health monitoring applications. One of the primary challenges with these devices is designing them for maximized productivity under low frequency chaotic vibration cases representative of realistic driving conditions.

Structures that demonstrate non-linear behavior have been shown to respond quite efficiently over a broad range of driving frequencies [1–3]. Pairing these types of nonlinear structures with a piezoelectric material or other energy production mechanism (electrostatics, etc.) allows for energy production that is based on strains induced within the composite system under transient deformations. Nonlinear behavior can be produced through a dual energy well arrangement using magnetic fields [2–5], or mechanically by using a bi-stable buckled structure [6–11]. For the mechanically bi-stable buckled structures, the transition between buckled states, or "snap-through", can induce very large deformations which generate a spike in electrical output. Thus, the mechanical stability state switching conditions are critical for optimizing power production.

In this work, the influence of geometrical parameters is explored for a recently development bi-stable energy harvesting structure. Unlike most bi-stable structures which use a Mode 1 buckled state, this device utilizes a quasi-pinned beam that symmetrically displaces the supports to induce a Mode 2, "S" shaped buckled beam. As shown schematically in Fig. 6.1, this design also employs two torque arms which transfer a bending moment to the center beam when transverse accelerations are applied (i.e., vibrations of the base). This torque is used to induce switching between the two stable buckled configurations of

D.A. Porter (✉) • T.A. Berfield
Department of Mechanical Engineering, University of Louisville, 200 Sackett Hall, Louisville, KY 40292, USA
e-mail: daport02@louisville.edu

N. Sottos et al. (eds.), *Experimental and Applied Mechanics, Volume 6: Proceedings of the 2014 Annual Conference on Experimental and Applied Mechanics*, Conference Proceedings of the Society for Experimental Mechanics Series,
DOI 10.1007/978-3-319-06989-0_6,

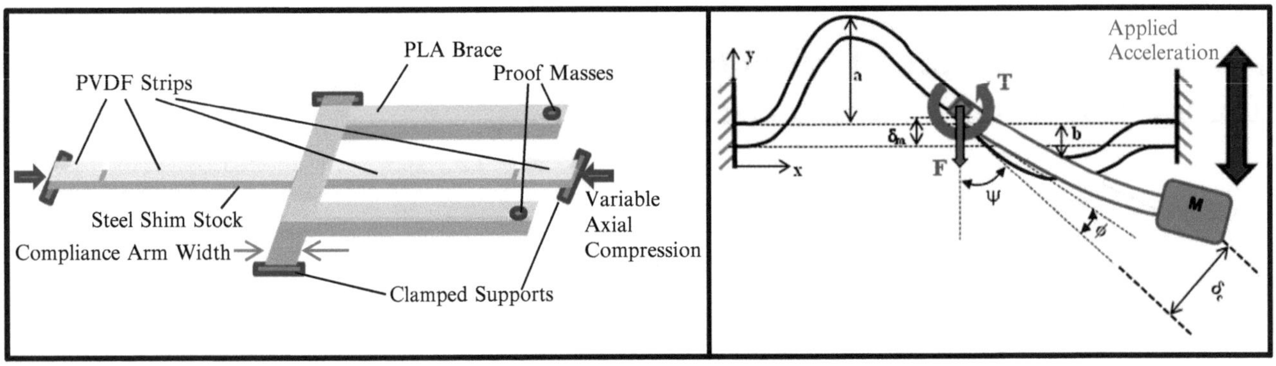

Fig. 6.1 Schematics showing the stacked layers within buckled device (*left*) and the idealized loading and quantifiable deformations of the center buckled beam (*right*)

the structure. The threshold for inducing "snap-though" events is explored with respect to structure dimensions. Specifically, the effects of compliance arm width on bi-stability will be investigated under a variety of vibrational loading conditions.

6.2 Experimental Procedure

6.2.1 Energy Harvesting Device Fabrication

The bi-stable energy harvesting devices were constructed using 100 μm thick steel shim stock as the base mechanical layer, as shown in Fig. 6.1. The steel shim stock was cut to specifications, and then mounted with the piezoelectric polyvinylidene-fluoride (PVDF) layer using adhesive strips. The aluminum clad piezoelectric PVDF sheets were purchased commercially from Goodfellow Corp., with a specified thickness of 110 μm and piezoelectric coefficients of d_{33} = ~20 pC/N and d_{31} = ~18 pC/N. For each of the devices, four strips of PVDF were bonded to the steel center beam as shown in Fig. 6.1. The size and placement of these strips was determined based on the curvature (and, therefore, induced strain) of the center beam in its "S"-shaped buckled configuration. Using an ANSYS finite element analysis model, the beam and constraint conditions were simulated to determine the transverse deflections and axial strains shown in Fig. 6.2. The labeling scheme for the electrodes is shown on the actual device in Fig. 6.3a.

Connected to the cantilevered arms holding the proof masses, a 1 mm thick 3D printed polylactic acid (PLA) brace was bonded to add structural rigidity. This stiffened section provides more efficient torque transfer to the center buckled beam than using the steel layer alone. At the end of each torque arm, a 0.17 g proof mass was attached. Initially, three variations of the bi-stable buckled energy harvesting device were constructed, each using a different compliance arm width (3.0, 5.0, or 7.0 mm). All other dimensional parameters, such as the center beam length, compliance arm length, and torque arm length were kept constant.

The PVDF strips bonded on the energy harvesting devices were electrically connected to the data acquisition system using silver epoxy and fine gage wires. The devices were then fitted into the custom constructed constraint base, which was mounted with a 3-axis accelerometer. The constraint base features adjustable grips and rigidly clamps an energy harvesting device at four support locations: once at each end of the center beam and once at the end of each compliance arm (Fig. 6.3). The center beam support can be translated inwards, applying a compressive axial force to the center beam. Symmetrically compressing the center beam induces the "S" shaped buckled configuration shown in Fig. 6.3b.

For each beam, the supports were compressed just enough so that the center beam buckled and demonstrated bi-stability as shown in Fig. 6.3b. The axial displacement required to reach this state varied with the compliance arm width, which is torsionally deformed when the center beam buckles. This compression was found to be 0.47 mm for the specimen with the 3 mm compliance arms, 0.36 mm for the 5 mm compliance arm device, and 0.17 mm for the 7 mm compliance arm case. For comparison purposes, the devices were also tested in vibration loading sweeps in a state of no axial compression (completely uncompressed).

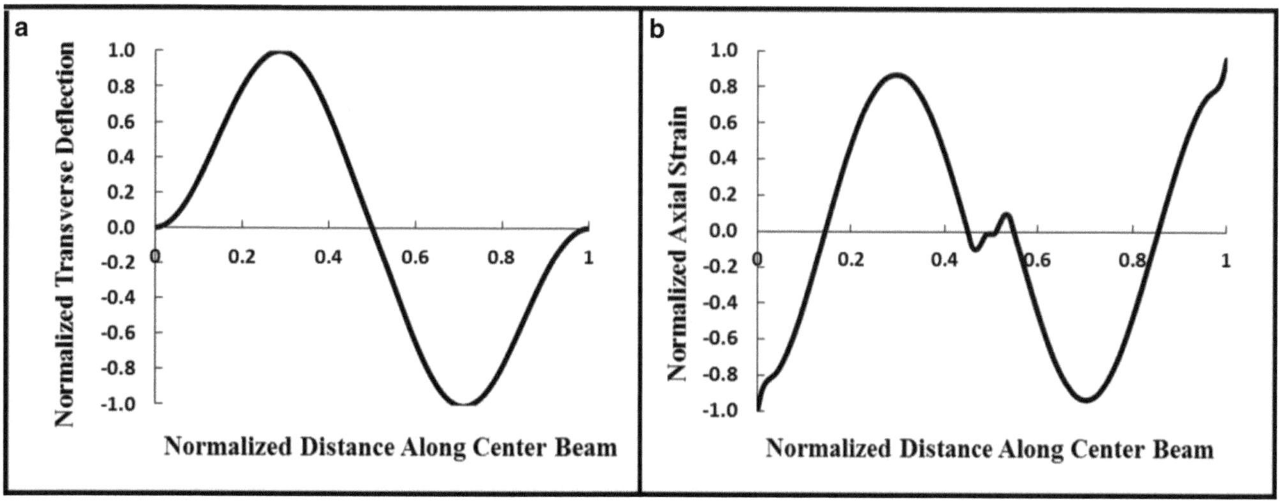

Fig. 6.2 Finite element analysis of a buckled center beam showing the normalized transverse deflection profile (**a**), and the corresponding axial strains (**b**). Results were used for PVDF strip placement to minimize charge neutralization

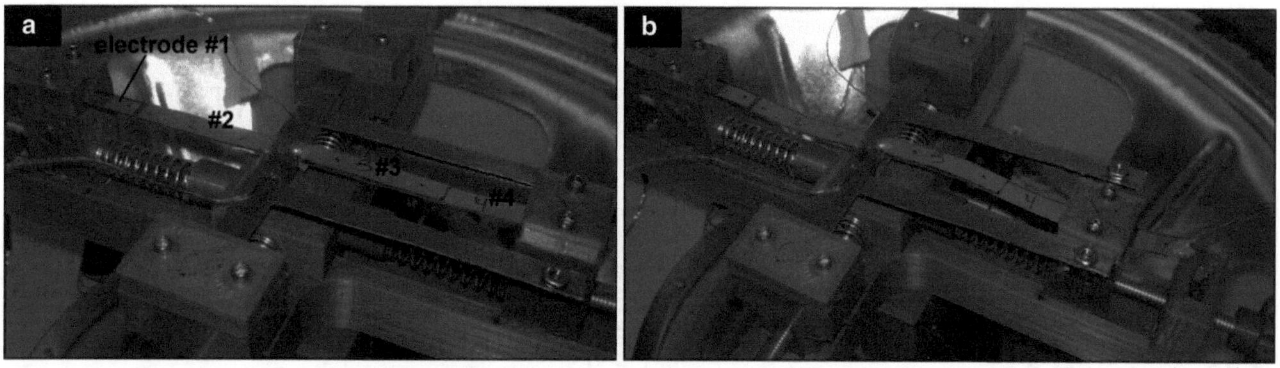

Fig. 6.3 Bi-stable buckled energy harvesting device with 7 mm torque arms fixed in the constraint base in the (**a**) uncompressed and the (**b**) 0.17 mm axial compression configurations. PVDF electrode numbering scheme is shown in (**a**)

6.2.2 Vibration Loading Frequency Sweeps

The constraint base in which the buckled energy harvester devices were mounted and buckled was rigidly fixed to a custom made variable amplitude, variable frequency shaker stage. This system uses a LabVIEW program to integrate feedback from the 3-axis accelerometer mounted in the shaker base to adjust the driving signal to match target values. Frequency sweeps were run in the forward (increasing frequency) and backwards (decreasing frequency) directions over the range of 16–40 Hz. The applied acceleration range was tested in increments of 0.2 g over the range from 0.5 up to 1.3 g.

During a single frequency sweep, a constant applied acceleration case was maintained. After both the target acceleration and frequency values are met and maintained for 5 s, data was logged for that particular vibration case and the next frequency target iteration was set. Power was calculated using a 3.3 MOhm resistive load which was constant for all devices. This, though, was not the optimal load resistance for any of the devices and because of proximity of the optimal load from previous experiments on other similar devices.

6.3 Results

The typical frequency sweep resulted in a data channel for each electroded section of PVDF. In the forward sweep direction, as shown in Fig. 6.4, the voltage would peak at the natural frequency and slowly decrease as the frequency swept higher. The same trend is observed in reverse for the backwards frequency sweeps. It should be noted that electrodes #2 and #3,

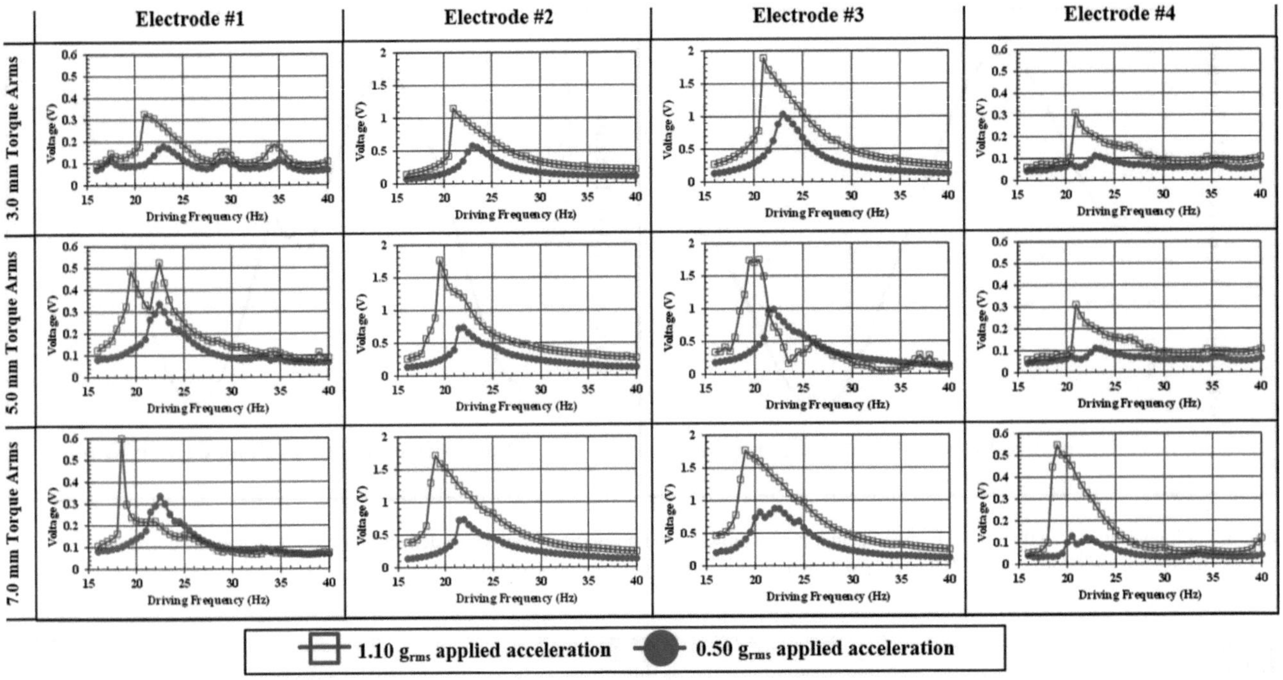

Fig. 6.4 Voltage output for the different electrode locations during forward frequency sweeps at both 0.5 g and 1.1 g applied accelerations on all the different compliance arm cases in the compressed configuration

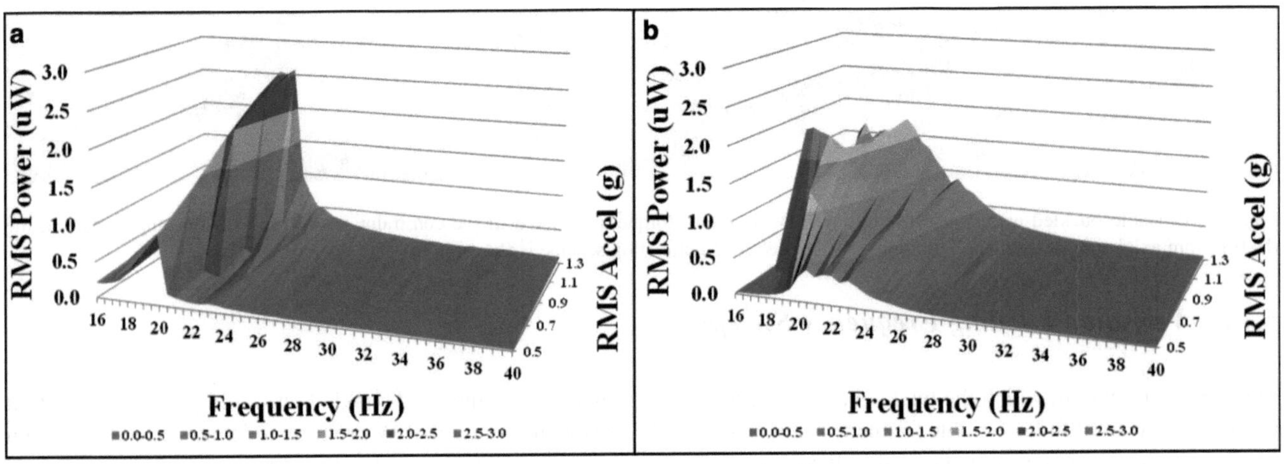

Fig. 6.5 Full power output sweeps in the forward (**a**) and backwards (**b**) directions for the 0.17 mm compressed specimen with 7.0 mm width compliance arms

consistently produced the highest voltages. This is expected as these center two PVDF strips are larger and experience more substantial strains during "snap-though" events than the outer electrode areas (#1 and #4).

Also observed during the frequency sweeps was the dependence of the natural frequency on the applied acceleration magnitude. When the devices are in their unbuckled state, an increase in sinusoidal acceleration input magnitude cause an increase in natural frequency. However, as the acceleration magnitude increases for devices in the "just bi-stable" buckled configuration, the peak output power occurred at lower frequencies. This frequency correlation was independent on sweep direction.

Full-width half-max (FWHM) measurements were a maximum when the devices were swept in reverse. Forward sweeps tended to give a higher RMS power output at the natural frequency. Snap-through during oscillation broadened the FWHM values and also gave higher peak-peak voltage values due to higher strains. RMS power output for the 7 mm compliance arm device in the buckled state is shown in Fig. 6.5 above.

Table 6.1 Maximum power output and frequency conditions

	Forward sweep			Reverse sweep		
Device configuration	Resonant frequency (Hz)	Max. RMS power (μW)	FWHM (Hz)	Resonant frequency (Hz)	Max. RMS power (μW)	FWHM (Hz)
3 mm, uncomprressed	26.5	5.71	2.84	25.5	4.60	2.50
3 mm, 0.47 mm compressed	20.5	1.60	3.64	19.0	1.93	4.64
5 mm, uncompressed	24.5	5.31	2.99	24.0	4.60	2.51
5 mm, 0.36 mm compressed	19.5	2.85	2.47	18.0	3.59	3.28
7 mm, uncompressed	23.0	3.03	3.07	22.5	2.57	3.32
7 mm, 0.17 mm compressed	20.5	1.68	5.72	20.5	1.75	6.59

Power production values under the maximum output driving frequency are listed in Table 6.1. The structures were tested under both uncompressed and compressed states, with the maximum power and FWHM values reported. Maximum RMS power for each channel was calculated by

$$P_{rms} = \frac{V_{rms}^2}{R_{load}} \tag{6.1}$$

where V_{rms} is the RMS voltage for the channel and R_{load} is the resistance load of 3.3 MOhm. Total device power was then derived by summing each of the four channels.

6.4 Discussion

While uncompressed beams do demonstrate a higher maximum power output, this output is limited to a very narrow frequency range (see FWHM values from Table 6.1). The advantage of the bi-stable configuration is the broadening of the viable energy harvesting range, though coming at the expense of peak output. In practice, this trade-off is quite beneficial as most "real-life" energy harvesting source produce non-regular vibration profiles. The width of the peak power range in Fig. 6.5 clearly shows the significance of this effect.

The 7 mm compliance arm device had a very shallow energy well, meaning that it easily crossed between the two bi-stable states with minimal force and exhibited snap-through for the majority of the experiments. Snap through contributed to high peak-peak voltage values and was easily seen and heard during the experiment. The snap-through behavior is seen as a decaying super-harmonic within the typical sinusoidal response voltage. Maintaining high energy orbitals during actuation helps to achieve high deflections thus giving higher voltage outputs.

It is important to note that the load resistance used for each channel/device was not determined to be the optimal load resistance which can change given different configurations. Thus this work is constrained to a single resistance load value. Future works will include an optimal load resistance sweep using digital pots; which will help determine the optimal load for all channels on a given device. Currently it is still feasible for the buckled devices to outperform their unbuckled counterparts (RMS power output) given an optimal design and setup. Also of particular interest is how these devices perform when being driven by Gaussian white noise frequencies, at super/sub-harmonics of the natural frequency, and at their behavior given different geometries.

6.5 Conclusions

The compliance arm width was found to have a substantial effect on the bi-stability behavior of the buckled energy harvesting device tested. Larger, stiffer compliance arms greatly reduced the compression required to induce a bi-stable buckled state and lower the acceleration threshold needed to switching between buckled states. In addition, buckled configurations provided broadened ranges of excitation frequencies viable for vibration-based energy harvesting. Unbuckled devices had their natural frequencies meander upward with an increasing acceleration. Buckled devices had their natural frequencies lowered with an increase in acceleration.

Acknowledgements The authors would like to acknowledge student support provided through the National Science Foundation, award number #1130528. In addition, the authors would like to thank Tim Varon and the University of Louisville, Department of Mechanical Engineering for their assistance in preparing this work.

References

1. Cottone F, Vocca H, Gammaitoni L (2009) Nonlinear energy harvesting. Phys Rev Lett 102(8):080601
2. Stanton SC, McGehee CC, Mann BP (2010) Nonlinear dynamics for broadband energy harvesting: investigation of a bistable piezoelectric inertial generator. Physica D 239(10):640–653
3. Mann BP, Owens BA (2010) Investigations of a nonlinear energy harvester with a bistable potential well. J Sound Vib 329(9):1215–1226
4. Lin J-T, Lee B, Alphenaar B (2010) The magnetic coupling of a piezoelectric cantilever for enhanced energy harvesting efficiency. Smart Mater Struct 19(4):045012
5. Tang L, Yang Y (2012) A nonlinear piezoelectric energy harvester with magnetic oscillator. Appl Phys Lett 101(9):094102
6. Harne RL, Wang KW (2013) A review of the recent research on vibration energy harvesting via bistable systems. Smart Mater Struct 22(2):023001
7. Ando B, Bagilo S, Bulsara AR, Marletta V, Medico I, Medico S (2013) A double piezo- snap through buckling device for energy harvesting. In: 2013 transducers & eurosensors XXVII: 17th international conference on solid-state sensors, actuators and microsystems (TRANSDUCERS & EUROSENSORS XXVII), Barcelona, pp 43–45
8. Cottone F et al (2012) Piezoelectric buckled beams for random vibration energy harvesting. Smart Mater Struct 21(3):035021
9. Giannopoulos G, Monreal J, Vantomme J (2007) Snap-through buckling behavior of piezoelectric bimorph beams: I. Analytical and numerical modeling. Smart Mater Struct 16(4):1148–1157
10. Jung S-M, Yun K-S (2010) Energy-harvesting device with mechanical frequency-up conversion mechanism for increased power efficiency and wideband operation. Appl Phys Lett 96(11):111906
11. Berfield TA, Porter DA (2014) Bi-stable energy harvesting device actuated via torque arm. Smart Mater Struct 23 075003 doi:10.1088/0964-1726/23/7/075003

Chapter 7
Validating FSI Simulations in LS-DYNA 971 R7

Kevin A. Gardner, Jeremy D. Seidt, and Amos Gilat

Abstract The latest release of LS-DYNA includes a multi-physics solver that combines computational fluid dynamics and structural solvers. This capability is a new computational design tool for the automotive industry. One of the simplest ways to reduce weight and increase fuel efficiency is to trim unnecessary weight from the body panels which comprise the vehicle. However, body panels that are made too thin are susceptible to a phenomenon known as oil-canning under loads such as those encountered from typical automotive air dryers. Oil-canning is a complex phenomenon that can result in permanent deformation or the panel can snap back. Oil-canning is to be avoided, even if temporary, for customer satisfaction reasons. An experimental program is presented where automotive roof panels are placed in a custom test rig and loaded with a high velocity air jet to replicate the oil-canning phenomenon. Flow characterization is performed using an array of piezo-electric pressure sensors. Panel deformation is measured using three-dimensional digital image correlation. Experimental data will be used to determine the validity of the multi-physics solver as an engineering design tool.

Keywords Fluid–structure interaction • Oil-canning • Digital image correlation • Numerical methods • Automotive

7.1 Introduction

LS-DYNA 971 R7 [1] includes both incompressible and compressible computational fluid dynamics solvers that are integrated with the solid mechanics solver allowing fluid–structure interaction (FSI) problems to be analyzed. This new capability presents a powerful design tool for the automotive industry that can be used to help reduce the overall weight of automobiles and improve fuel economy. The body panels of automobiles are typically stamped from sheet steel and can represent a large portion of the vehicle's overall weight. Significant weight savings can be had by thinning the body panels or using lighter materials such as aluminum. Thinning the steel used to fabricate the roof panel can give a large weight reduction, however if the panel is thinned too much a phenomenon known as roof panel oil-canning can occur. This phenomenon occurs when an automobile exits an automatic car wash. The high velocity air dryers can cause temporary or permanent deformation of the roof panel. Small deformations can be temporary as the panel will simply snap back to its original shaped after the loading is removed. However, if the dryer velocity is high enough or the panel thin enough the deformation can be permanent as the panel will remain deformed even after the loading is removed. It is important that the oil-canning phenomenon be avoided, even if the deformation is temporary, since it leads to customer dissatisfaction. Results from fluid–structure interaction simulations would be a useful tool to minimize the roof panel's weight while ensuring that oil-canning is avoided for a typical automotive dryer.

Many researchers have studied fluid–structure interactions over a wide range of applications. The deformation of composite structural elements has been investigated with underwater impulsive loading using an experimental setup that recreates fluid–structure interactions that are commonly encountered [2–4]. Similarly, air blast loading tests have been performed on carbon composite panels to determine the response of the material and the influence plate curvature has on the fluid–structure interactions [5–7]. Fluid–structure interaction has also been studied to optimize the design of steel panel roofing systems so that the uplift created by high winds can be minimized [8].

K.A. Gardner (✉) • J.D. Seidt • A. Gilat
Scott Laboratory, Department of Mechanical and Aerospace Engineering, The Ohio State University, 201W 19th Ave, Columbus, OH 43210, USA
e-mail: gardner.796@osu.edu

N. Sottos et al. (eds.), *Experimental and Applied Mechanics, Volume 6: Proceedings of the 2014 Annual Conference on Experimental and Applied Mechanics*, Conference Proceedings of the Society for Experimental Mechanics Series, DOI 10.1007/978-3-319-06989-0_7,

7.2 Experimental

Roof panel oil-canning is a complex phenomenon that is dependent upon many different properties including the sample material and dimensions, the boundary conditions, and the ratio of the jet area to the panel area. In order to give the most accurate representation of the phenomenon on a production vehicle it was decided that tests would be performed on actual automotive roof panels. A compact sedan was chosen for the project and the provided panels had a nominal thickness of 0.65 mm. In order to match the boundary conditions of the experiment to that of the real car the test panels are constrained along their edges by matching forms that clamp the panels in place within the test rig shown in Fig. 7.1. The matching forms were fabricated from medium density polyethylene blocks and match the contours of the panel edges exactly.

The test rig consists of a clamping inner enclosure that is supported by the outer frame by two turnbuckles. The panels and matching forms are positioned within the clamping frame and the forms are clamped in place, thus also locking the panel in place. The clamping inner enclosure is then hoisted in place to be supported by the two turnbuckles. The turnbuckles support the full weight of the inner enclosure. Three load cells are placed between the inner enclosure and the support frame to measure the total force on the panel generated by the air jet. The load cells provide an accurate air jet force measurement since the entire weight of the inner enclosure is supported by the turnbuckles.

The high velocity air jet is generated using an Exair model 1117 industrial air nozzle. The nozzle is supported by the outer frame of the test rig on an adjustable arm that allows the nozzle to be accurately positioned in three directions. The nozzle generates approximately 10 lb of force when operating at 80 psi at a distance of 12 in. from the nozzle. At this standoff distance, the flow acts on a circular area approximately 10 in. in diameter. The nozzle is fed from a 120 gal Ingersoll Rand air compressor that provides constant pressure during a test and adjustable operating pressures which in turn allows the total force on the panel to be varied.

The automotive roof panel was tested using an operating pressure of 100 psi. This pressure resulted in a total average force of 12 lb as shown in Fig. 7.2. The total force was determined by summing the force measurements from each of the three load cells. The test was performed with the nozzle located approximately at the center of the panel. Three-dimensional digital image correlation was used to directly measure displacements on the surface of the panel. A high contrast random

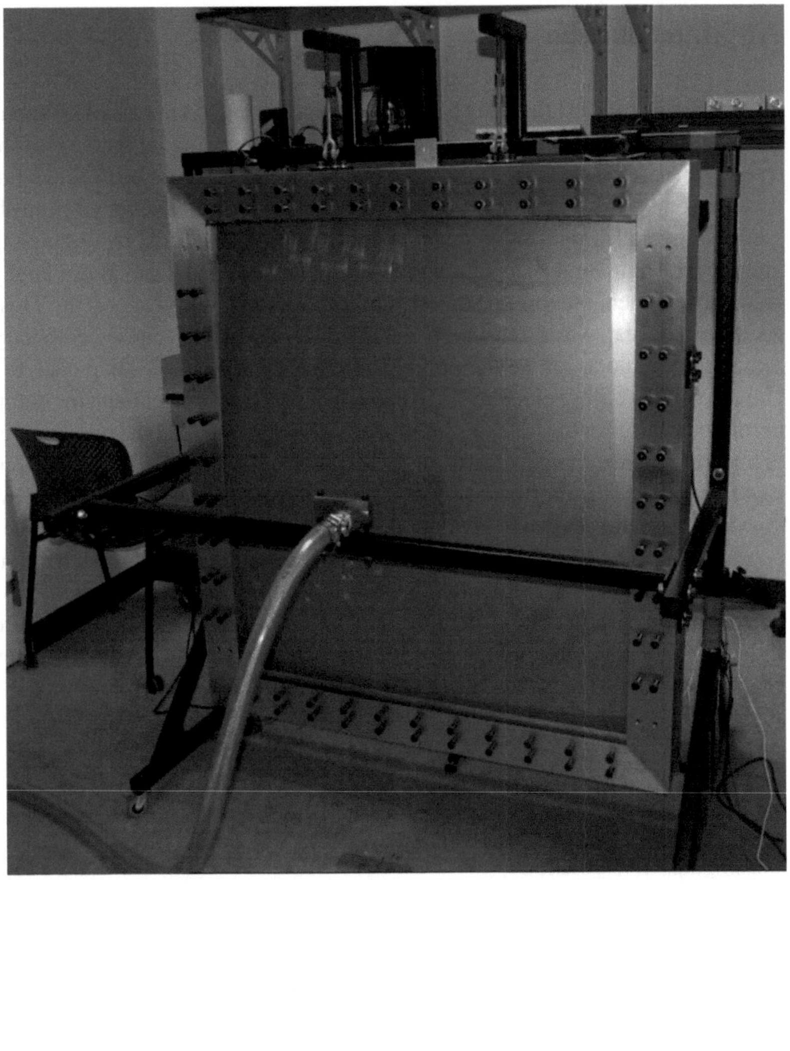

Fig. 7.1 Test rig

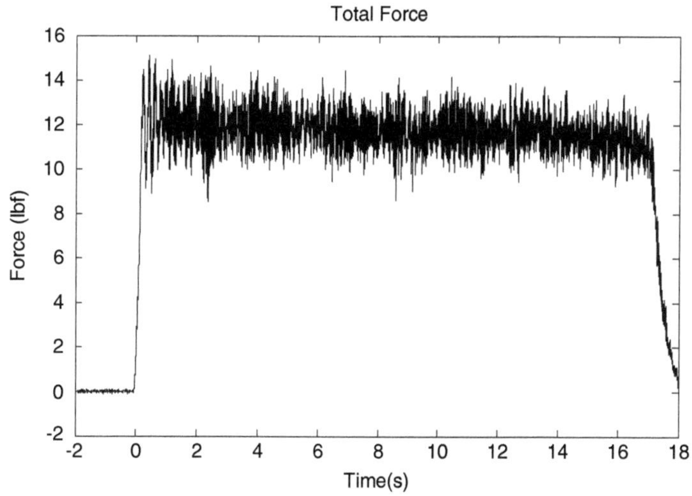

Fig. 7.2 Force vs. time, force determined from summation of three load cell values

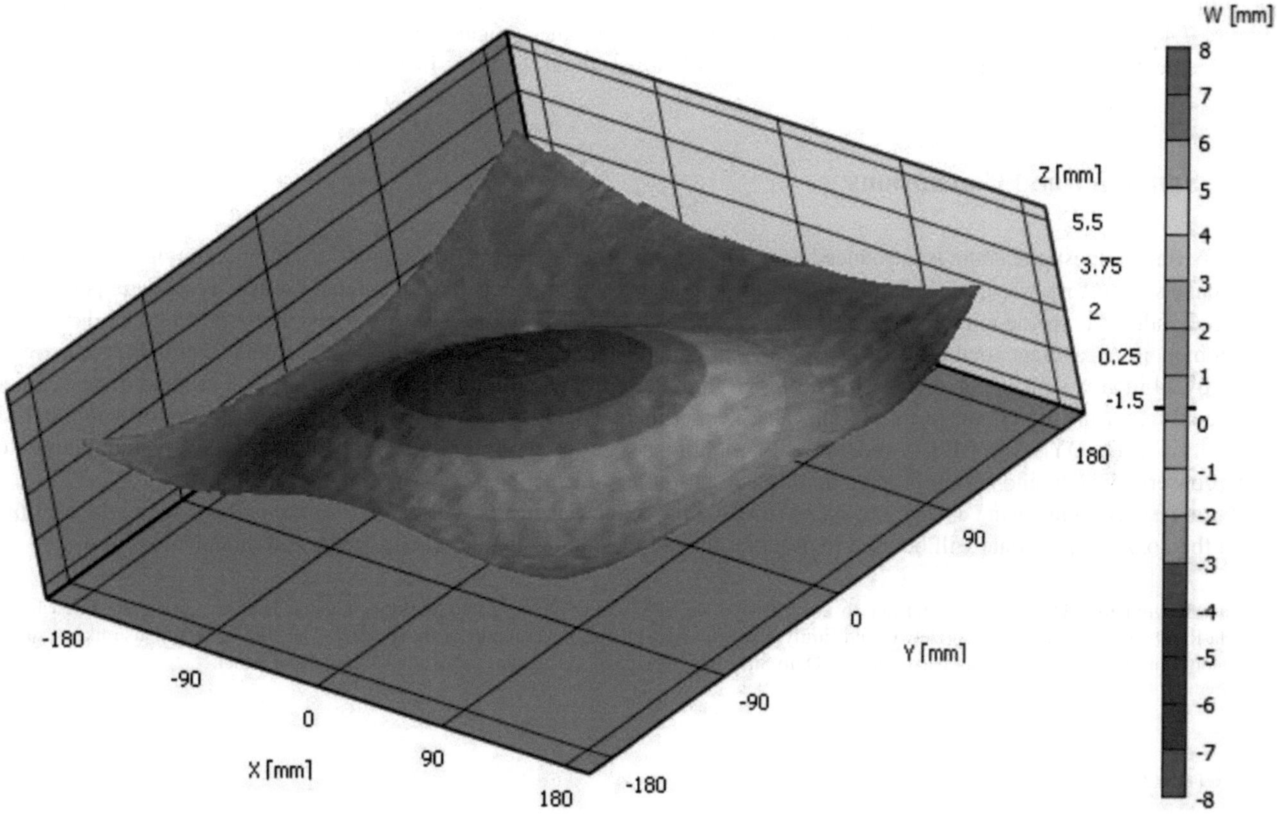

Fig. 7.3 Displacement contour from digital image correlation

speckle pattern was applied, by hand, to a roughly 24 in by 18 in area in the center of the primed roof panel. Images were captured using two Photron Fastcam SA1.1 cameras operating at 500 frames per second at 1024 × 1024 resolution. Correlated Solutions Vic 3D software was used to calculate the displacements on the surface of the panel. Figure 7.3 shows a characteristic plot of displacement calculated from DIC. The maximum displacement occurs at the center of the loading and has a magnitude of approximately 0.28 in. The displacement occurs over an area measuring approximately 14 in. by 14 in. No significant strain is observed in the panel during a test which confirms the notion that oil-canning is a reversible process that does no damage to a vehicle in most situations.

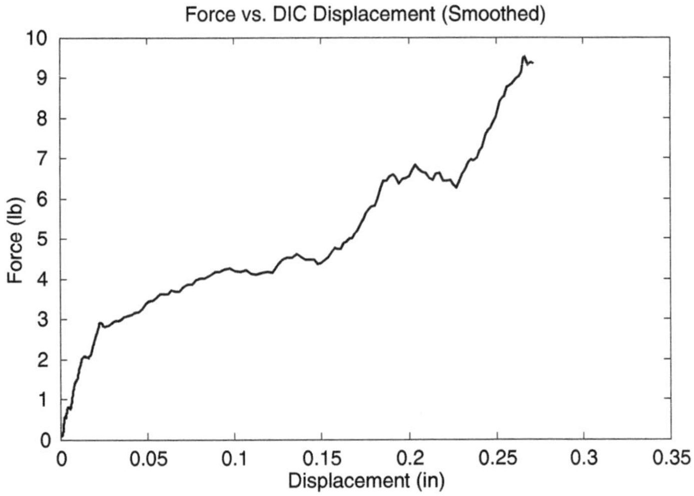

Fig. 7.4 Force vs. displacement

Oil-canning is typically accompanied by a snap through response of the panel. During the snap through a large amount of deformation occurs at nearly constant load. Once the snap through is complete the load on the panel begins to increase. Figure 7.4 shows a plot of force vs. displacement which exhibits the typical snap through behavior. A large amount of displacement is developed at a load of approximately 4 lb after which the force increases to the average value of 12 lb.

7.3 Summary and Conclusions

An experimental test procedure is introduced to validate fluid–structure interaction simulations of oil-canning in automotive roof panels. A test rig was designed to conduct full scale tests on roof panels from a compact sedan. Tests were performed using an industrial nozzle to replicate the loading conditions created by automotive air dryers and replicate the oil-canning phenomenon. An average total force of 12 lb was applied to the panel and out of plane displacements were observed over a 14 in by 14 in area with a peak displacement of 0.28 in directly in line with the center of the air jet. Full field displacement was measured using digital image correlation. These experimental data will be used as validation metrics for oil-canning FSI simulation in LS-DYNA 971 R7. Validated FSI simulations would be useful design tool for automotive manufacturers to both conserve weight and ensure product quality and customer satisfaction.

Future experiments using an array of piezo-resistive pressure transducers will be performed to characterize the air flow from the nozzle. These data will be used to describe the air jet loading characteristics in the FSI model.

Acknowledgements The authors would like to thank Duane Detwiler, Andy Fields, Allen Sheldon, Thomas Ramsay, and Yong-Bae Cho of Honda R&D Americas, Inc. and Ed Helwig and Emily Nutwell of LSTC for their support of this project. The authors would also like to thank Michael Tomaso of the architecture department at Ohio State who machined the custom forms to clamp the roof panel.

References

1. Pin FD (2010) Advances on the incompressible CFD solver in LS-DYNA. In: 11th international LS-DYNA users conference
2. Espinosa HD, Lee S, Moldovan N (2006) A novel fluid structure interaction experiment to investigate deformation of structural elements subjected to impulsive loading. Exp Mech 46:805–824
3. Wei X, Tran P, Vaucorbeil A, Ramaswamy RB, Latourte F, Espinosa HD (2013) Three-dimensional numerical modeling of composite panels subjected to underwater blast. J Mech Phys Solids 61:1319–1336
4. Wei X, Vaucorbeil A, Tran P, Espinosa HD (2013) A new rate-dependent unidirectional composite model - application to panels subjected to underwater blast. J Mech Phys Solids 61:1305–1318
5. Kumar P, Shukla A (2011) Dynamic response of glass panels subjected to shock loading. J Non-Cryst Solids 357:3917–3923
6. Kumar P, Stargel DS, Shukla A (2013) Effect of plate curvature on blast response of carbon composite panels. Compos Struct 99:19–30
7. Wang E, Wright J, Shukla A (2011) Analytical and experimental study on the fluid structure interaction during air blast loading. J Appl Phys 110:114901
8. He WX, Hong HP (2012) Probabilistic characterization of roof panel uplift capacity under wind loading. Can J Civ Eng 39:1285–1296

Chapter 8
Fundamental Frequencies of Slender Beams Subject to Imposed Axial End Displacements

G. Piana, A. Manuello, R. Malvano, and A. Carpinteri

Abstract The dependence of the fundamental frequency on the axial load in slender beams subject to imposed axial end displacements was experimentally investigated. The considered specimens presented different geometrical imperfections (initial curvatures), and were tested in two different constraint conditions (hinged–hinged and hinged-clamped). The natural frequencies were extracted in both conditions of forced and free vibration, using an electromagnet and a laser displacement transducer. In addition, the responses observed during the experiments, for the hinged–hinged case, were reproduced by numerical simulations, obtaining a good agreement between numerical and experimental results.

Keywords Experimental test • Numerical simulation • Fundamental frequency • Imposed displacement • Buckling

8.1 Introduction

As is well known, the influence of applied compressive axial loads on the fundamental vibration frequency is strictly connected with the stability analysis of elastic slender beams [1]. For this reason, the correct evaluation of the fundamental frequency is of primary importance in designing new structures and components, as well as in monitoring existing ones. At the same time, if an internal axial load arises in a slender element as the consequence of an imposed (static) axial end displacement, then a different dynamic structural response is encountered respect to the case in which a beam end is free to slide, during transverse vibration, and a (constant) axial load is applied externally. This difference is due to the change in the axial boundary condition. Moreover, the presence of an initial curvature of the beam axis may significantly affect the aforesaid response. The latter situation—less infrequent than we might think—may appear when slender structures are subjected to external moving constraints. For example, for bridges and large structures the reduction in bearing capacity and the sudden collapse of compressed elements can be the result of constraint settlements or structural subsidence due to seismic events. Also, a situation analogous to that of a prescribed axial displacement in a slender beam is obtained when, for example, a uniform thermal expansion is restrained.

In case of prescribed axial displacement, the internal axial load varies (oscillates) during the beam vibration, due to the change in axial strain; consequently, the variation of the frequency with the axial load will differ from that of the case in which a beam end is free to slide and a (constant) axial load is applied from outside [2, 3]. Furthermore, when both ends are axially restrained, the presence of an initial curvature contributes to the stiffness of the beam which behaves like a pinned shallow arch [4–7].

In the present contribution, we deal with an experimental investigation on the fundamental frequency evolution in slender steel beams subjected to axial end displacements [8]. Different geometrical imperfections, in terms of initial curvature, were considered in order to understand their influence on the evolution of the first natural frequency. In addition, each specimen

G. Piana (✉) • A. Manuello • A. Carpinteri
Department of Structural, Geotechnical and Building Engineering, Politecnico di Torino, Corso Duca degli Abruzzi 24, 10129 Torino, Italy
e-mail: gianfranco.piana@polito.it

R. Malvano
Department of Mechanical and Aerospace Engineering, Fluid Dynamics Unit C/o Politecnico di Torino, National Institute of Metrological Research - INRIM, Corso Duca degli Abruzzi 24, 10129 Torino, Italy

N. Sottos et al. (eds.), *Experimental and Applied Mechanics, Volume 6: Proceedings of the 2014 Annual Conference on Experimental and Applied Mechanics*, Conference Proceedings of the Society for Experimental Mechanics Series, DOI 10.1007/978-3-319-06989-0_8,

was tested in two different constraint conditions. The tests were conducted using a servo-controlled testing machine under displacement-controlled condition. The experimental apparatus was realized according to metrological requirements and was fine tuned at the National Institute of Metrological Research (INRIM), while the experiments were carried out at the Experimental Laboratory of Materials and Structures of the Politecnico di Torino (MASTRLAB). For each specimen, the fundamental frequencies corresponding to different imposed displacements—and, therefore, to different axial loads—were evaluated analyzing both the free and forced responses: an electromagnet and a laser displacement transducer with suitable characteristics were adopted. Hence, the fundamental frequency vs. axial load curves were obtained, therefore showing a transition from a beam-like behavior to an arch-like behavior, with an increase in the beam stiffness as the load increases [5, 7]. This transition may be used to discriminate between two different phases and to evaluate the stability of structural elements subjected to external conditions such as constraint subsidence: the achievement of particular values of the natural frequency could be used as a monitoring indicator in existing structures. In addition, numerical simulations were performed, getting a valid interpretation of the experimental results and contributing to understand the phenomenon under investigation.

The present work distinguishes from the cited references on the subject [4–7], in addition to the peculiarities of the experimental setup and test procedure, mainly for the profile of the initial geometrical imperfection. Also, the hinged-clamped constraint condition has not been analyzed in the abovementioned references. Moreover, in this paper we show how the phenomenon under investigation can be numerically simulated in a multi-body framework, which can be conveniently applied as a valid alternative to the finite element method when an analytical solution is not available.

8.2 Experimental Setup

The specimens used during the experimental tests consisted of three steel bars of rectangular cross-section ($15 \times 30\ \text{mm}^2$) and length ($l$) equal to 1,000 mm (Fig. 8.1a). Specimen B1 was a rectified beam, with a maximum offset (e_0) from the rectilinear configuration no larger than 0.2 mm ($e_0/l \leq 0.2/1,000$). For the other two specimens, different geometrical deviations were preliminarily introduced by imposing a prescribed offset from the rectilinear configuration. The profile of the initial deflection from straightness was linear in the end portions for a length equal to 15 cm, while it was parabolic in the central part. In particular, specimens B2 and B3 presented a maximum offset at mid-length equal to 1 mm ($e_0/l = 1/1{,}000$) and 2 mm ($e_0/l = 2/1{,}000$), respectively (Fig. 8.1a). At both the ends of each specimen a spherical cavity was created, with a radius of 8 mm and a maximum depth of 2.5 mm, and centered with a tolerance of ± 0.1 mm (Fig. 8.1b). During the tests, two steel spheres having a diameter of 15 mm were positioned in said cavities in order to facilitate the specimen centering and to realize the external constraints at the same time, as it will be specified later on.

Figure 8.2a shows a general view of the experimental setup. The experimental tests were conducted using a servo-controlled machine (MTS) with a closed-loop electronic control, having a maximum capacity of 100 kN. The specimen was positioned vertically and connected to the testing machine by two end joints that realized the external constraints. Two horizontal transverses were positioned at mid-height of the tested beam: one supported a laser displacement transducer (measuring device) and the other supported an electromagnet (forcing device). Both the electromagnet and the measuring device acted without any contact with the specimen. The complete experimental apparatus was constituted by a wave generator, a signal amplifier, an oscilloscope, and a data acquisition system (see Fig. 8.2a).

Figure 8.2b shows a schematic representation of the two end joints. These joints were rigidly connected through a double-threaded pin to the testing machine, with steel rings interposed to guarantee a correct transmission of the load. While the upper joint, once mounted, remained in a fixed position, the lower one could be adjusted in the horizontal plane to allow the correct positioning therefore ensuring the vertical alignment. More details about the specimens centering can be found in [8].

Two constraint conditions were realized during the experiments: hinged–hinged and hinged-clamped. In the former case, both the lower and the upper joints worked, through the steel spheres, as spherical hinges that react only in compression (unilateral spherical hinges). In the hinged-clamped case, two steel plates were opportunely introduced and fixed into the upper joint in order to prevent, at least partially, end flexural rotations, while the lower joint remained unchanged.

a

b
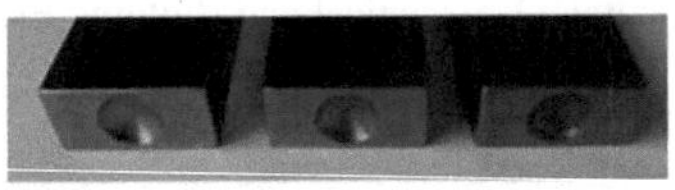

Fig. 8.1 (**a**) Specimens with different initial curvatures and (**b**) detail of end cross-sections with spherical cavities

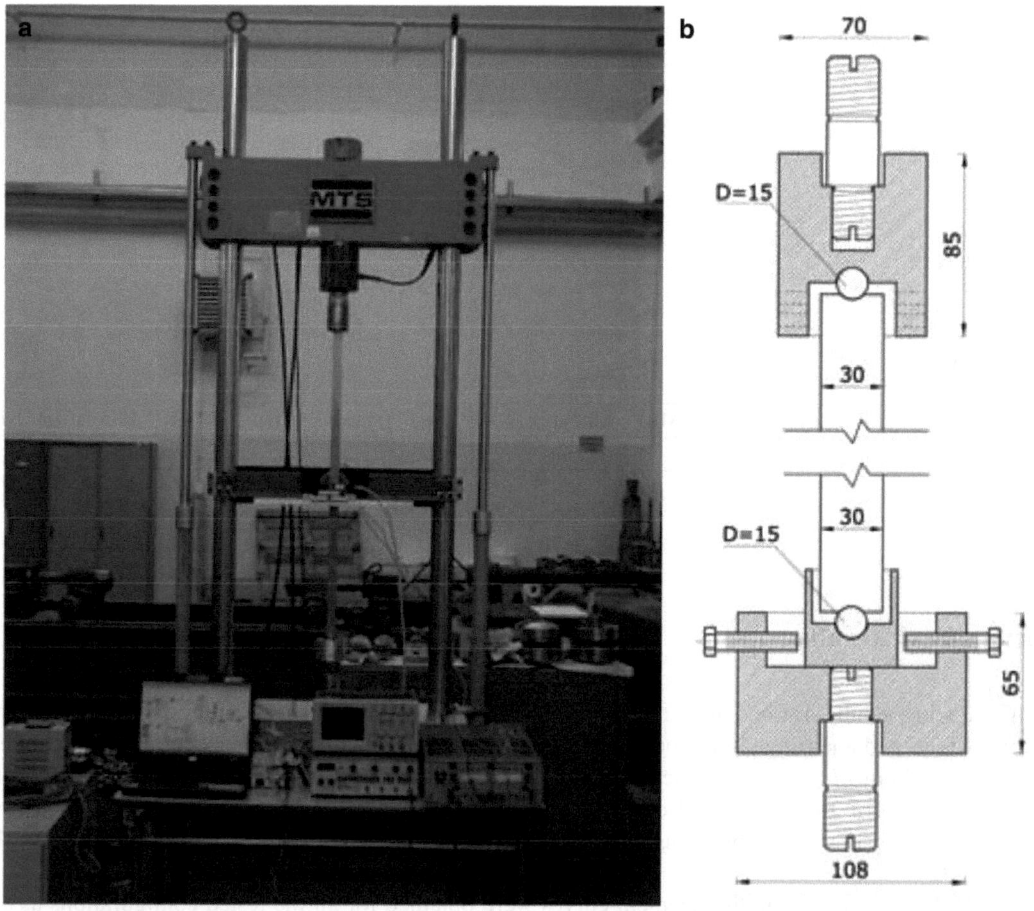

Fig. 8.2 (**a**) Experimental setup and (**b**) schemes of upper and lower end joints

8.3 Experimental Equipment and Procedure

The experimental equipment is shown in Fig. 8.3a. It consisted of two transverses connected each other by two aluminum circular joints that served for connection to the testing machine frame. On the transversal element presenting a C-shaped cross-section the electromagnet was fastened with a bolt in middle position, while the second transversal element was used to allocate the laser sensor on a doubly-adjustable guide. Thus, during the tests, the laser sensor was positioned in front of the electromagnet on the other side of the monitored beam (Fig. 8.3b). The adopted electromagnet was the T100 produced by C. E.I. s.r.l., while the laser sensor was the optoNCDT 1302–20 produced by Micro-Epsilon, whose main characteristics are listed in Table 8.1. During the experiments, the laser sensor was coupled with a data acquisition device NI USB-6009 produced by National Instruments, and the default parameters were adopted.

The tests were performed controlling the vertical displacement transmitted by the linear actuator to the beam. The input was constituted by several steps of the imposed displacement and by the electromagnet forcing. The output was constituted by the time history of the beam's midpoint transversal displacement, measured with the laser sensor. Hence, for each imposed displacement the fundamental vibration frequency was obtained under both conditions of forced and free vibrations. In the case of forced vibrations, the electromagnet excited the beam with a sinusoidal force, controlled in frequency and amplitude by the wave generator. In the case of free vibrations, the electromagnet was turned off after 10 s and the free response was measured. Three independent methods were used to analyze the response, as it will be described in Sect. 8.4. Acquisition and analysis of the measured signals were carried out through computer programs that were specially developed using the software LabVIEW.

Fig. 8.3 (**a**) Experimental equipment and (**b**) testing configuration with specimen B1 undergoing post-critical deflections

Table 8.1 Main characteristics of the laser sensor

Measuring range (MR)	Start of measuring range (SMR)	Midrange (MR)	End of measuring range (EMR)	Resolution (dynamic 750 Hz)
20 mm	30 mm	40 mm	50 mm	10 μm

8.4 Experimental Results

A total of six testing configurations was realized by analyzing each specimen in the following two constraint conditions: hinged–hinged and hinged-clamped. In the former case, the theoretical buckling load calculated according to Euler's formula is about 17.5 kN, while the theoretical fundamental flexural frequency calculated according to linear dynamics is about 35.2 Hz. For the hinged-clamped specimen, the Euler's critical load is about 37.2 kN, while the fundamental frequency is 55 Hz. The previous theoretical values are not affected by imperfections and so they hold for all specimens B1-3.

First, the axial load vs. transversal displacement curves were obtained for all the tested configurations as a preliminary phase (Fig. 8.4). The axial load was intended as the specimen reaction corresponding to the axial displacement imposed to the lower beam end by the testing machine actuator, therefore its value came from the MTS. The transversal displacement was intended as the midpoint deflection of the specimen: its value was measured with the laser sensor.

Figure 8.4a shows the axial load vs. transversal displacement curves for specimens B1-3 when hinged at both ends. While specimen B1 showed a bifurcation point almost corresponding to the Euler's buckling load (17.5 kN), specimens B2 and B3 showed the typical response of beams with initial geometric imperfection. Figure 8.4b shows the analogous curves for the case in which specimens B1-3 were hinged at one end and clamped at the other. In this case larger differences between the expected and the measured results were obtained. In fact, the graph corresponding to specimen B1 showed a bifurcation in correspondence of a load equal to about 30 kN, while the theoretical Euler's load was equal to 37.2 kN. This behavior can be interpreted as an inefficiency of the constraint condition at the upper joint, with a condition of not-perfectly-clamped end.

For each testing configuration, the fundamental vibration frequency was determined under different values of the mean axial force obtained imposing different vertical displacements to the beam end. The fundamental natural frequencies were obtained by applying three different and independent procedures: (1) by identifying the resonance condition in the case of forced oscillation, i.e. analyzing the response corresponding to different frequencies of the harmonic excitation transmitted by the electromagnet; by applying the Logarithmic Decrement Method to the free response signal induced by opportune initial conditions; and, lastly, by performing the Fast Fourier Transform of the free response signal. In all the analyzed cases there was a perfect correspondence between the results given by the three adopted methods.

Figure 8.5 shows the fundamental frequency vs. axial load curves for specimens B1-3, in both the analyzed constraint conditions: Fig. 8.5a refers to the hinged–hinged constraint condition, while the results related to the hinged-clamped constraint condition are collected in Fig. 8.5b. First of all, we observe that the zero-frequency condition was never reached, due to the combined effects of the geometric and material imperfections (the rectified beam was non a perfect beam in reality) and of the corresponding non-trivial equilibrium path associated to the imposed displacement. As can easily be seen, in all the analyzed cases two different trends of the frequencies with respect to the axial load were observed. In fact, the evolution of the fundamental frequency showed a decrease down to

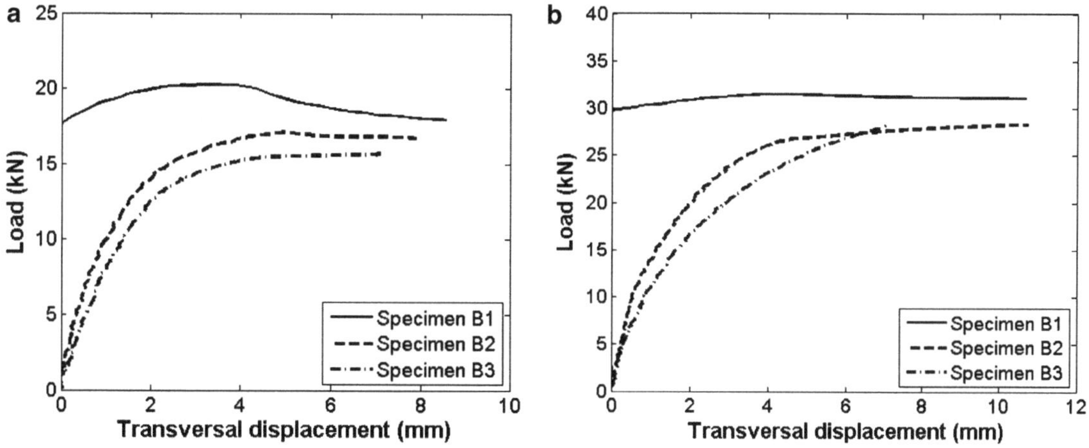

Fig. 8.4 Axial load vs. midpoint transversal displacement curves for specimens B1-3: (**a**) hinged–hinged and (**b**) hinged-clamped constraint condition

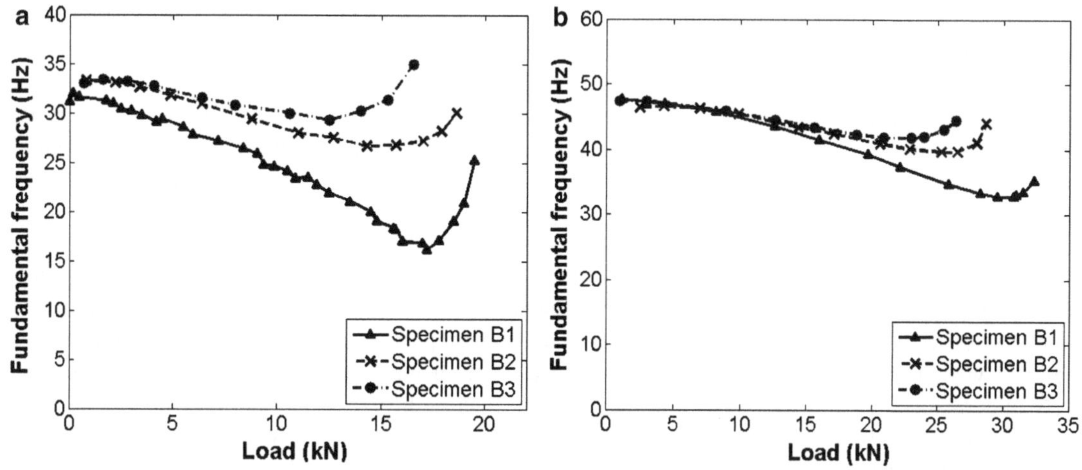

Fig. 8.5 Fundamental frequency vs. axial load curves for specimens B1-3: (**a**) hinged–hinged and (**b**) hinged-clamped constraint condition

a minimum in the curve, after which the natural frequency started to increase sharply, thus denoting a stiffening of the specimen. In this case of displacement controlled experiments the response was always stable, but after a certain value of the transversal displacement the tendency was reversed. In the first phase, the specimens followed a straight beam-like behavior, showing a reduction in the fundamental frequency as the axial compression load increased. During the second phase, the specimens exhibited an arch-like (curved beam-like) behavior, showing a stiffening leading to an increase in the fundamental frequency.

Also, an important influence of the initial imperfections on the natural frequency evolution was observed. In fact, while for the rectified beam the transition from the first to the second phase was particularly sharp, specimens B2 and B3 showed wider transition intervals: the greater the initial geometrical deviations the smaller the slope in the first branch of the frequency trend and the smoother the transition phase (see Fig. 8.5). Moreover, the presence of the initial curvature favored the transition from the first to the second phase (see Fig. 8.5). Although the behaviors shown in Fig. 8.5a, b are similar, we observe that in the case of hinged-clamped specimens the three curves are closer to each other than in the case of hinged–hinged specimens: this can be read as an effect due to the additional stiffness given by the fixed joint, which induced the three specimens to behave in a more similar way than in the hinged–hinged case.

8.5 Multi-body Dynamic Simulations

Numerical models were implemented in order to reproduce the experimental tests conducted on specimens B1, B2 and B3 in the hinged–hinged constraint condition. The simulations were conducted using Working Model 2D, a multi-body dynamic simulator. Figure 8.6a shows a general view of the model reproducing the rectified beam. The model was constituted by nine masses, ten massless connection units, ten axial springs, eighteen torsional springs, two end constraints, and, lastly, a rod element that simulated the effect of the hydraulic jack. The model concentrated the mass of the specimen into nine squared elements, each one being 2 cm long and having a mass equal to one ninth of the specimen's total mass; in particular, the value was set up equal to 3.09×10^{-3} kg. These masses were connected to each other through connection units that were massless and infinitely rigid in flexure, but at the same time they allowed axial displacements (Fig. 8.6b). The axial stiffness of the specimen was concentrated in ten axial springs, placed inside the connection units. These springs presented a relaxed length of 2 cm and a linear stiffness equal to 9.45×10^{8} N/m. The flexural stiffness of the specimen was concentrated into the torsional springs connecting the masses to the rigid massless elements (see Fig. 8.6b). Each torsional spring had a linear stiffness equal to 6.0×10^{4} Nm/rad. The external constraints were represented by two hinges, one of which was fixed and simulated the upper joint, while the other one could slide on a guide in order to simulate the lower joint (see Fig. 8.2a). The hydraulic jack was simulated by means of an infinitely rigid rod element, whose length could be adjusted in order to reproduce the imposed displacements obtained experimentally.

The numerical models allowed the simulation of the response of the three specimens, with their geometrical imperfections. Specimen B1 (rectified beam) was considered perfectly straight, and its initial configuration was assumed as rectilinear. For specimens B2 and B3 an originally curved shape was generated to reproduce the imperfect geometry of the two specimens. This approximation was very close to the experimental conditions realized in the laboratory and it allowed us to obtain a good numerical reproduction of the experimental results. Therefore, under each imposed displacement, the fundamental vibration frequency was obtained by analyzing the time history of the transversal displacement of the central mass. The initial condition was represented by a deformed shape similar to the first vibration mode of the beam, with zero-velocity. The maximum displacement imposed in the initial condition was equal to 2 mm (deflection of the central mass), in all the simulations. Further details concerning the numerical procedure and the simulation accuracy adopted for the solution can be found in [8].

Figure 8.7 shows the comparison between the experimental results and those obtained by the numerical simulations. It is possible to observe that fundamental frequency vs. load behaviors obtained during the tests were well reproduced by the numerical models. In the case of specimen B1, for instance, the frequency trend obtained by increasing the imposed displacement was well reproduced, showing an abrupt transition from the first to the second phase: the cusp localized approximately in correspondence of 16 kN is even more evident than in the experimental plot (see Fig. 8.7a). Figure 8.7d collects all the fundamental frequency vs. load curves obtained by the numerical simulations. From those graphs it is possible to observe that all the aspects originally observed from the experimental results were confirmed and reproduced by the numerical simulations.

An improvement of the closeness of the numerical values to the experimental ones could be obtained by refining the models. A larger number of elements, as well as a better fitting of the numerical values attributed to mass and stiffness of each component should induce a better reproduction of the experiments. Finally, models like those implemented in the present study may be useful to predict, by simulating laboratory tests, the behavior of structural elements to obtain indications in designing new structures and components or to offer a valid interpretation to experimental measurements carried out on existing structures subjected to imposed displacements or constraint settlements.

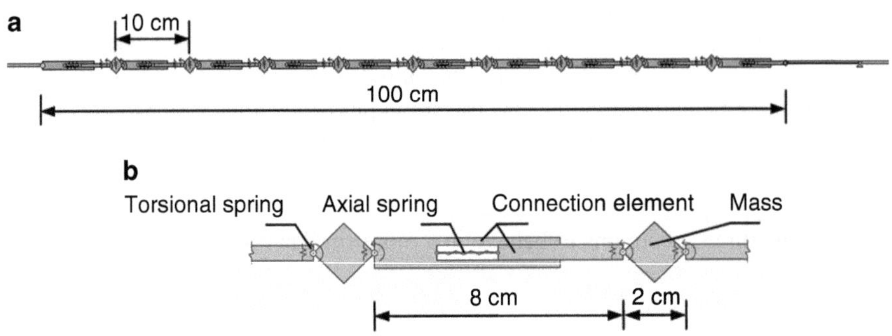

Fig. 8.6 (**a**) General view of the multi-body model implemented and (**b**) detail of a portion

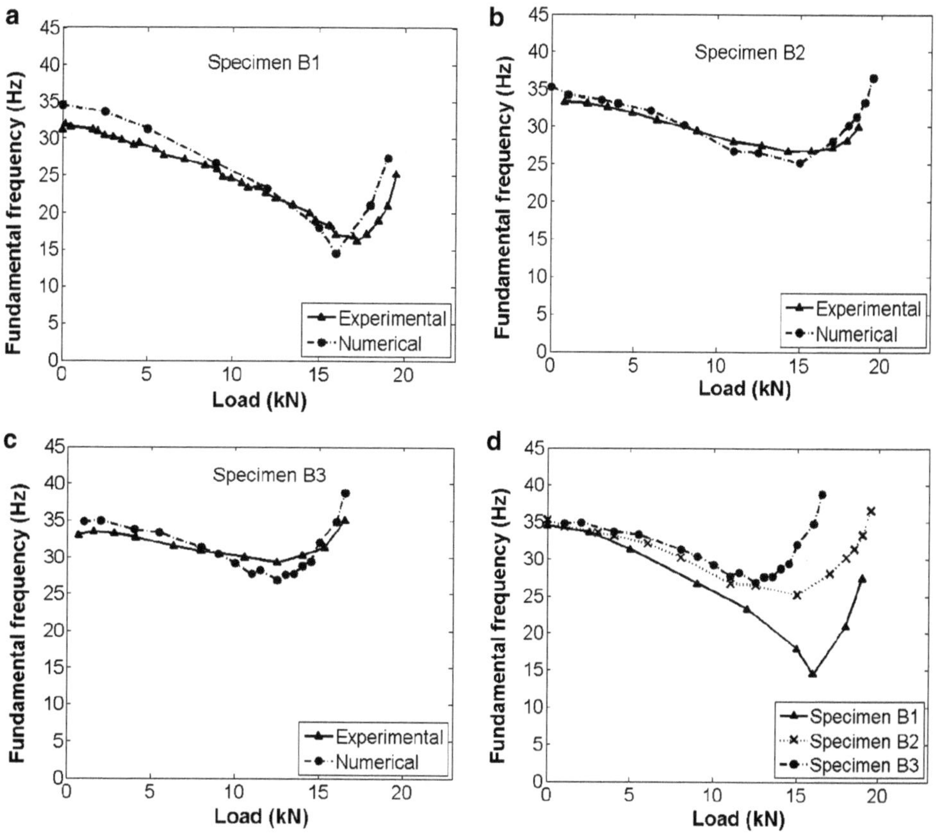

Fig. 8.7 (**a–c**) Comparison between experimental and numerical fundamental frequency vs. axial load curves for specimens B1-3 and (**d**) collection of numerical results for specimens B1-3

8.6 Conclusions

The evolution of the fundamental frequency in slender beams subject to prescribed axial end displacements was experimentally investigated: three elements presenting different initial curvatures were tested in the hinged–hinged and hinged-clamped constraint conditions. For all the analyzed cases, it can be noted that a first phase, where the fundamental frequency decreases with the axial load, is followed by a stiffening one, where the trend is reversed. The effect of the geometric imperfection (initial curvature) is to reduce the negative slope of the first branch, therefore anticipating the transition point and making this transition smoother at the same time. This type of behavior, analytically predicted by previous studies [5, 7] for other shapes of the initial imperfection and end conditions, must be well taken into account when designing new structures as well as in monitoring existing ones. For example, the possibility to foresee the transition point, where the minimum frequency (minimum stiffness) is localized, could find application in monitoring in-service structures and components in order to plan consolidating interventions. Further, numerical simulations can be helpful in interpreting in-situ measurements, as well as they may allow us to perform virtual laboratory tests.

References

1. Virgin LN (2007) Vibration of axially loaded structures. Cambridge University Press, New York
2. Woinowsky-Krieger S (1950) The effect of an axial force on the vibration of hinged bars. J Appl Mech 17:35–36
3. Burgreen D (1951) Free vibrations of a pin-ended column with constant distance between pin ends. J Appl Mech 18:135–139
4. Dickinson SM (1980) The lateral vibration of slightly bent slender beams subject to prescribed axial end displacement. J Sound Vib 68(4):507–514

5. Kim CS, Dickinson SM (1986) The flexural vibration of slightly curved slender beams subject to axial end displacement. J Sound Vib 104(1):170–175
6. Yamaki N, Mori A (1980) Non-linear vibrations of a clamped beam with initial deflection and initial axial displacement. Part I: theory. J Sound Vib 71(3):333–346
7. Ilanko S (1990) The vibration behaviour of initially imperfect simply supported beams subject to axial loading. J Sound Vib 142(2):355–359
8. Carpinteri A, Malvano R, Manuello A, Piana G (2014) Fundamental frequency evolution in slender beams subjected to imposed axial displacements. J Sound Vib 333:2390–2403

Chapter 9
Characterization of a Heating and Quenching Apparatus for Microgravity Testing

Anthony S. Torres, Jeff Ganley, and Arup Maji

Abstract Microgravity experimentation enables new materials to be developed and traditional materials to be improved, which can't be completed under terrestrial conditions. Recent developments on Heavy Metal Fluoride Glasses (HMFG's) have shown that, when heated, there is a crystallization dependency on gravity. HMFG's have the potential for optical transmission from 0.3 μm in the UV to 7 μm in the IR region, enabling fiber optic applications such as fiber amplifiers, radiometry, and mid-IR laser technology for surgery, drilling and cutting. The problem of devitrification from heat processing prevents this material from achieving its theoretical transmission range. Past researchers have shown that crystallization of HMFG's is suppressed in microgravity and enhanced in hyper-gravity, however further investigation is still needed for a determination of this phenomenon. In this study, a HMFG heating and quenching testing apparatus was characterized and developed for microgravity and hyper-gravity testing. The testing apparatus was developed and characterized for use on a parabolic aircraft that provides a microgravity and hyper-gravity environment for experimental testing. The apparatus was successful in processing HMFG's, which produced crystalline and non-crystalline glasses for future studies.

Keywords Heavy metal fluoride glasses (HMFG) • Characterization • Heating and quenching testing • Microgravity • Crystallization

9.1 Introduction

Heavy Metal Fluoride Glasses (HMFG)s are fluoride based optical fibers that have the potential for very low optical attenuation especially in the infrared region (2–7 μm), however these glasses are extremely susceptible to crystallization.

Past researchers have shown that crystallization of HMFG's is suppressed in microgravity and enhanced in hyper-gravity, showing a crystallization dependency on gravity. The specific material used for this research is a fluorozirconate glass, a subset of the HMFG family known as ZBLAN (ZrF_4-BaF_2-LaF_3-AlF_3-NaF). The optical transmission spectrum for ZBLAN is from 0.3 μm in the Ultraviolet (UV) to 7 μm in the Infrared (IR) region [1]. Many researchers have conducted studies on crystal growth under microgravity [2–7]. While these researchers have studied ZBLAN, protein crystals, and semiconductor growth in microgravity, all have noticed that crystallization processes are suppressed in microgravity. These studies used sounding rockets, parabolic flight aircraft and even a drop tower to achieve free fall. The authors also noticed that crystal growth was increased during times of high accelerations [6]. When using the parabolic flight aircraft there are periods of approximately 2-g accelerations. Varma et al. detected twice as many crystals formed during the period of high acceleration as terrestrial processed fibers. Most of these experimental programs used an automated glass heating apparatus and have limited detail to the exact science and build of the experimentation. This study focuses on the characterization of a HMFG

A.S. Torres (✉)
Department of Engineering Technology, Texas State University, San Marcos, TX 78666, USA
e-mail: ast36@txstate.edu

J. Ganley
Air Force Research Lab/Space Vehicles Directorate, Kirtland AFB, NM 87117, USA

A. Maji
Department of Civil Engineering, University of New Mexico, Albuquerque, TX 87106, USA

N. Sottos et al. (eds.), *Experimental and Applied Mechanics, Volume 6: Proceedings of the 2014 Annual Conference on Experimental and Applied Mechanics*, Conference Proceedings of the Society for Experimental Mechanics Series, DOI 10.1007/978-3-319-06989-0_9,

heating and quenching testing apparatus for microgravity and hyper-gravity testing. The testing apparatus was characterized for use on a parabolic aircraft that provides a microgravity and hyper-gravity environment for experimental testing. The apparatus was ultimately successful in processing HMFG's, which produced crystalline and non-crystalline glasses for future studies.

9.2 Microgravity Experimental Testing Apparatus

The heating and quenching testing apparatus used for this study is a two-stage heating apparatus. The apparatus is also known as the "Quencher" and is used to describe the assembly throughout this paper. A photo of the Quencher can be seen in Fig. 9.1.

The two-stage furnace and pushrod system was originally designed by Tucker et al. [8] in 1997 and was donated to this study for use on the parabolic aircraft. This research focuses on the experimental technique, operating procedure, and updated developments required for the Quencher. The furnaces, pushrod, and electrical system were re-used and all other components and modifications were completed as part of this study. The Quencher assembly consists of two parts: the actual Quencher apparatus; ~22 in. L × 10 in. W × 12 in. H; ~25 lb (~55.88 cm L × 25.4 cm W × 30.48 cm; ~11.34 kg) and a table on which the Quencher sits, ~26 in. L × 22 in. W × 30 in. H; ~30 lb (66.04 cm L × 55.88 cm W × 76.2 cm; 13.6 kg). The Quencher table is constructed from 80/20 aluminum (6105-T5) framing material (www.8020.com), and has an approximately 1 in. (2.54 cm) thick piece of foam between the table and the Quencher for vibration damping/isolation. The two different furnaces were used as a pre-heating step and the annealing step of the experiment. Due to the limited μ-g time of the aircraft, it is essential that the sample reached the desired temperature during the controlled time frame. Thus a pre-heat furnace would reduce the temperature differential of the testing temperature. The two furnaces are encased in a Plexiglass box that holds the electronic controllers and also prevents any debris from entering the furnace area. During operation, a single quartz ampoule of average length 1.25 in. × 0.11 in. (3.2 cm by 3 mm) in diameter, containing two ZBLAN samples were placed in the end of the aluminum pushrod. The aluminum pushrod mounted the sample and then translated the sample to each furnace and then out the other end of the Quencher. Each ampoule contained two 0.11 in. × 0.04 in. (3 mm by 1.03 mm) diameter samples of the same type. A small piece of silica glass was used to separate the two ZBLAN samples to ensure they don't come in contact with each other. Eliminating sample-to-sample contact prevents any potential (heterogeneous) crystal nucleation [9]. The ZBLAN samples are free to contact the silica glass separator and the ampoule walls. The silica glass has a very stable molecular structure and a glass transition temperature (Tg) of 1,175 °C [1]. Hence it does not physically or chemically interact with the

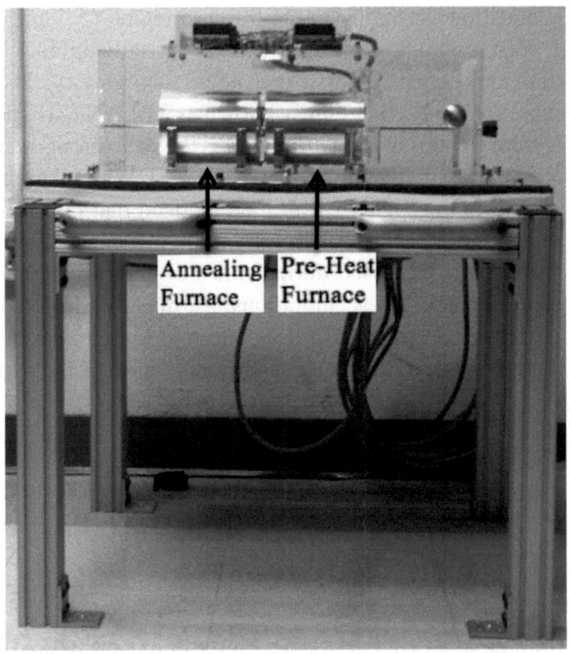

Fig. 9.1 The Quencher (heating and quenching apparatus)

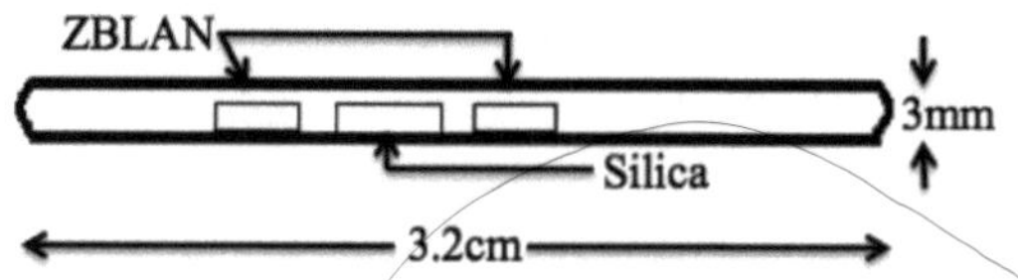

Fig. 9.2 Schematic of ampoule housing w/ZBLAN samples and Silica separator glass

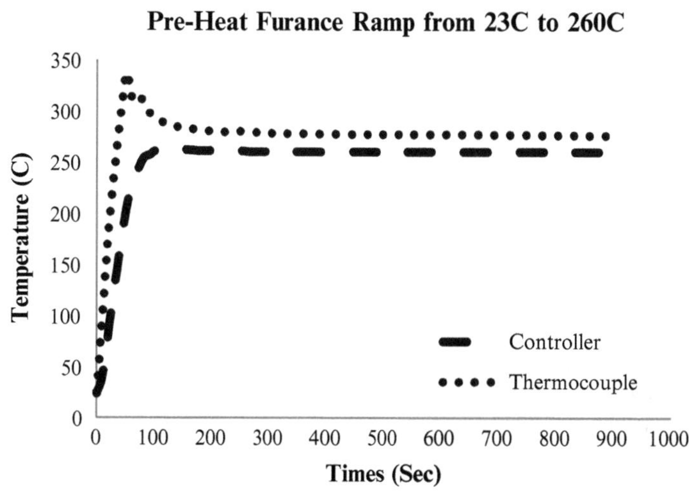

Fig. 9.3 Pre-heat furnace ramp time from 23 to 260 °C

ZBLAN during the experiment [9]. No alteration to the silica glass was observed during experimentation; the two sides of the ZBLAN samples did not demonstrate any observable differences, irrespective of their proximity to the Silica glass. Any effect of moisture was removed by sealing the ampoules under vacuum. The ampoule and sample configuration can be seen in Fig. 9.2.

Prior to any testing the Quencher was initially characterized. The purpose of this was to gain a better understanding of how the Quencher operates, how to manually control the Quencher, the temperature ramp up/cool down time, and any other critical information required to obtain accurate results. The characterization process was done with the furnace controller output and a type-K thermocouple with a diameter of 0.031 in. (0.787 mm).

9.2.1 *Pre-heat Furnace*

The purpose of this furnace is to increase the temperature of the sample so that the sample reaches the annealing temperature quicker. Due to the μ-g/hyper-g transitions there is limited time to raise the sample temperature to the annealing temperature. Therefore, the Pre-Heat furnace was required to reach the annealing temperature in an appropriate timeframe (~22 s). For example, one temperature test point is 350 °C (annealing furnace) where the Pre-Heat furnace will be set at 250 °C, which is below the crystallization temperature (360 °C) and doesn't affect experimental results. The Pre-Heat furnace reduces the temperature differential to 100 °C, in this example, as opposed to a ~320 °C difference from room temperature (~22 °C). The main purpose of the characterization process is to compare the controller read out temperature to that of the thermocouple. Thus, the operators know a more precise temperature range inside the furnace. For all characterization tests, the thermocouple was placed directly in the center of the furnace, unless otherwise stated. Figure 9.3 shows the temperature ramp from near room temperature to a final temperature of 260 °C. This test will ultimately show how fast it will take the furnace to stabilize at its set point, as well as how far the actual temperature over-shoots.

It can be seen (a 3) that within 1 min the thermocouple jumps about 100° over the set point on the controller. However, the temperature drops back to the to within the set point within 2 min. The controller also goes above the temperature set point but only by approximately10°, and stabilizes within 2 min. Also represented on the graph is that after both the controller and thermocouple have stabilized, the thermocouple still reads at a higher temperature. The exact readings are 260 °C on the controller and 276 °C on the thermocouple. Figure 9.4 shows the time required to increase the temperature by 10° (from a previous stable temperature, 260 °C).

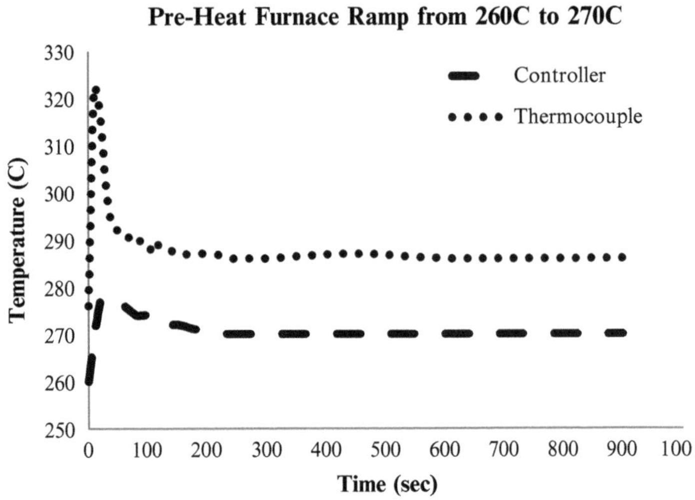

Fig. 9.4 Pre-heat furnace ramp from 260 to 270 °C

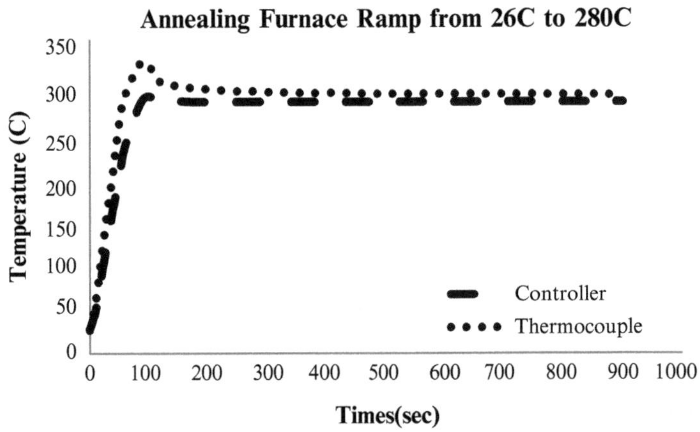

Fig. 9.5 Annealing furnace ramp time from 26 to 280 °C

Figure 9.4 shows an immediate jump for both the controller and the thermocouple. The thermocouple maxes out at 322 °C within 10 s, and the controller at 278 °C within 30 s. Both the thermocouple and controller stabilize within 3 min. As previously noted, the thermocouple stabilizes at a higher temperature (286 °C) than the controller (270 °C). This test was completed to determine the time required for the pre-heat furnace to stabilize if the operator chose to adjust the temperature by at least 10°.

9.2.2 Annealing Furnace

The purpose of the annealing furnace is to raise and hold the temperature of the sample to the applicable testing temperature while in μ-g or hyper-g depending upon the test matrix. The sample will be in this furnace for the duration of the desired gravity level (μ-g/hyper-g). The annealing furnace was characterized in the same manner as the pre-heat furnace. Figure 9.5 shows the annealing furnace ramp from room temperature to a set point of 280 °C.

As seen in Fig. 9.5, the temperature quickly jumps up to above the set point, but eventually levels out to the desired temperature. The thermocouple reads a higher temperature initially (325 °C) but eventually settles down to just above the set point (288 °C). The controller reaches the set point within 3 min for the thermocouple; the set point is never reached but stabilizes within 6 min. The next step in characterizing the annealing furnace was to increase the temperature by 20°. This would show how quickly the furnace would adjust to a small temperature change. Figure 9.6 shows the annealing furnace ramp from 280 °C to the new set point of 300 °C.

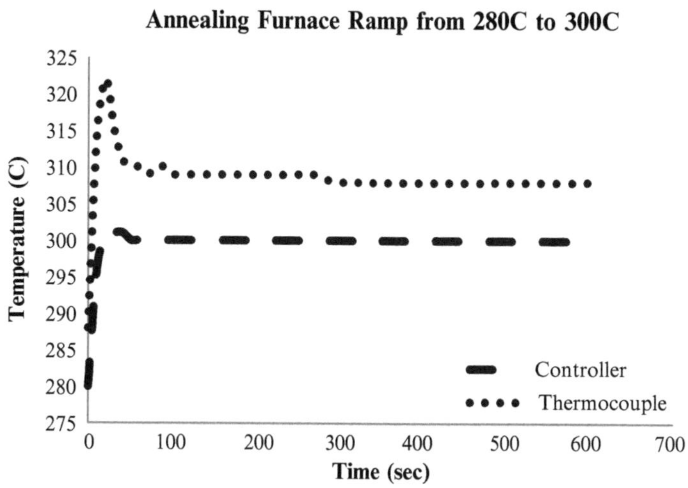

Fig. 9.6 Annealing furnace ramp time from 280 to 300 °C

As depicted in Fig. 9.6, the thermocouple reading spikes to a much higher temperature than the controller reading does. Similar to findings with the pre-heat furnace characterization, the thermocouple read a few degrees higher than the controller. The thermocouple read an initial high of 310 °C within 90 s of recording while the controller peaked at 301 °C and leveled off within 60 s. The thermocouple leveled off at a higher temperature of 308 °C. The 8° difference was also noticed in previous characterization testing.

9.3 Experimental Program

The majority of the experimentation was completed on board the Zero-G Corporation's parabolic aircraft sponsored by the Air Force Space Test Program (STP) and the National Aeronautics and Space Administration (NASA). The Zero-G Corporation stipulates a pristine 0.00 ± 0.05 g environment of approximately 22 s. The transition between the microgravity portion of the flight puts the aircraft in a hyper-g environment of approximately 1.2–1.8 g. The hyper-g environment was also used as an experimental variable for testing.

The in-flight Quencher operation requires two operators, one to load the samples and operate the controllers and the second to quench and store the samples. The operation consisted of the following steps, depending upon the desired temperature set point. Once the ampoule is loaded into the end of the pushrod the pushrod is manually translated to the center of preheat furnace for the duration of either the μ-g or the hyper-g section of the flight. After the sample has been pre-heated the pushrod is relocated to the center of the annealing furnace for the duration of the gravity level under testing. Lastly, the pushrod is extended out the end of the annealing furnace and the glass ampoule is grasped with a damp sponge and removed from the pushrod. This final process simultaneously quenches the sample as the operator removes the ampoule from the pushrod. This entire process will take place during the μ-g/hyper-g sections of the parabolic flight. The test matrix for both flights was developed based on a temperature scheme that spanned just below the glass transition temperature (270 °C) to well above the crystallization temperature (650 °C). Both the microgravity and hyper-g portions of the parabolas were used as variables in the experiment. Lower set point temperatures (270–350 °C) were studied in the hyper-g section to determine if crystallization can be initiated below the recorded (1-g) crystallization temperature of 360 °C. The second set of tests (360–650 °C) were established to determine if crystallization can be suppressed in a μ-g environment and to determine at what temperature will crystallization occur regardless of the μ-g environment.

9.4 Analysis

Optical microscopy was used as the primary investigation technique as the transmitting light through the sample will illuminate any transmission inclusions (crystals). A single temperature was selected to represent each technique. The temperature focused on was 360 °C, in μ-g and 1-g, as this is the reported crystallization temperature of the ZBLAN

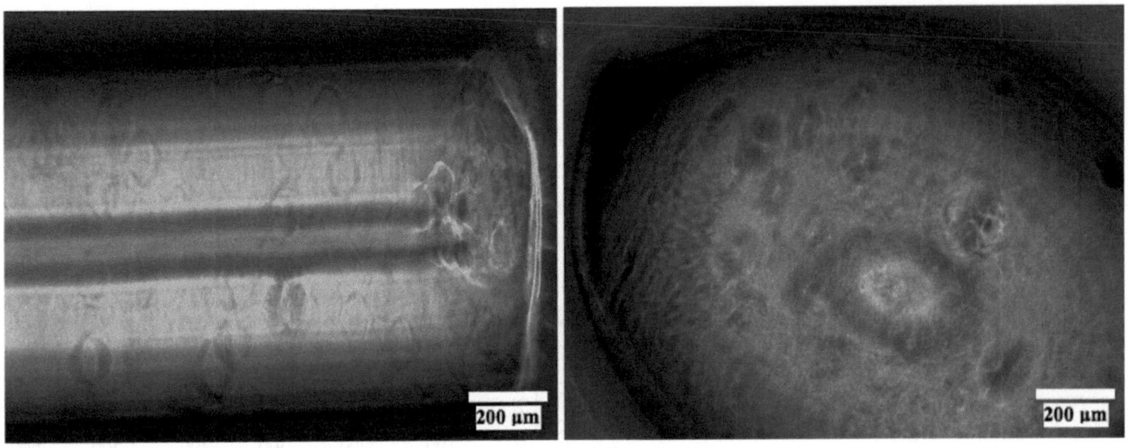

Fig. 9.7 Phase contrast optical microscopy (×100): μ-g (*left*) and 1-g (*right*)

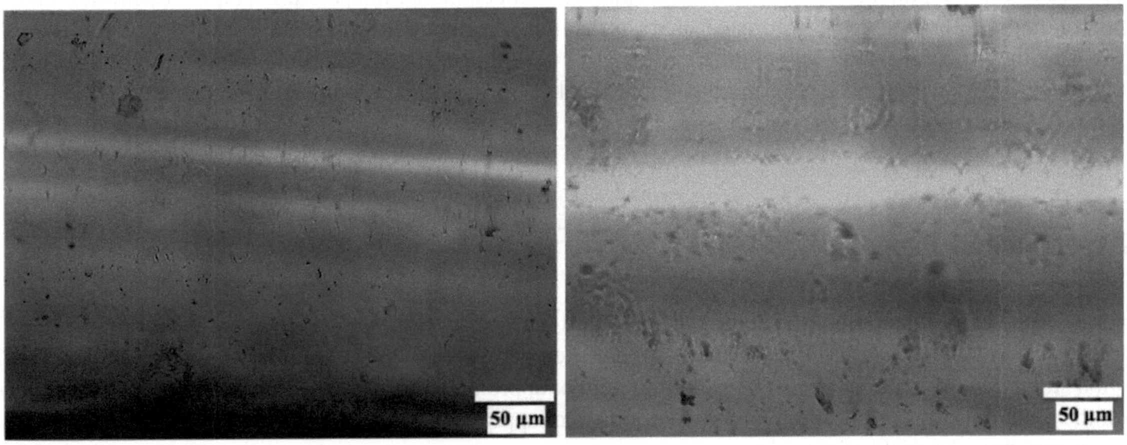

Fig. 9.8 Phase contrast optical microscopy (×400): μ-g (*left*) and 1-g (*right*)

material. One micrograph (μ-g/1-g) per analysis technique is sufficient to demonstrate the processing technique and capability of each analysis method. For all techniques described, the ampoule (sample housing) was opened and the samples were removed, placed lengthwise on microscope platform and imaged. Standard optical microscopy didn't reveal any essential features, thus phase contrast optical microscopy was used. Phase contrast microscopy works by introducing a stained slide, which enhances the contrast of the image. This process highlights essential features such as crystals and other cellular structures not apparent in standard microscopy. Each micrograph was taken with a constant light intensity and constant focus, to minimize sample-to-sample image variability. The microscope used was an Olympus BX51, which has a built-in stained slide for phase contrast microscopy. Three different magnifications, 100×, 400×, and 1,000×, were taken using this method; the micrographs can be seen in Figs. 9.7, 9.8, and 9.9, respectively.

As shown in the above figures, phase contrast microscopy readily shows the crystalline features of the samples. It is also shown that the μ-g (left) consistently has fewer inclusions than the 1-g (right). The higher magnification (400×/1,000×) was overall beneficial to this process, however it's difficult to use the small area to represent the sample in its entirety. The higher magnification levels should primarily be used to focus on a particular area for detailed study or to determine the approximate size of smaller inclusions not apparent in lower magnification. Further analysis such as, polarized optical microscopy and Scanning Electron Microscopy were completed on these samples and will be reported in future reports on this topic. Further processing techniques that could be utilized include X-Ray Diffraction (XRD), Transmission Electron Microscopy (TEM) and Atomic Force Microscopy (AFM). These methods well aid in capturing the crystal type and other specific crystallinity issues related to the sample characterization.

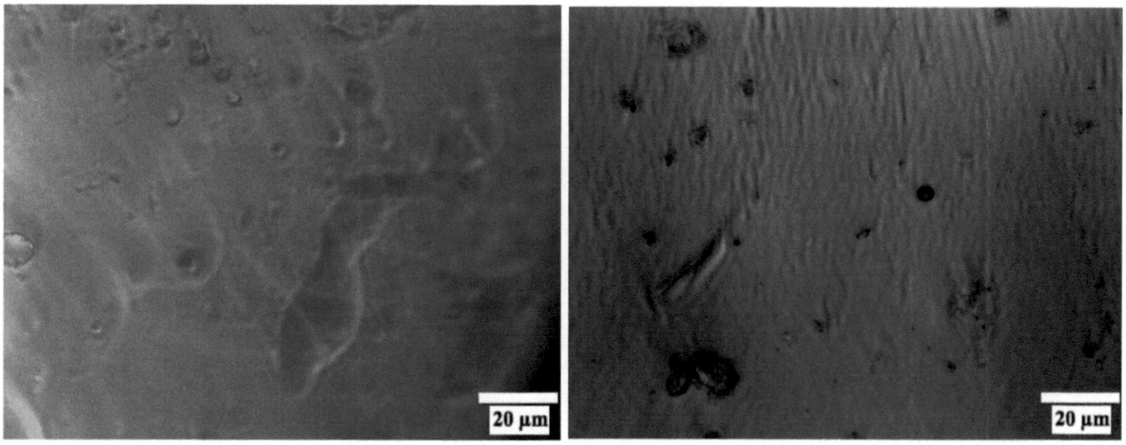

Fig. 9.9 Phase contrast optical microscopy (×1,000): µ-g (*left*) and 1-g (*right*)

9.5 Conclusions

This research reports on the experimental techniques and processing methods for ZBLAN crystallization suppression studies in microgravity. This experimental and processing campaign can be utilized to analyze similar materials that exhibit similar behavior, such as chalcogenide glasses and semiconductor materials that can be processed in 1-g/µ-g. Prior to this report, there were few researchers working in this area [2, 6, 7, 10, 11, 12] with minimal reporting on the experimentation, but mostly reporting that the phenomenon of ZBLAN crystallization suppression in µ-g exists. Therefore, a more detailed experimental and analytical method is needed. The experimental program included in this study describes a heating and quenching testing apparatus, known as the 'Quencher'. The Quencher was flown on the Zero-G Corporation's parabolic aircraft, which provided a µ-g and hyper-g environment, 1-g based tested was also completed for comparison. A detailed description of the in-flight operation was described as well as the analysis (imaging) technique used to elucidate the degree of crystallinity in the sample. The samples were analyzed using phase contrast optical microscopy. This technique confirms that crystallization is suppressed in µ-g at 360 °C by showing the degree of crystallinity between each sample. In addition to the previously mentioned techniques, supplementary analysis techniques were also recommended for future studies including polarized optical microscopy, SEM, XRD, TEM, and AFM. Future work to be completed by Torres et al. is to determine the effect gravity plays on the crystallization suppression of ZBLAN glass.

Acknowledgements The author's of this work would like to thank Dr. Dennis Tucker, the Air Force Space Test Program, and NASA for their generous support of this research.

References

1. Harrington JA (2007) Infrared fiber optics. In: M. Bass, J.M. Enoch, E.W. Van Stryland, and W.L. Wolfe, (eds) Handbook of Optics: Fiber and Integrated Optics, Vol. 4, Mcgraw-Hill, New York, pp 12.1–12.14
2. Tucker D (2004) Effects of gravity on ZBLAN glass crystallization. Annal NY Acad Sci 1027:129–137
3. Kundrot C (2001) Microgravity and macromolecular crystallography. Cryst Growth Design V1(1):87–99
4. Regel LL (1990) Experiments on crystallization of semiconductor materials, eutectic alloys and crystal growth from water solution in microgravity. Acta Astronaut 21(5):331–348
5. Voloshin AE (2002) Perfection and homogeneity of space-grown GaSb:Te crystals. Crystallogr Rep 47(suppl 1):S136–S148
6. Varma S (2001) Effect of gravity on crystallization in heavy metal fluoride glasses processed on the T-33 parabolic flight aircraft. J Mater Sci V36:4551–4559
7. Dunkley RI (2004) The study of devitrification processes in heavy-metal fluoride glasses. Annal NY Acad Sci V1027:150–157
8. Tucker D (1997) Effects of gravity on processing heavy metal fluoride fibers. J Mater Res V12(9):2223–2225
9. Tucker D, Ethridge E (2001) Explanation of the effects of gravity on crystallization of ZrF_2-BaF_2-LaF_3-AlF_3-NaF glass. J Mater Res 16(11):3027–3029
10. Tran DC, Siegel GH, Bendow B (1984) Heavy metal fluoride glasses and fibers: a review. J Lightwave Technol 2(5):121–138
11. Wilson SJ (1985) Observations of crystallization in fluorozirconate glasses. Mater Sci Forum 5:269–274
12. Workman G, Smith G, O'Brien S, Adcock L (1995) ZBLAN microgravity study. In: Final report submitted to national aeronautics and space administration George C. Marshall space flight center

Chapter 10
Phase Unwrapping Work Done via Graphic Processing Unit

M.J. Huang and Y.C. Liu

Abstract Phase-shifting technique is widely used in phase detection field. With the aid of this technique, hundreds to thousands of wavelength accuracy result can be achieved. Further calculating the grabbed phase-stepping frames yields the phase map, which is in wrapped format and has to be further treated to convert into a continuous (i.e., an unwrapped) mode. Different algorithms—path-dependent or path-independent, temporal or spatial, point-by-point or regional, noise-sensitive or noise-immune, etc., have been proposed for solving problems in different applications. In present study, graphic processing unit (GPU), which has been developing rapidly in recent years, is utilized in addition to shorten and accelerate the processing time needed. By the aid of GPU parallel processing technique, the retrieving work is much more time effective. Since temporal phase unwrapping deals wrapped data from load-stepping or wavelength-stepping basis, it can be easily benefit from the parallel processing of GPU. Experimental work of photoelasticity is verified by the present study.

Keywords Photoelasticity • Temporal phase unwrapping • Parallel processing • Isoclinic • Isochromatic

10.1 Introduction

GPU consists of thousands of cores which can handle multiple tasks simultaneously, comparing with CPU (central processing unit) which has only a few cores to deal with serial processing tasks. On usage of GPU, the processing time can be dramatically reduced, especially while temporal wrapped data are treated. Temporal phase unwrapping is an effective way of phase retrieving method, like load stepping method [1] and different wavelength method [2, 3]. This technique later on is applied on the isoclinic as well as isochromatic wrapped phase retrieving and gets quite well results. Kihara [4] proposed a coincidence method with three wavelengths (RGB) to retrieve the phase jump $2\pi N$ for wrapped isochromatic, which can get unwrapped isochromatic correctly. This algorithm is implemented via parallel pixels processing and can reduce the consumed time significantly.

10.2 Parallel Processing

Parallel processing is the ability to handle multiple tasks simultaneously, and GPU has highly parallel structure that makes it more effective than CPU for algorithms where processing of an amount of data is done in parallel. CUDA (the Compute Unified Device Architecture), is a parallel computing platform created by NVIDIA, and carried out by the GPUs that they produce. In CUDA, the data must be copy to the memory of GPUs for parallel processing, and after that, the result will be

M.J. Huang (✉) • Y.C. Liu
Department of Mechanical Engineering, National Chung Hsing University, 250, Kuo-Kuang Road, Taichung, Taiwan 40227, Republic of China
e-mail: mjhuang@dragon.nchu.edu.tw

N. Sottos et al. (eds.), *Experimental and Applied Mechanics, Volume 6: Proceedings of the 2014 Annual Conference on Experimental and Applied Mechanics*, Conference Proceedings of the Society for Experimental Mechanics Series,
DOI 10.1007/978-3-319-06989-0_10,

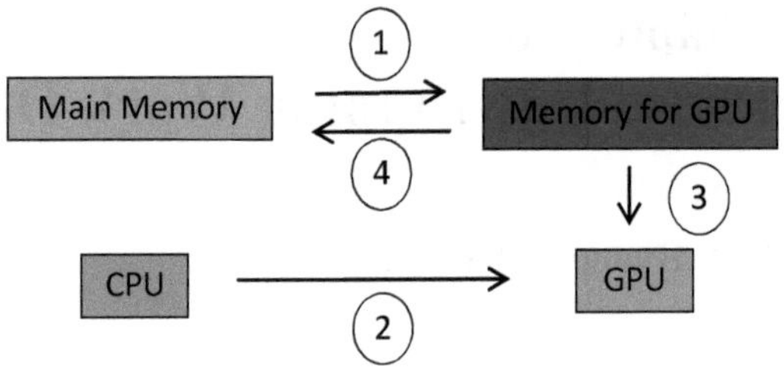

Fig. 10.1 CUDA processing flow (*1* Copy data, *2* Instruct the tasks, *3* Execute parallel in each core, *4* Copy the result)

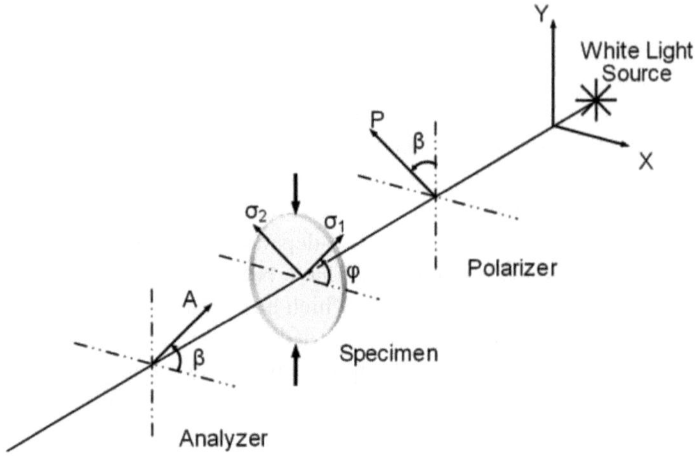

Fig. 10.2 The optical arrangement of plane polariscope

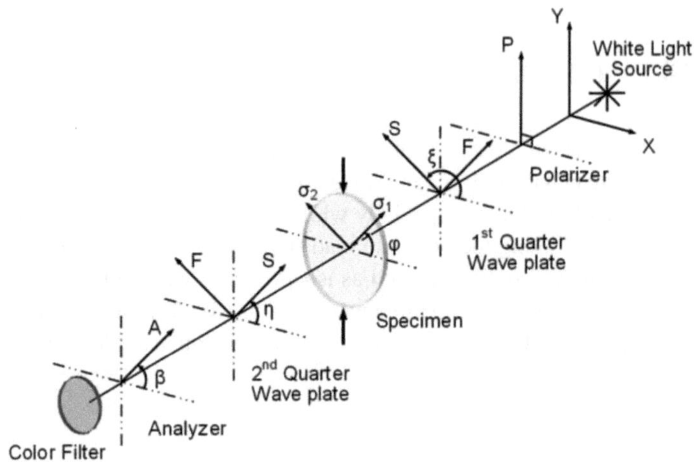

Fig. 10.3 The optical arrangement of circular polariscope

coping to the main memory to make use of it, the processing flow is shown in Fig. 10.1. The CUDA platform is accessible to industry-standard programming languages, including C, C++ and FORTRAN, which can adapt to various develop environment.

10.3 Phase Stepping Photoelasticity

A general optical arrangement of a plane polariscope and a circular polariscope are shown in Figs. 10.2 and 10.3, respectively.

White light is used as the light source and a general digital camera with RGB filters is used as the fringes recorder. Table 10.1 summarizes the used phase stepping parameters and their corresponding intensity results. The isoclinic φ_w and isochromatic δ_w parameters can be obtained as follows.

Table 10.1 Polariscope arrangement and the intensity results for phase stepping algorithm

ξ	η	β	Intensity equation
–	–	0	$I_1 = I_B + \frac{1}{2}I_A \sin^2\frac{1}{2}\delta(1 - \cos 4\alpha)$
–	–	π/8	$I_2 = I_B + \frac{1}{2}I_A \sin^2\frac{1}{2}\delta(1 - \sin 4\alpha)$
–	–	π/4	$I_3 = I_B + \frac{1}{2}I_A \sin^2\frac{1}{2}\delta(1 + \cos 4\alpha)$
–	–	3π/8	$I_4 = I_B + \frac{1}{2}I_A \sin^2\frac{1}{2}\delta(1 + \sin 4\alpha)$
3π/4	π/4	π/2	$I_5 = I_b + \frac{1}{2}I_a(1 + \cos\delta)$
3π/4	π/4	0	$I_6 = I_b + \frac{1}{2}I_a(1 - \cos\delta)$
3π/4	0	0	$I_7 = I_b + \frac{1}{2}I_a(1 - \sin 2\varphi \sin\delta)$
3π/4	π/4	π/4	$I_8 = I_b + \frac{1}{2}I_a(1 + \cos 2\varphi \sin\delta)$
π/4	0	0	$I_9 = I_b + \frac{1}{2}I_a(1 + \sin 2\varphi \sin\delta)$
π/4	3π/4	π/4	$I_{10} = I_b + \frac{1}{2}I_a(1 - \cos 2\varphi \sin\delta)$

$$Ø_w = \frac{1}{4}\tan^{-1}\left\{\frac{I_4 - I_2}{I_3 - I_1}\right\} = \frac{1}{4}\tan^{-1}\left\{\frac{\sin Ø}{\cos Ø}\right\} \tag{10.1}$$

$$Ø = W^{-1}\{Ø_w\} \tag{10.2}$$

$$\delta_w = \tan^{-1}\left\{\frac{(I_9 - I_7)\sin 2Ø + (I_8 - I_{10})\cos 2Ø}{I_5 - I_6}\right\} = \tan^{-1}\left\{\frac{I_a \sin\delta}{I_a \cos\delta}\right\} \tag{10.3}$$

$W^{-1}\{\}$ represents the unwrapping operator, and $\tan^{-1}\{\}$ of Eqs. 10.1 and 10.3 is arctan2 function with their results ranging between −π/4 ~ π/4 and −π ~ π, respectively. Since the isochromatic data is dependent on the isoclinic data, the isoclinic data should be correctly unwrapped and ranged in the range of −π/2 ~ π/2 before substituting into Eq. 10.3. Provided the substituted isoclinic data are correct, the isochomatic calculation is ambiguity free and that makes the following retrieval work of the isochromatic data extremely easy.

10.4 The Proposed Method

The coincidence method proposed by Kihara [4] is used herein to solve the unwrapped phases of each pixel in a parallel manner (see Eqs. 10.4 and 10.5).

$$F_i = \left(\frac{\lambda_i}{C_i}\right)(\delta_{w,i} + 2\pi N_i) = 2\pi t(\sigma_1 - \sigma_2) \;\; (i = 1, 2, 3; N_i = 0, \pm 1, \pm 2, \pm 3, \ldots) \tag{10.4}$$

$$\mathrm{E} = \left[(F_1 - F_2)^2 + (F_2 - F_3)^2 + (F_3 - F_1)^2\right]^2 \tag{10.5}$$

where $(\sigma_1 - \sigma_2)$ is the difference of principal stresses of the model, t is the thickness, and C_i is the stress optical coefficient of the model at λ_i. N_1, N_2, N_3 can be find with the coincidence method using Eq. 10.4 with $\delta_{w,1}$, $\delta_{w,2}$, and $\delta_{w,3}$. The integer N_2 of the photoelastic model is given in a predefine range and N_2 is retrieved from the integral value that minimum of the error E in Eq. 10.5. Therefore, unwrapped isochromatic at λ_i is obtained from Eq. 10.6.

$$\delta_i = \delta_{w,i} + 2\pi N_i (N_i = 0, \pm 1, \pm 2, \pm 3, \ldots) \tag{10.6}$$

With the aid of the parallel processing, the algorithm described above can be accelerated while the calculation of each pixel is independent, compare with the serial processing is calculated pixel by pixel, the speed of unwrap is improved significantly.

10.5 Conclusions

The coincidence method proposed by Kihara is used herein to solve the unwrapped phases of each pixel in a parallel manner. GPU technique is used for the unwrapping of the isoclinic and isochromatic data. The consumed time is much less than one without GPU processor. The present results indicate that the on-line unwrapping work could be possible if the GPU processing technique can be used for the retrieving work of the temporal retrieval.

Acknowledgments The authors would like to thank the National Science Council of the Republic of China for financially supporting this research under contract Number NSC 102-2221-E-005-019.

References

1. Nurse AD (2002) Load-stepping photoelasticity. Opt Lasers Eng 38:57–70
2. Ramesh K, Deshmukh SS (1996) Three fringe photoelasticity – use of color image processing hardware to automate ordering of isochromatics. Strain 32:79–86
3. Chen TY (1997) Digital determination of photoelastic birefringence using two wavelengths. Exp Mech 37:232–236
4. Kihara T (2003) An arctangent unwrapping technique of photoelasticity using linearly polarized light at three wavelengths. Strain 39:65–71

Chapter 11
Classification of Low Velocity Impact Using Spiral Sensing Technique

Chijioke Agbasi and Sourav Banerjee

Abstract In this paper, the non-linear elastodynamics of a flat plate subjected to a low velocity foreign body impact is studied experimentally. The work is based on a central hypothesis that in addition to identifying the impact locations, the material properties of the foreign objects can be classified using acousto-ultrasonic signals. A novel cluster of thin piezoelectric sensors is proposed and a carefully formulated dominant frequency approach is studied to investigate the nonlinearities. Such nonlinearities with their highest resolution are quantified with the proposed Theodorus spiral configuration of the sensors (TSSC). It is found that the frequency and speed of the guided wave generated in the plate can be quantized based on the impactor's relationship with the plate, i.e. the wave speed and the impactor's mechanical properties are coupled. In this work, in order to characterize the impact location and mechanical properties of impactors, nonlinear transient phenomenon is empirically studied to decouple the understanding using the dominant frequency band (DFB) and lag coefficients of the acousto-ultrasonic signals through TSSC. Next the understanding was correlated with the elastic modulus of the impactor to predict transmitted force histories.

Keywords Impactor • Theodorus spiral • Acousto-ultrasonic • Elastodynamic • Dissipation

11.1 Introduction

Over the years, studying impact events and its effect on the host structure have become an important topic, especially in the aerospace industries. Space operating vehicles and aircrafts often receive impacts from the space debris present in the lower and greater earth orbit. However, to estimate the degree of damage in the real time, no comprehensive method exists that can characterize the transmitted force and the elastic modulus of the impactors during the impact, accurately. In this work, we show that this can be achieved through a data driven modified model. A key fact to note is that the coupled nature of the mechanical properties of the impactor and the structure influences the transmitted force in the structure and hence the degree of damage. Transmitted acousto-ultrasonic signals (AUSs) from the impact event can be processed to extract quantitative features that can characterize the physical properties by implementing ad-hoc signal processing algorithms [1, 2]. To localize the foreign impacts, generally, the aero structures are idealized to a flat plate in the laboratories and are considered in this work too. Boundary conditions of the plate, the geometry and the mechanical properties of the impactors and the plate, the impact velocity, etc. influence the nature of transmitted energy in the plate in the form of guided waves. Transient displacements, contact area and transmitted stresses are unknowns and the physical events are simultaneously influenced by the impactor and plate's response [3]. *Most analysis separates the problem into a local contact problem at the impact location and a global problem of the plate's response, which does not capture the true phenomena.* Here, we study the influence of spherical impactors with different mechanical properties on the impact force history and the transient elastodynamics of an isotropic plate undergoing low velocity impact, empirically. To provide sufficient data for the empirical model, the impact events were sensed using a Theodorus spiral configuration of piezoelectric sensors (TSSC) in passive mode. Conventionally, the Hertzian contact law is implemented for the

C. Agbasi (✉) • S. Banerjee
Department of Mechanical Engineering, University of South Carolina, Columbia, SC 29208, USA
e-mail: agbasi@email.sc.edu; BANERJES@cec.sc.edu

N. Sottos et al. (eds.), *Experimental and Applied Mechanics, Volume 6: Proceedings of the 2014 Annual Conference on Experimental and Applied Mechanics*, Conference Proceedings of the Society for Experimental Mechanics Series,
DOI 10.1007/978-3-319-06989-0_11,

analysis of the local response from which the contact force can be theoretically derived. Yang J and Chun [4], Sun and Yang S [5], Suemasu et al [6], Liu and Swaddiwudhipong [7], Olsson [8], Zheng and Binienda [9], employed the Hertzian contact law in their formulations for understanding the impact events. Clerence Zener (1941) proposed an analytical solution for an isotropic plate impacted by spherical impactor based on infinite Hertzian and Kirchhoff-Love theory [10]. Olsson (2000) showed that small mass and large mass impactors of identical impact energy initiate different plate responses [11]. Lee et al. [3, 12] studied the transmission of energy flow in the plate from a structural intensity approach. However, we took a different approach and implement two feature extraction models that analyze the transmitted AUSs. We redefined the analytical solutions using constructed indices and designed an algorithm to predict the force history generated from an impact more accurately.

11.2 Analysis

For a small displacement in a flat plate, we use the unified particle motion equation given by

$$\mu\nabla^2\mathbf{u} + (\lambda+\mu)\nabla(\nabla\cdot\mathbf{u}) + \mathbf{f} = \rho_p\ddot{\mathbf{u}} + \mathbf{F}, \tag{11.1}$$

where, $\mathbf{u}$ is the particle displacement vector, λ and μ are the Lame parameters, ρ_p is the density of the plate, $\mathbf{f}$ is the body force, and $\mathbf{F}$ is the externally transmitted force [13, 14]. The homogenous solution of the Eq. 11.1 (i.e. $\mathbf{F} = \mathbf{f} = 0$) yields the solution of shear horizontal wave and coupled Lamb waves. However, in this analysis it is important to note that the externally applied load is non-uniform and is transmitted with a natural intensity factor e that is governed by the unique coupled properties of the impactor and the plate,

$$\mathbf{F} = e\left(E_{1,2},\rho_{1,2},\sigma_{1,2},h,R\right)\int_v^{\square}\rho_2\ddot{\mathbf{u}}(X,Y,Z,t)dV, \tag{11.2}$$

where, $E_{p,i}$ is the elastic modulus of the plate (p) and the impactor (i), likewise ρ is the density, σ is the Poisson ratio, w is the plate thickness and R is the radius of the impactor. Our objective is twofold (1) accurately estimate the material properties of the impactors and (2) experimentally investigate $\mathbf{F}$ and correlate with the TSSC sensor signals. Here we use the strength of the signal at a distance d from the impact, after degeneration of the displacement due to damped wave propagation and the geometric spreading of the energy using the following equation, where b is the function of coupled material properties.

$$\ddot{\mathbf{u}}_\mathbf{d} = \frac{1}{\sqrt{d}}\mathrm{e}^{-bd}\ddot{\mathbf{u}}. \tag{11.3}$$

The displacement and force history at the point of impact can be analytically approximated using Clarence Zener's (CZ) theory for hard material impactors, while for soft materials like Teflon, there is huge discrepancy. Consequently, data driven indices are constructed for more accurate results. Accurate results are needed for damage prediction which depends on the force transmitted and dissipated in the structure as wave propagates particularly due to the coupled plate-impactor properties. The theory accords with the fact that force history of the impact event is dependent on coupled physical properties of the impactor and the plate, from which a dimensionless parameter λ can be obtained. We express all parameters in their respective domain, keeping the plate parameters constant in order to isolate impactor parameters. The theoretical solution is applicable to the impact events, whose effects are concluded before the arrival of the reflected waves from the plate boundaries. It is safe to assume that the impact events are elastic and complete after the impactor to plate collision, pressure exertion and retraction. However, the displacement on the plate at the point of impact remains perfectly inelastic until the boundary reflections return to the origin [10]. The first equation is the acceleration of the impactor modeled as a spherical body and is given by

$$m\frac{d^2s}{dt^2} = -F, \tag{11.4}$$

where s is taken as the displacement of the center of the sphere in contact with the plate, m is its mass, and F is the plate reaction. Displacement of the mid-plane of the plate at the point of impact is said to be directly proportional to impulse and is expressed as

$$Z = \left(\frac{3\rho_p}{E'_p}\right)^{0.5} \frac{P}{16\rho_p w^2}, \qquad E' = \frac{E}{1-\mu^2}, \tag{11.5}$$

$$Z = \alpha \int F dt, \tag{11.6}$$

where Z is the point on the mid-plane of the plate underneath the impact point, P is impulse, and α is a constant of proportionality determined from Kirchhoff-Love theory by imposing fixed boundary and initial conditions of a square plate. It is then sought to express the plate reaction as a function of the relative displacement U between the center of the sphere and the mid-plane of the plate in the form

$$U = s - Z. \tag{11.7}$$

Equation 11.7 is differentiated twice [10] and subtracted from Eq. 11.4 to obtain the governing equation of motion of the plate impact event.

$$m\frac{d^2U}{dt^2} + \alpha\frac{dF(U)}{dt} + F(U) = 0, \qquad \text{given } at\ t = 0 \begin{cases} U = 0 \\ \dfrac{dU}{dt} = v_0 \end{cases} \tag{11.8}$$

where v_0 is the initial velocity of the impactor. The governing Eq. 11.8 is a nonlinear ordinary differential equation. The interaction of plate and sphere is nonlinear and is approximated explicitly by classical Hertzian contact theory for a half-spaced impacted solid:

$$F(U) = kU^{1.5} \tag{11.9}$$

$$k = \left(\frac{4}{3}\right) r^{\frac{1}{2}} \left(\frac{E'_p E'_i}{E'_p + E'_i}\right). \tag{11.10}$$

Here, k is the contact stiffness that depends on the elastic parameters E_p' and E_i' of the plate and impactor respectively, which in turn depends on Young's modulus E and Poisson ratio μ. As mentioned in [9], Eq. 11.9 can be modified to account for finite thickness of the plate by rewriting as

$$F(U) = \varsigma k U^{\varepsilon} \tag{11.11}$$

where ς is a constant that accounts for contact force reduction in the plates due to finite thickness and the exponent ε takes the limits $1 < \varepsilon < 3/2$. Substituting Eq. 11.9 into 11.10 and then into Eq. 11.8, the resulting equation can be integrated numerically to obtain the contact force history of the impact event at the point of impact. Force history plots can be obtained for any range of impactor and plate materials. For all force history plots in this work, the plate properties were kept constant and the impact energy H was approximated by

$$H = mgh = \frac{1}{2}mv_0^2. \tag{11.12}$$

H was also kept constant, as well as the effective diameter of the impactor (i.e. the diameter of the sphere). As a result, solutions obtained were a function of the mechanical properties of the impactor only. Here, g is acceleration due to gravity and h is the drop height of the sphere which was varied to equalize the impact energy and nullify the energy effect.

11.3 Modified Clerence Zener Theory

After conducting several experiments in our laboratory of idealized impact events on a plate, it was found that the theory developed by Clerence Zener is not sufficient to accurately capture the transient force profiles particular to the soft impactors, especially in the time domain. Transient characteristics of the force profiles due to the impacts are important to understand the nonlinear plate response and further necessary for predicting the damage occurrence more accurately, because of the effect of wave-controlled impact [9] or energy dissipation as opposed to plastic shear and deformation [15]. Short duration impacts have compact force profiles that transmit through-thickness waves which generate significant longitudinal waves aka p-wave, found to be a characteristic of the hard material impactors. On the other hand, long duration impacts are dominated by shear and flexural waves aka s-wave, found to be a characteristic of the soft material impactors. Consequently, we propose two feature extraction techniques, experimentally and numerically determined, to be capable of characterizing impactor elastic modulus from plate AUSs. From this information, impactor type can be identified and subsequently, the transmitted force history can be accurately estimated. Although previous authors have proposed iterating ς or ε defined in Eq. 11.11 in order to provide more accurate results, we have found that these parameters only influence the force peaks and do not reconstruct the transient profile of the transmitted force accurately. We introduce two new parameters: β and γ in the non-dimensional equation to control and account for numerical contact time discrepancy and viscoelastic dissipation, respectively. In order to introduce these two parameters we make the transformation

$$\sigma = \frac{s}{v_0 T}; \qquad \tau = t\left(\frac{\beta k v_0^{0.5}}{m}\right)^{-0.4}, \tag{11.13}$$

$$\lambda = \frac{\pi^{0.6}}{3^{0.5}}\left(\frac{r}{2w}\right)^2\left(\frac{\rho_i}{\rho_p}\right)^{0.6}\left(\frac{v_0}{v'}\right)^{0.2}\left(\frac{E_i'}{E_p'+E_i'}\right)^{0.4}, \tag{11.14}$$

where v' is velocity defined by

$$v' = \left(\frac{E_p'}{\rho_p}\right)^{0.5}. \tag{11.15}$$

Hence, the motion equation (Eq. 11.8) can be modified to account for contact time discrepancy and viscoelastic dissipation according to

$$\frac{d^2\sigma}{d\tau^2} + (\lambda+\gamma)\frac{d\sigma^{1.5}}{d\tau} + \sigma^{1.5} = 0, \qquad \text{given } at\ \tau = 0 \begin{cases} \sigma = 0 \\ \dfrac{d\sigma}{d\tau} = 1 \end{cases} \tag{11.16}$$

11.4 Experiment

To verify the modified theory, we have conducted experiments where six different impactors of different materials (High-Alumina, Silicon Nitride, Stainless Steel, Titanium, Delrin Acetal, and Teflon) were impacted on an Aluminum plate (610 $\times$ 610 $\times$ 1 mm) embedded with TSSC at its center. Table 11.1 reports the vendor specified material properties of the specimens used in the experiments. The impactor materials were chosen to represent hard, intermediate, and soft impactors. Their impact energies were equalized by varying the impact velocities calibrated by mass according to Eq. 11.12. To impose a constant impact point and constant impact energy imparted by the impactors (not to be confused with actual transmitted energy into the plate), the drop heights of the impactors were controlled through the calibrated and surface lubricated guide rail system shown in Fig. 11.1a. The exploded view to the top right of Fig. 11.1b shows a close shot of TSSC in situ for AUS reception and the denoted sequential numbering of the discrete sensors. The boundaries of the plate were fixed. To study the impact force by

Table 11.1 Material properties of the specimens

Material	Elastic modulus (GPa)	Density (kg/m^3)	Poisson ratio
High-alumina	372	3,875	0.24
Silicon nitride	310	3,322	0.27
Stainless steel	209	7,750	0.29
Titanium	103	4,512	0.37
Aluminum	69	2,712	0.34
Delrin acetal	3.1	1,412	0.35
Teflon	1	2,214	0.46

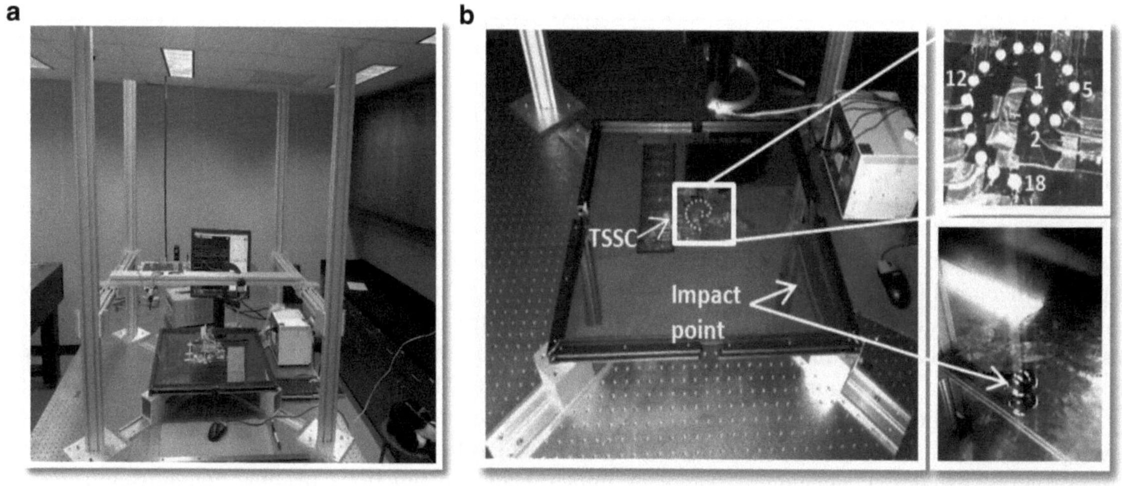

Fig. 11.1 Experiment setup of idealized impact event. (**a**) Guide rail system, (**b**) Magnified views of plate, TSSC, and impact event

the different impactors, a force sensor was also embedded underneath the plate at the impact point. We used the novel sensor cluster, TSSC, because TSSC provides multiple possibilities with the strategic placement of "discrete sensors" that enhance wave mode and frequency sensitivity [16]. AUSs obtained from the TSSC are mutually exclusive due to special geometric configuration; thereby, assuring efficient use of the discrete sensors and sensitivity with improved resolution on the type of impactors. Although mutually exclusive signals can be acquired by random placement of sensors, TSSC facilitates an organized and clustered layout which in turn helps wiring and connection of the sensors to hardware for practical applications. When comparing TSSC to conventional sensor cluster geometries like circle or square, it is easy to observe that in circle or square clusters, an inherent symmetry and overlap with respect to an incident wavefront exists, which makes half of the sensors redundant. This is similar for other linear sensor arrays. However, no symmetry or overlap among the discrete sensors exists in TSSC with respect to an incident wavefront from any direction, thus making the sensor array have higher sensitivity to AUSs. The TSSC implemented in this experiment consisted of 18 discrete APC-850 lead zirconate titanate (PZT) sensors, located on the vertices of the Theodorus spiral geometry. Due to piezoelectric effect, a polarized vector (i.e. the acquired AUSs as electric potentials) is generated in the thickness direction of the piezoelectric structures when internal stress is induced in them by the elastic waves propagating in the plate. The AUSs were acquired via NI PXI-5105 Module (8 channel, 60 MS/s digitizer). A trigger level was set to record acquisition at the slightest gain in the force sensor signal. Due to limited channels, signals from all sensors were collect in three batches. The first batch consisted of the force sensor and the first seven discrete sensors of the TSSC. The second batch consisted of the force sensor again and the next seven discrete sensors of the TSSC. The last batch consisted of the force sensor yet again and the remaining four discrete sensors in the cluster. The force sensor signal acquired in all batches served as a measure to ensure that the excitation level was kept the same during acquisition. Altogether, a total of 19 useful sensor signals were collected for every impact test: 18 from the discrete sensors in the TSSC and 1 from the force sensor. Surely, multiple repeats where conducted for an averaged and consistent data set. Figure 11.2 shows a typical AUS acquired from sensor 1 for the bounded time of data acquisition. This was acquired for all impactors and similarly for all other discrete sensors in the TSSC.

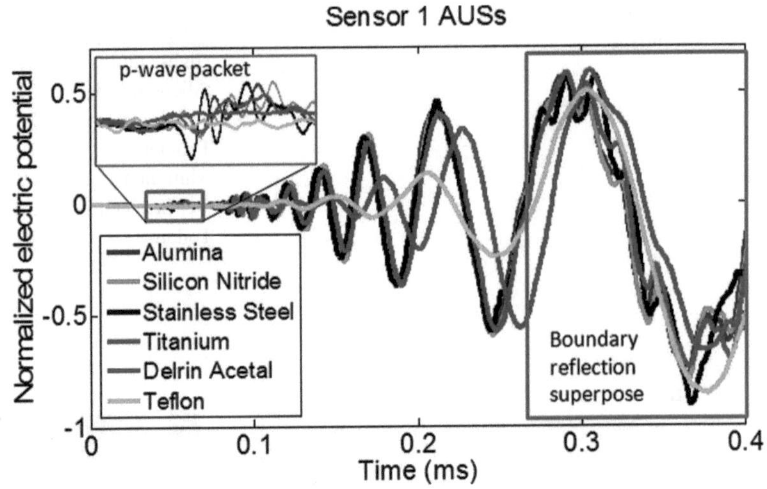

Fig. 11.2 Typical features of AUS excited by various impactor materials as acquired from discrete sensor 1

11.5 Feature Extraction Techniques

Generally, it is known that the impactors with higher elastic modulus although of the same size excite higher frequencies. This can be visualized in Fig. 11.2. Consequently, the modulus also regulates the wave velocities in the plate. However, we found many discrepancies or nonlinear effects e.g. wave velocity in the plate is a function of impact energy or the energy flow transmitting in the plate. Titanium impactor shows significant discrepancies and the nonlinear effect comes from the coupled behavior between Titanium impactor and the Aluminum plate. This was also confirmed with a spectrogram study. Similarly it was found that the nonlinearity in the experiments with steel impactors comes from the coupled density effect. In order to study these nonlinearities and manipulatively linearize them for the purpose of impactor-type classification, we adopted two feature extraction techniques that were implemented as functions of time-domain-segmented transient elastodynamic vibrations of the plate. The feature extraction techniques adopted are the lag index (LI) and dominant frequency band (DFB), constructed from the sensor cluster array. A Hann window was used for the time domain data segmentation.

LI. The lag index are directly related to the time delays in the AUSs transmitting between paired sensors in the cluster for the duration of the impact or succeeding the impact and can be further related to the elastic wave frequency and speeds excited in the plate during the impact event. From Lamb wave theory, we expect that lag coefficients are proportional to the speeds of the wave, given that the lower frequency AUSs generate bigger lag coefficients. LI is obtained as the cross-correlation function with mathematical expectation defined by

$$L = \sum_{i=1}^{n} 1 - x_t^r x_{t-\varphi}^j W(t), \tag{11.17}$$

where x_t^r are the AUSs of the reference discrete sensor, $x_{t-\varphi}^j$ are the AUSs of the correlated discrete sensors as a function of phase lag, φ and $W(t)$ is the windowing function [17, 18]. Figure 11.3 displays the Lag coefficients obtained from processing respective time-domain-segmented AUSs from all discrete sensors in TSSC as excited by the various impactor materials. It can be observed that the elastic moduli of the impactors are linearly classified most especially in the impactor transmission time window that spanned approximately 0–0.2 ms (see Fig. 11.3a). During this time, Delrin Acetal and Teflon impactors are distinctly classified. However, for the time window spanning 0.2–0.4 ms after the impact event is complete, the soft impactors are no longer distinctly classified and the order of the impactor materials with respect to increasing LI is reversed (see Fig. 11.3b).

DFB. Dominant frequency is a key feature for signal contraction of this scope [19, 20]. It captures the subtle deviations in the transient elastodynamics of the plate due to plate-impactor material coupling. The DFB processes the AUSs for dominant frequencies with a windowing function and sensor position accordingly. To do this, we take the classical Fourier transform of the transient AUSs and extract the peak magnitudes. We use these peak magnitudes to identify the transient dominant frequencies transmitting in the plate due to the impact energy flow. These dominant frequencies are then compiled for the entire discrete sensors in the cluster. The details are documented in this paper in a comprehensive manner. Figure 11.4a shows the extracted dominant frequency bands with respect to impactor type. It can be seen that different zones of TSSC are

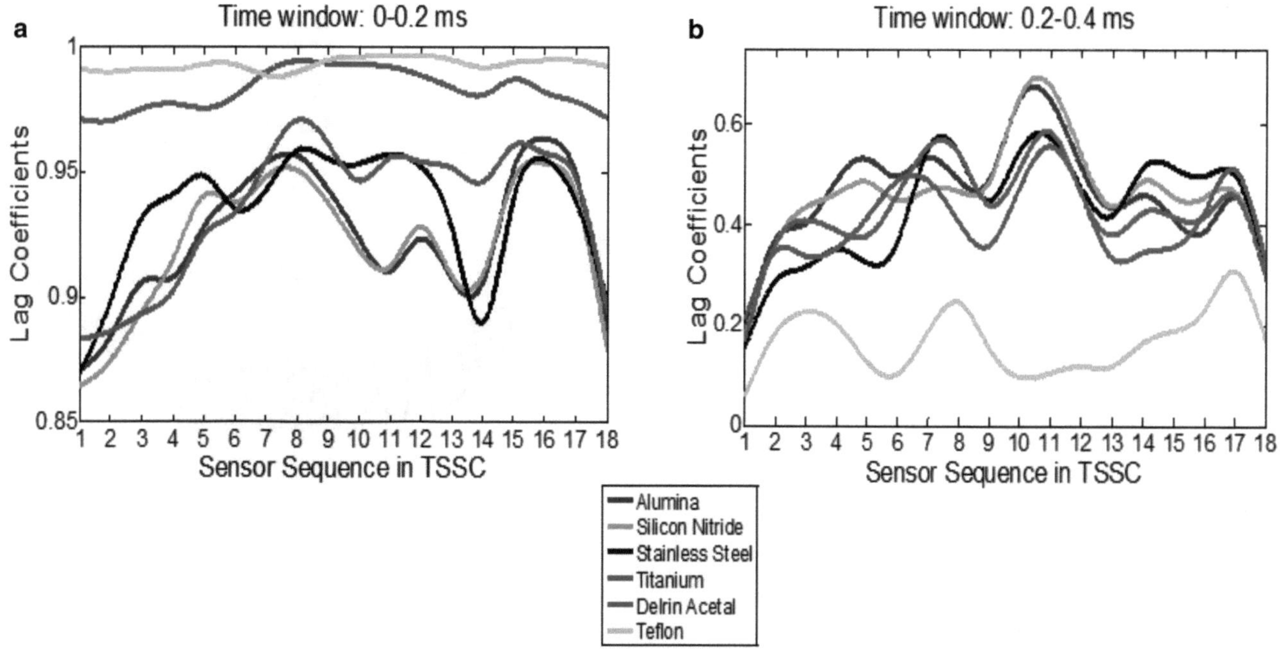

Fig. 11.3 Lag coefficients obtained from TSSC due to the range of impactor materials exciting AUSs for the time window of (**a**) 0–0.2 ms, (**b**) 0.2–0.4 ms

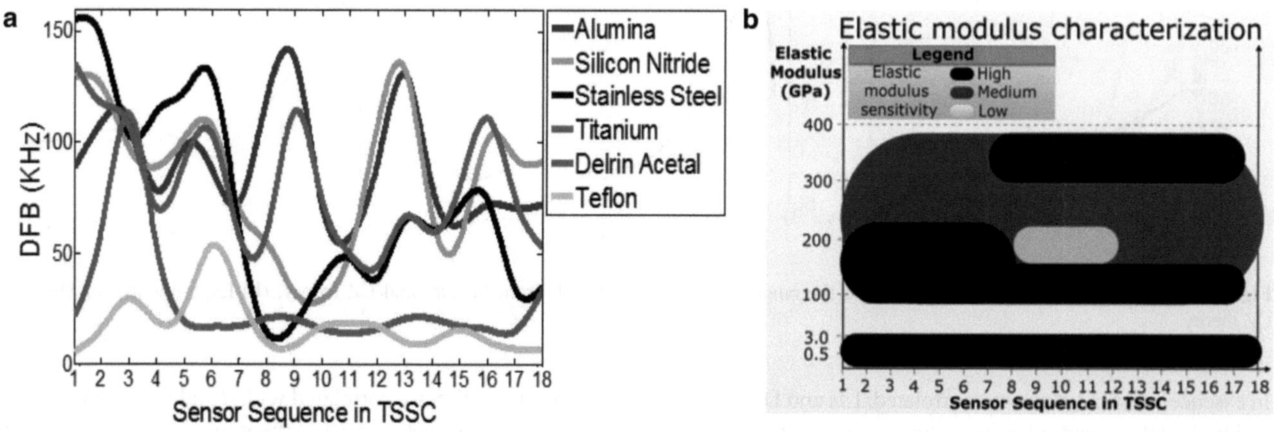

Fig. 11.4 Dominant frequency bands obtained from TSSC due to the range of impactor materials exciting AUSs. (**a**) DFB plots, (**b**) DFB sensitivity to impactor elastic modulus chart

capable of quantifying the material properties of the impactors. The soft materials in particular are easily classified as can be seen deduced in the elastic modulus sensitivity chart constructed in Fig. 11.4b. Sensitivity on the chart was determined by linearly correlating the impactor elastic modulus to DFBs and dominant nonlinear peaks.

11.6 Parametric Model Study

Figure 11.5 shows the first symmetric and anti-symmetric wave modes simulated in the plate and traveling towards TSSC. This wave packet visible in the simulation provided best results when performing LI and DFB analysis. In essence, the wave excited during the impact which spanned approximately 0–0.2 ms provided better results than subsequent waves. For our particular analysis, sensors 1–18, sequentially denoted in Fig. 11.1b, were cross-correlated with sensor 1 to obtain lag coefficients.

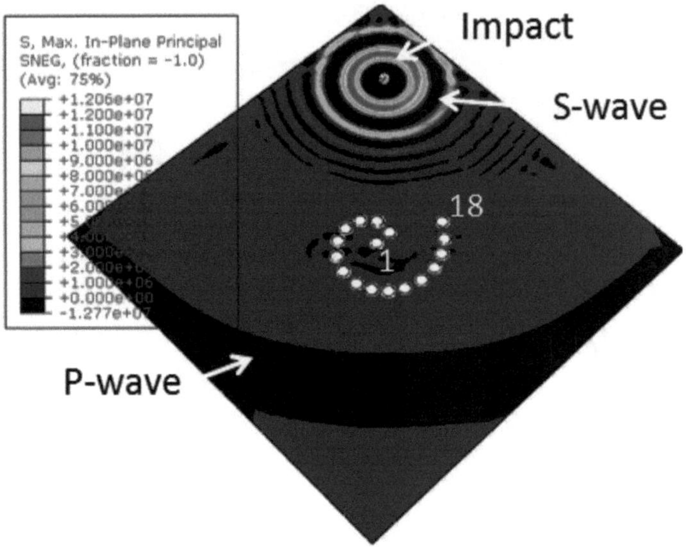

Fig. 11.5 Finite element simulation of an impact event capturing the boundary reflection free wave packet

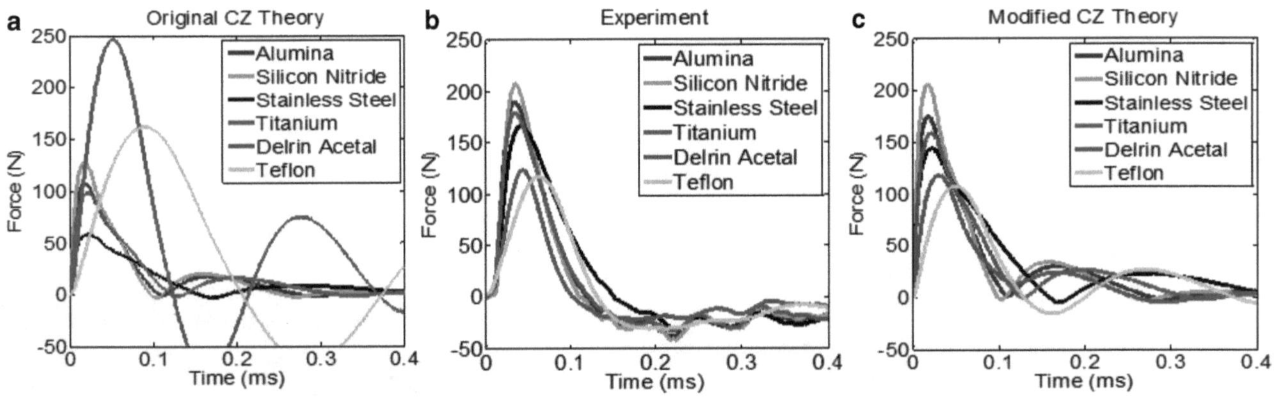

Fig. 11.6 Transient force histories obtained for the range of impactor materials from (**a**) Original CZ theory, (**b**) Experiments, (**c**) Modified CZ theory

In essence, sensor 1 was auto-correlated. LIs and DFBs from experiments were directly correlated with β and γ simultaneously, and both features served as impactor classifiers that activated the modification parameters in a CZ theoretical algorithm. Hence, solutions for the case of 'soft impactors' (i.e. Delrin Acetal and Teflon) where corrected. Highest LI and lowest DFB identify Teflon, and then appropriate weights on β and γ obtained from best fitting are activated. A similar correlation is done for Delrin Acetal. This enables better force history solutions for soft impactors, computed using CZ theory. Without this, the solutions for the soft impactors are exaggerated as can be seen in Fig. 11.6a.

The force history signals from experiment as voltage histories were converted to unit of force using conservation of momentum law and impulse equation, surmised to give

$$mv_0 = a\int_0^{t_p} V(t)dt, \tag{11.18}$$

where m and v_0 are the impactor mass and velocity at impact, respectively, a is a proportionality constant, t_p is the time at peak voltage and $V(t)$ is the voltage signal [21]. Although not the full momentum is translated to the sensor location, Eq. 11.18 serves as a fair approximation for force sensor calibration. By directly linking force histories from experiments to CZ theory using empirically derived indices from DFB and LI AUS processing techniques, we found that the values of β and γ for best and consistent fit in the time domain for Delrin Acetal impactor were $(\beta, \gamma) = (3.0 \pm 0.5, 2.50 \pm 0.05)$ and for

Teflon impactor were $(\beta, \gamma) = (4.0 \pm 0.5, 0.70 \pm 0.05)$. Force histories obtained from original CZ theory, experiment, and modified CZ theory are displayed side by side in Fig. 11.6. It can be seen that the exaggerated transient force histories obtain from original CZ theory (see Fig. 11.6a) for soft material impactors are modified (see Fig. 11.6c) based on the understanding obtained from experiments (see Fig. 11.6b). In addition, the force peaks of the profiles where modified using a universal ς and ε value found to be $(\varsigma, \varepsilon) = (1.50 \pm 0.10, 1.460 \pm 0.005)$ for best matched results. Notice that the dissipation parameter γ and solution time discrepancy parameter β depend on the impactor material. Delrin Acetal has a higher γ -parameter than Teflon because, as physically witness in our lab experiments although not measured, its restitution coefficient is significantly and visibly greater than Teflon's. Delrin Acetal impactor behaves like a tennis ball, restoring a significant amount of energy during impact which other impactor materials transmit to the plate. On the other hand, Teflon has greater β-parameter due to its soft polymeric nature that is quantized by its low elastic modulus. Inspecting Eqs. 11.10 and 11.13, it can be observed that a low elastic modulus will lead to divergence of the solution in the time domain, resulting in the need for the β-parameter to mitigate the mathematical bias.

11.7 Conclusion

In this work, we have successfully shown that CZ theory can be modified to obtain more accurate solutions by implementing a novel spiral sensor array called Theodorus spiral sensor cluster (TSSC) and two ad-hoc AUS processing techniques called the lag index (LI) and dominant frequency band (DFB) for impactor type, low velocity impact classification and the consequent application of empirically determined modification parameters. TSSC provided mutually exclusive data sets that enabled Lag coefficient and DFB to have robust sensitivity to impactor elastic modulus. This in turn was used to identify, most importantly, the soft impactors in order for the paramount and automated application of modification parameters for accurate solutions. Future studies on this topic will look into solution accuracy as a function of other impact parameters such as impact energy, impactor effective diameter, plate material properties and thickness, and the eventual correlation with experimentally determined damage indices.

References

1. Zhongqing S, Chao Z, Ming H, Li C, Qiang W (2013) Acousto-ultrasonics-based fatigue damage characterization: linear versus nonlinear signal features. J Mech Sys Signal Process 45:225–239
2. Davarcioğlu B (2010) The acousto-ultrasonic characterization of physical properties of human bones. J Appl Biol Sci 4(1):41–44
3. Xinlin QingeLiu Z (2003) Structural intensity study of plates under low-velocity impact. Int J Impact Eng 957–975
4. Yang JCS, Chun DS (1969) Application of the Hertz contact law to problems of impact in plates. United States Naval Ordnance Laboratory, White Oak, MD, NOLTR 69-152
5. Sun CT, Yang SH (1980) Contact law and impact response of laminated composite. NASA. CR-159884
6. Suemasu H, Kerth S, Maier M (1994) Indentation of spherical head indentors on transversely isotropic composit plates. J Compos Mater 28:1723–1739
7. Liu Z, Swaddiwudhipong S (1997) Response of plate and shell structures due to low velocity impact. J Eng Mech 123(12):1230–1237
8. Olsson R (2002) Engineering method for prediction of impact response and damage in sandwich panels. J Sandwich Struct Mater 4(1):83–95
9. Zheng D, Binienda WK (2009) Semianalytical solution of wave-controlled impact on composite laminates. J Aerosp Eng 22(3):318–323
10. Zener C (1941) The intrinsic inelasticity of large plates. Phys Rev 59:669–673
11. Olsson R (2002) Mass criterion for wave controlled impact response of composite plates. Composites 31(8):879–887
12. Lee HP, Lu C, Liu ZS, Xu XD (2003) The structural intensity analysis of plates under dynamic loading. Proceedings of the 16th nordic seminar on computational mechanics, Trondheim, Norway, pp 145–148
13. Victorov A (1967) Rayleigh and lamb waves. Plenum, New York, NY
14. Giurgiutiu V (2008) Structural health monitoring with piezo wafer active sensors. Elsevier, New York, NY
15. Carretero-Gonza´lez R, Khatri D, Porter MA, Kevrekidis PG, Daraio C (2009) Dissipative solitary waves in granular crystals. Phys Rev Lett 102:024102
16. Davis PJ (1993) Spirals: from theodorus to chaos. AK Peters, Wellesley, MA
17. Ljung L, Glad T (1994) Modeling of dynamic systems. Prentice Hall, Upper Saddle River, NJ
18. Orfanidis SJ (1996) Optimum signal processing: an introduction, 2nd edn. Prentice Hall, Englewood Cliffs, NJ
19. Morchen F (2003) Time series feature extraction for data mining using DWT and DFT. Data Bionics, Philipps-University Marburg, Hans-Meerwein-Strasse, 35032 Marburg, Germany
20. Anstey JS, Peters DK, Dawson C (2007) An improved feature extraction technique for high volume time series data. Proceedings of the 4th conference on IASTED International Conference: Signal Processing, Pattern Recognition, and Applications. pp 74–81
21. Daraio C, Nesterenko V, Herbold E, Jin S (2004) Strongly nonlinear waves in a chain of teflon beads. American Institute of Physics, AIP Conference Proceedings, 845. pp 1507–1510

Chapter 12
Residual Stress Measurements in Finite-Thickness Materials by Hole-Drilling

Gary S. Schajer and Colin Abraham

Abstract Hole-drilling measurements of residual stresses are traditionally made on materials that are either very thick or very thin compared with the hole diameter. The calibration constants needed to evaluate the local residual stresses from the measured strain data are well established for these two extreme cases. However, the calibration constants for a material with finite thickness between the extremes cannot be determined by simple interpolations because of the occurrence of local bending effects not present at either extreme. An analytical model is presented of the bending around a drilled hole in a finite thickness material and a practical procedure is proposed to evaluate the corresponding hole-drilling calibration constants.

Keywords Residual stress • Hole-drilling • Finite thickness • ASTM E837 • Calibration constants

12.1 Introduction

The hole-drilling method [1, 2, 3] is a well-established method for measuring residual stresses in materials. The method, formalized in the ASTM standard test procedure E837-13, is very versatile and can be applied to a wide range of material types and specimens. The measurement procedure involves drilling a small hole in the test specimen and measuring the deformation of the surrounding material caused by the removal of the stressed material within the hole. The stresses originally within the hole can then be evaluated from the surrounding deformations using calibration coefficients determined from prior numerical calibrations. For example, the ASTM procedure involves using specially designed strain gauge rosettes to measure the deformations, for which the document tabulates the numerical values of the needed coefficients. Alternative approaches using optical techniques such as Electronic Speckle Pattern Interferometry (ESPI) [4, 5] and Digital Image Correlation (DIC) [6, 7] require analogous calibrations of the surface deformations.

The ASTM document identifies two extreme cases, the first and most common, where the specimen material depth is much greater than the hole diameter, called a "thick" material. The hole depth reaches about half the hole diameter, thus producing a "blind hole". Under these circumstances, the specimen is effectively "infinitely" deep. The second extreme case occurs in plate specimens where the material depth is much less than hole diameter, called a "thin" material. In this case the hole goes through the entire specimen depth. Both extreme cases allow significant analytical economies to be made. For example, the calibration coefficients required for the residual stress evaluation need only be tabulated for different hole diameters and for the "thick" case, also for different hole depths. In neither case does specimen depth influence the response, so this factor need not be included and the needed tables of calibration constants can be kept compact.

The gap between the thickness of "thin" and "thick" specimens defined in ASTM E837-13 is significant. For a typical strain gauge rosette that uses a 2 mm diameter hole, the unavailable range is from 1.03 mm to 5.13 mm [8]. This is a substantial interval and many practical specimens will have thicknesses within it. ASTM E837-13 does not directly address this finite thickness case, apart from suggesting that the residual stress can be approximately estimated by comparisons with the tabulated "thick" and "thin" calibration coefficients. The nature of the required evaluation is not specified.

Hole-drilling in a finite-thickness material brings a further important factor into play beyond the need to enlarge the calibration tables required for the "thick" and "thin" material cases. Hole-drilling into a finite-thickness material creates a

G.S. Schajer (✉) • C. Abraham
Department of Mechanical Engineering, University of British Columbia, Vancouver, BC, Canada
e-mail: schajer@mech.ubc.ca

N. Sottos et al. (eds.), *Experimental and Applied Mechanics, Volume 6: Proceedings of the 2014 Annual Conference on Experimental and Applied Mechanics*, Conference Proceedings of the Society for Experimental Mechanics Series, DOI 10.1007/978-3-319-06989-0_12,

one-sided load on a flexible structure and so causes the specimen to bend locally. In the "thick" case, the stiffness of the thick specimen inhibits such bending. In the "thin" case, the hole goes right through, so there is no non-symmetry, thus no bending effect. However, in the finite-thickness case, substantial bending can occur and it contributes substantially to the hole-drilling response. Since this bending is not contained in either extreme case, a simple interpolation between them will not reveal this important response. Thus, a more detailed analysis is required to describe the hole-drilling response in the finite-thickness range. This is done in this paper for the case of residual stresses that are uniform through the depth of the drilled hole, first using finite element calculations and then using an analytical model. Finally, a hybrid model is proposed that is aimed to extend the use of the existing tables of "thick" and "thin" calibration constants so as to provide a practical interpolation scheme for the finite-thickness range between them.

12.2 Mathematical Background

The hole-drilling method is typically used on specimens whose in-plane dimensions are significantly larger than the diameter of the drilled hole, commonly with outside boundaries more than 3–4 hole diameters away from the hole center [8]. Under these conditions the relieved radial strain in the material surrounding the drilled hole has the trigonometric form:

$$\varepsilon(\mathrm{r},\theta) = \left(\sigma_x+\sigma_y\right)\ \mathrm{A(r)} + \left(\sigma_x-\sigma_y\right)\ \mathrm{B(r)}\ \cos 2\theta \tag{12.1}$$

where x and y are the directions of the in-plane principal stresses σ_x and σ_y, θ is the counter-clockwise angle from the x-direction, and A(r) and B(r) are numerical constants that depend on the radial position and the material elastic properties. For a through hole in a thin plate, the numerical constants can be determined analytically [9] and for a blind hole in a thick material they can be determined by finite element calculations [10]. When doing hole-drilling measurements according to the ASTM Standard Test Method E837-13 [8], the use of strain gauge rosettes of standardized dimensions is specified. To allow application to general materials, the corresponding A(r) and B(r) are replaced by dimensionless quantities $\bar{a}$ and $\bar{b}$ whose numerical values for the standardized rosettes are tabulated within E837-13.

$$\varepsilon(\theta) = \frac{\sigma_x+\sigma_y}{2}\,\frac{(1+\nu)\,\bar{a}}{E} + \frac{\sigma_x-\sigma_y}{2}\,\frac{\bar{b}}{E}\,\cos 2\theta \tag{12.2}$$

The division by E in Eq. 12.2 represents the effect of variation in material Young's modulus. The effect of Poisson's ratio is complex and is approximately represented by the $(1 + \nu)$ term shown.

The first stress term $(\sigma_x + \sigma_y)/2$ of Eqs. 12.1 and 12.2 corresponds to an in-plane hydrostatic stress. This term is axisymmetric and hence does not have angular dependence. The second stress term $(\sigma_x - \sigma_y)/2$ corresponds to a shear stress at 45° to the x-y axes and has a cos 2θ angular dependence. The trigonometric format of the equations is general and it applies to all mechanics quantities around a drilled hole, for example, stresses and in-plane and out-of-plane displacements. These examples differ only in the associated numerical values for the "A" and "B" quantities. This division of loading and response into "A" (isotropic) and "B" (shearing) components perpetuates throughout the discussions in this paper.

12.3 Finite Element Analysis

During hole-drilling residual stress measurements, the removal of the material containing residual stresses perturbs and redistributes the associated stresses in the surrounding material and causes the surface deformations that are measured. These deformations are measured by strain gauges when using the ASTM E837-13 technique. Alternatively, non-contact optical techniques can be used. The Bueckner method [11] provides a convenient way to evaluate the numerical constants needed in Eqs. 12.1 or 12.2. Using this approach it may be shown that the response due to hole drilling into a material containing residual stresses can be evaluated by setting up a model of the specimen material and loading the hole boundary with the residual stresses with reversed sign. Figure 12.1 shows a cross-sectional view of such a model. For a through hole in a thin plate, an analytical solution [9] can be used, but for the general case of a blind hole in a thicker material, numerical solutions are required, typically using finite element calculations [10, 11].

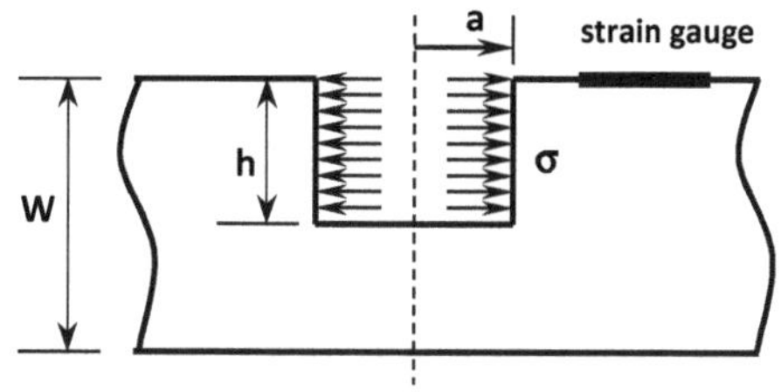

Fig. 12.1 Cross-sectional view of a hole of radius a and depth h, drilled into a material of finite thickness W

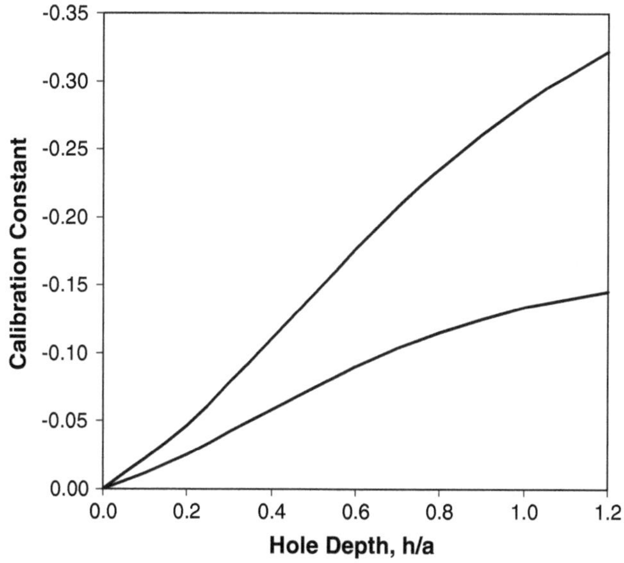

Fig. 12.2 Dimensionless calibration constants $\bar{a}$ and $\bar{b}$ for an E837-13 Type A rosette for hole drilling in a "thick" material with a hole of radius 0.36 x the mean radius of the rosette

Finite element analysis of the cross-sectional model shown in Fig. 12.1 involves dividing the material area into a finite element mesh and then completing a numerical analysis for the indicated loading. Significant modeling simplification and economy can be achieved by noting that the hole has circular shape and so the geometry of the material around the hole is axisymmetric. The "A" term in Eq. 12.1 or 12.2 represents a loading that is also axisymmetric, so the corresponding material response can be evaluated using a 2-D axisymmetric analysis instead of a 3-D general analysis. It turns out that the trigonometric character of the "B" term also allows an analogous approach to be taken for this load component. The case corresponds to a harmonic load varying circumferentially with a cos nθ distribution around a structure of axisymmetric geometry. For a linear material, it may be shown that all associated deformations have cos nθ or sin nθ distributions [12]. Specialized "axi-harmonic" finite elements are available for such calculations [13], here the $n = 2$ case is needed. It can be seen that the axisymmetric case represented by the "A" term follows the same format, with $n = 0$. This division into $n = 0$ (axisymmetric) and $n = 2$ (axiharmonic) components not only provides computational economy, it also provides an important conceptual simplification into isotropic and shearing loadings, each of which have distinctive physical characteristics. It should be noted that the use of the two 2-D finite element analyses with $n = 0$ and $n = 2$ does not involve any analytical approximation, and so a full 3-D finite element analysis offers no advantage. To the contrary, a 3-D analysis would be massively larger and so would likely involve coarser meshing and thus inferior numerical accuracy.

Figure 12.2 shows the strain vs. hole depth response of a ASTM E837-13 style strain gauge rosette as a hole is drilled into a "thick" material containing uniform residual stresses. The curves are determined from finite element calculations using the Bueckner method [10]. The numerical values are negative because they correspond to stress unloading.

A convenient way to evaluate the response of the strain gauge in the non-uniform strain field around the drilled hole is to observe that the measured strain is the integral of the strains existing over the strain gauge area, and that the integral of strain is displacement. Therefore, the strain integral can be directly evaluated as the difference of the in-place displacements at each end of the strain gauge:

$$\varepsilon_{\text{strain gauge}} = \frac{1}{r_2 - r_1} \int_{r_1}^{r_2} \varepsilon(r)\, dr = \frac{\delta_2 - \delta_1}{r_2 - r_1} \tag{12.3}$$

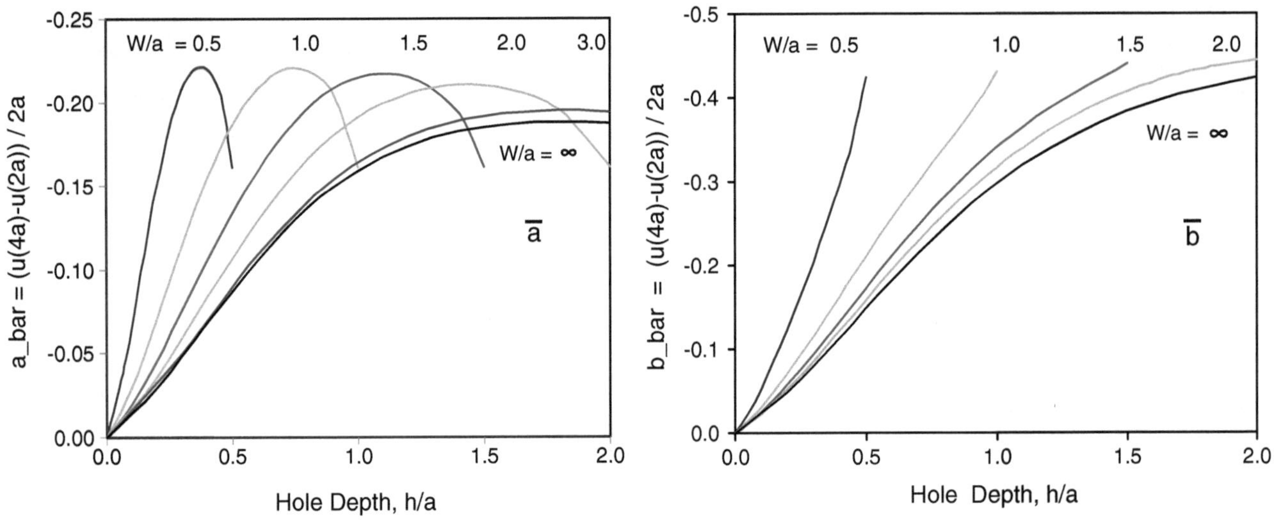

Fig. 12.3 Dimensionless calibration constants $\bar{a}$ and $\bar{b}$ for hole-drilling into finite thickness specimens

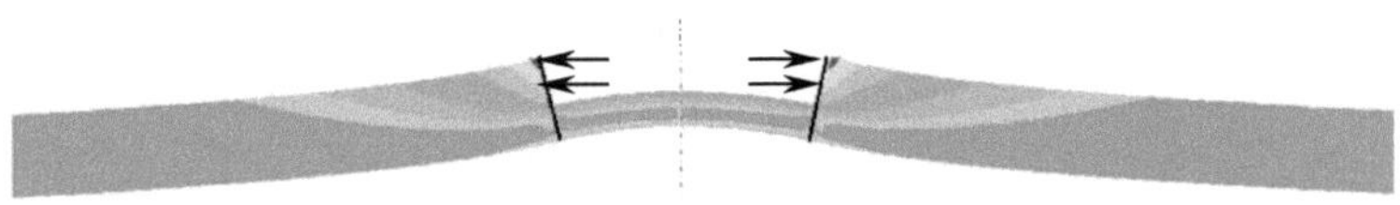

Fig. 12.4 Deformation of a finite thickness specimen due to residual stress loads on the hole boundary, calculated using finite element analysis. The deformations are greatly exaggerated for pictorial clarity

where δ_1 and δ_2 are the radial displacements at the inner and outer radii of the strain gauge, r_1 and r_2 respectively. Equation 12.3 applies only to the center strand of a strain gauge and in general the equation must be augmented by similar quantities representing all the parallel strands [14]. This has been done for the results shown in Fig. 12.2. However, when making size comparisons, a realistic indication of the strain gauge response can be obtained simply from Eq. 12.3. This approach is taken here to allow conceptual and computational simplicity.

Figure 12.3 shows the dimensionless calibration constants $\bar{a}$ and $\bar{b}$ for hole-drilling into finite thickness specimens. These results are computed using finite element calculations [13] using axi-symmetric elements for the $\bar{a}$ results and axi-harmonic elements with $n = 2$ for the $\bar{b}$ results. For conceptual simplicity, Eq. 12.3 has been used to represent the strains that could be measured if strain gauges were used. The chosen radii are $r_1 = 2a$ and $r_2 = 4a$. For an ASTM E837-13 standard strain gauge rosette, this corresponds to a hole radius ≈ 0.36 times the mean rosette radius. To generalize the graphical presentation, the hole depth h and specimen thickness W are both normalized by the hole radius a. In Fig. 12.3, the responses of a very thick material are represented by the line $W/a = \infty$. Both lines start from zero and reach a plateau at a hole depth of approximately one diameter (=2a). Both graphs show that the response generally increases with reduction in material thickness, with increasingly rapid response for very thin specimens. This behavior is to be expected because reduction in specimen thickness reduces the amount of material immediately below the hole and therefore reduces the constraint on the surrounding material. This constraint reduction in turn allows the deformation response to increase more rapidly. All response curves reach the same terminal value, corresponding to the through-hole case when the drilled hole reaches the lower

A prominent feature of the $\bar{a}$ curves in Fig. 12.3a is the peak that appears for finite thickness specimens. This occurs because, as shown in Fig. 12.4, a blind hole in a finite thickness specimen creates an unsymmetrical in-plane load on the specimen that causes local bending of the specimen and hence enlarged surface deformations. As the hole reaches the lower surface the load becomes symmetrical, the bending effect disappears and the deformation response reduces from its peak. For the axisymmetric case corresponding to $\bar{a}$, the cross-section shape shown in Fig. 12.4 corresponds to a circular hill with the hole at the top. For the shear stress case corresponding to $\bar{b}$, the surface is saddle shaped, with one diameter folded down and the perpendicular diameter folded up. The reversed curvatures of the two diameters tend to oppose each other, thereby reducing local bending effects and inhibiting the growth of bending peaks in the $\bar{b}$ curves in Fig. 12.3b.

12.4 Analytical Model

The deformed specimen in Fig. 12.4 suggests an analytical model that can be useful for explaining and predicting the material behavior around a drilled hole. Figure 12.5 schematically illustrates a cross-section of the model, which is axisymmetric with center axis at the left. The model consists of a circular disk of thickness W-h on the left, joined to an outer annular ring of thickness W, inner radius a, and "infinite" outer radius on the right. The two parts share a common lower surface. The drilled hole of depth h is located above the disk. The residual stresses relieved by hole cutting act on the hole surface as shown in Fig. 12.1, creating a radial force per unit angle $S = \sigma$ h a. An internal force F and moment M are created between the inner disk and outer ring. They combine to create a net radial force $C = S - F$ and a net moment per unit angle $Q = S\,(W-h)/2 + F\,h/2 - M$ on the outer ring.

From consideration of equilibrium of the forces and moments between the inner disk and outer ring, and of compatibility of displacement and rotation of the boundary between them, it may be shown [15] that:

$$Q = S\left[\frac{2hk_1K_2 + 2K_2(k_1+k_2)(W-h)}{h^2k_1k_2 + 4(k_1+k_2)(K_1+K_2)}\right] \tag{12.4}$$

$$C = S\left[\frac{4k_2(K_1+K_2) + hk_1k_2W)}{h^2k_1k_2 + 4(k_1+k_2)(K_1+K_2)}\right] \tag{12.5}$$

where the in-plane stiffnesses k_1 and k_2 and the bending stiffnesses K_1 and K_2 are for the axisymmetric "A" case:

$$k_1 = \frac{F}{V_r(a)} = \frac{(W-h)\,E}{1-\nu} \qquad k_2 = \frac{C}{V_r(a)} = \frac{W\,E}{1+\nu} \tag{12.6}$$

$$K_1 = \frac{M}{\theta(a)} = \frac{(W-h)^3\,E}{12(1-\nu)} \qquad K_2 = \frac{Q}{\theta(a)} = \frac{W^3\,E}{12(1+\nu)} \tag{12.7}$$

where the subscripts 1 and 2 respectively refer to the inner disk and outer ring. The displacements Ur(r) at radius r on the surface of the outer ring combine contributions from the centerline radial displacement Vr(r) and the centerline rotation θ(r). Following the method of Vinson [16] the combination of the two contributions is:

$$U_r(r) = \frac{Ca}{EWr}(1+\nu) + \frac{W}{2}\left[\frac{Qa^2 12(1+\nu)}{EW^2r}\right] \tag{12.8}$$

Similarly, for the anticlastic "B" case:

$$k_1 = \frac{F}{U_r(a)} = \frac{(W-h)\,E}{1+\nu} \qquad k_2 = \frac{C}{U_r(a)} = \frac{W\,E}{3-\nu} \tag{12.9}$$

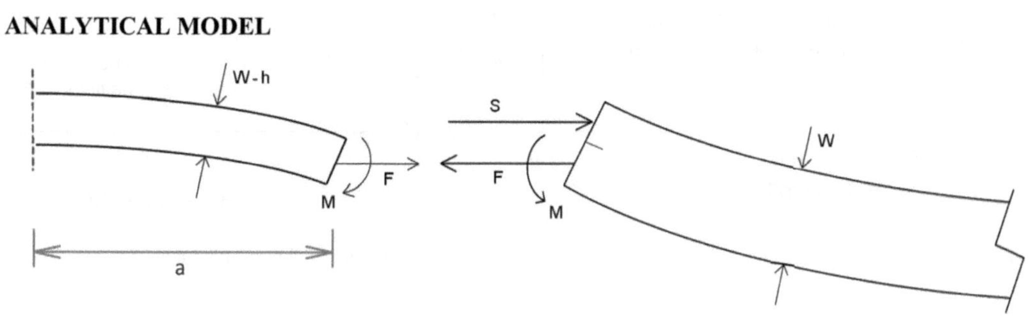

Fig. 12.5 Analytical model of a hole in a finite thickness specimen

$$K_1 = \frac{M}{\theta(a)} = \frac{(W-h)^3 E}{12(1+\nu)} \quad K_2 = \frac{Q}{\theta(a)} = \frac{W^3 E}{4(1+\nu)} \tag{12.10}$$

Substituting these results into the solution for shear stress loading found following Vinson [16], the deformations on the upper surface of the outer ring are:

$$U_r(r,\theta) = \left(\frac{Cr}{aEW}\left[4\frac{a^2}{r^2} - (1+\nu)\frac{a^4}{r^4}\right] + \frac{2Q(1+\nu)}{EW^2}\frac{a^3}{r^3}\right)\cos 2\theta \tag{12.11}$$

$$U_\theta(r,\theta) = \left(\frac{-Cr}{aEW}\left[2(1-\nu)\frac{a^2}{r^2} + (1+\nu)\frac{a^4}{r^4}\right] + \frac{2Q(1+\nu)}{EW^2}\frac{a^3}{r^3}\right)\sin 2\theta \tag{12.12}$$

12.5 Results

The dashed lines in Fig. 12.6 show the dimensionless calibration constants $\bar{a}$ and $\bar{b}$ from the analytical model, superimposed on the finite element values from Fig. 12.3, shown as solid lines. It can be seen that the analytical model gives results close to the finite element solutions for small specimen thickness up to $W/a = 0.5$, but with increased deviation at greater specimen thicknesses. This is particularly evident with the $\bar{b}$ results. The deviation occurs because the analytical model is based on the assumption that the boundary line between the inner disk and outer ring remains straight, as shown in Figs. 12.4 and 12.5. This is reasonable for modest specimen thicknesses, but it is increasingly unrealistic for large depths. In addition, the stress concentrations that occur at the lower corners of the drilled hole further curve the boundary line. This is a relatively modest effect for the $\bar{a}$ curves, where the stress concentration factor due to the hole is 2, but is a much greater effect for the $\bar{b}$ curves, where the corresponding stress concentration factor is 4.

Figure 12.6 shows that the specimen thickness has the greatest effect on the calibration constants for small specimen thicknesses. For larger specimen thicknesses the effect reduces, disappearing beyond about $W/a = 3$. Thus, the "thick" solution in ASTM E837-13 can be used at these greater specimen thicknesses. An interpolation scheme is proposed here to bridge the use of the analytical solution at small specimen thicknesses and the "thick" solution at large specimen thicknesses. The scheme, which is designed to mimic the finite element results in Fig. 12.3, is as follows:

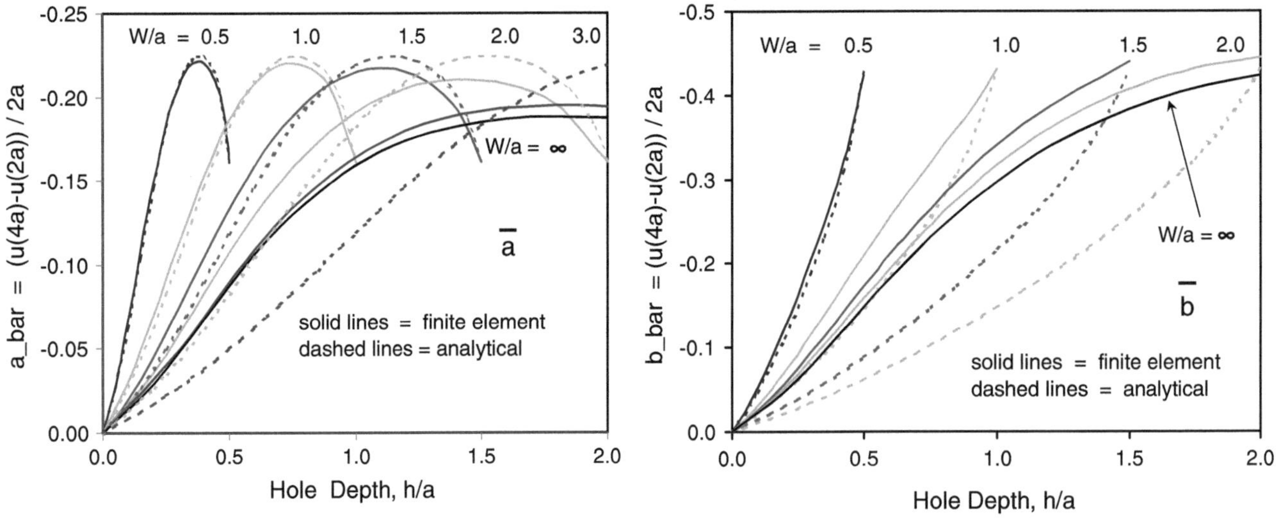

Fig. 12.6 Analytical calibration constants $\bar{a}$ and $\bar{b}$ vs. hole depth for hole-drilling into finite thickness specimens

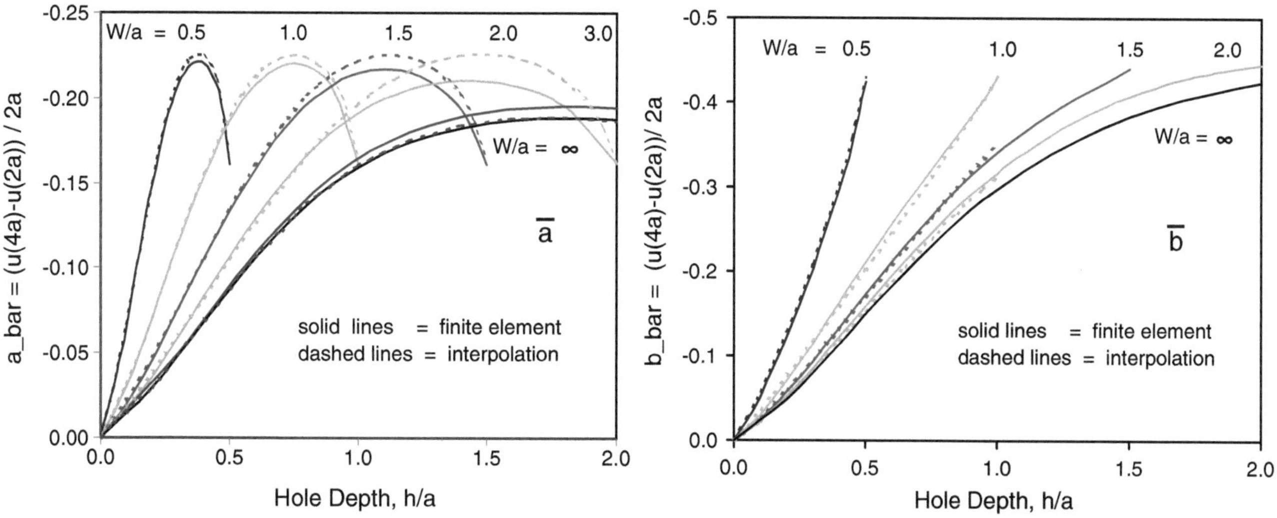

Fig. 12.7 Interpolated calibration constants $\bar{a}$ and $\bar{b}$ vs. hole depth for hole-drilling into finite thickness specimens

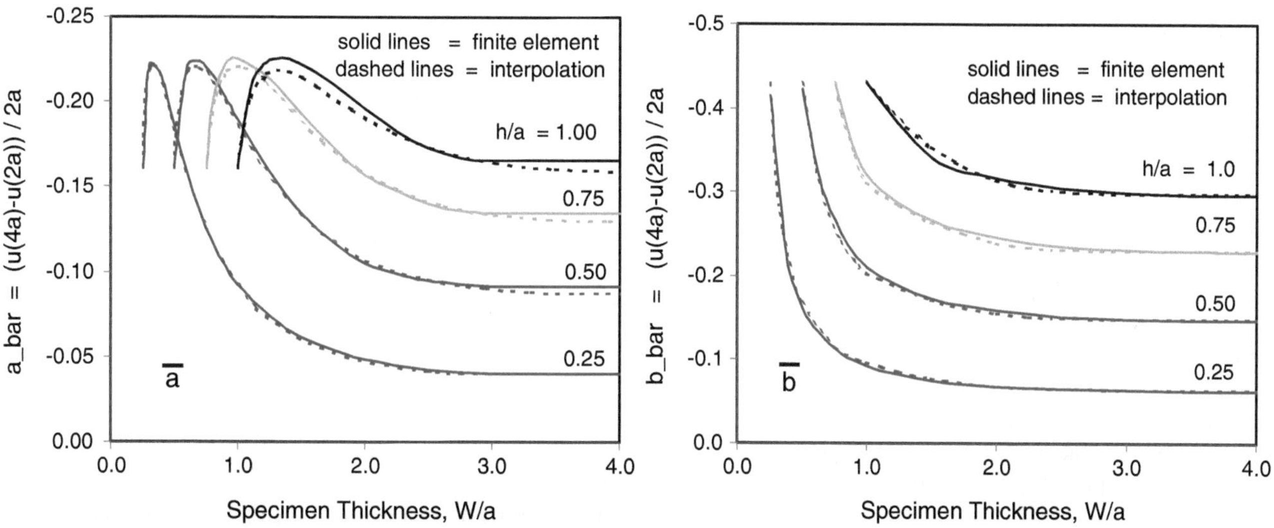

Fig. 12.8 Interpolated calibration constants $\bar{a}$ and $\bar{b}$ vs. specimen thickness for hole-drilling into finite thickness specimens

For $\bar{a}$: For $W/a \leq 2$ use the analytical solution with K_1 reduced by the factor $(1\text{–}0.3\ W/a)$ with minimum $= 0$
For $W/a \geq 3$ use the "thick" solution $\bar{a}(\infty)$
For intermediate W/a compute an interpolation factor $\alpha = (3\text{-}W/a)^2$

$$\text{and interpolate } \bar{a}(W/a) = \alpha\bar{a}(2) + (1\text{-}\alpha)\bar{a}(\infty) \tag{12.13}$$

For $\bar{b}$: For $W/a \leq 1$ use the analytical solution with k_1 reduced by the factor $(1 - 0.8\ W/a)$ with minimum $= 0$
For $W/a \geq 3$ use the "thick" solution $\bar{b}(\infty)$
For intermediate W/a compute an interpolation factor $\beta = ((3 - W/a)/2)^3$

$$\text{and interpolate } \bar{b}(W/a) = \beta\bar{b}(2) + (1\text{-}\beta)\bar{b}(\infty) \tag{12.14}$$

The reductions in the stiffness quantities K_1 and k_1 are to account for the stress concentration effects at the lower corner of the drilled hole. They were determined empirically to give a good fit to the finite element results. Figure 12.7 illustrates the interpolation results.

Figure 12.8 illustrates the calibration constants vs. specimen thickness for holes of various depths. As in Fig. 12.7, the solid lines show the finite element results and the dashed lines the interpolations using Eqs. 12.13 and 12.14. The right sides of these graphs show the "thick" case and the flatness of the lines confirm that the deformation response has largely reached the "infinite" thickness case at $W/a = 3$. The left sides of the graphs show the "thin" case, where the analytical solution applies. The peaks in the $\bar{a}$ curves correspond to the local bending also observed in Fig. 12.7. The interpolation defined in Eqs. 12.13 and 12.14 connects the two extreme cases. It can be seen that the interpolated results closely match the finite element results.

12.6 Discussion

It can be seen that the initial finite element results in Fig. 12.3 are sufficient to provide the needed calibration data for hole-drilling measurements in finite thickness materials. Thus, the subsequent analytical model and interpolation are not essential requirements. However, the analytical model is useful because it explains the local bending behavior that occurs in finite thickness specimens. In addition, the interpolation removes the need for a custom finite element calculation to be done for every combination of hole diameter, depth and specimen thickness. The need to do such calculations would be a substantial burden on a practical user, whose objectives are experimental, not theoretical. Computer codes already exist that contain the "infinite" thickness case described in ASTM E837-13. The theoretical model and interpolation allow those codes to be extended in a compact and practical way to model finite thickness specimens without having to store finite element data for numerous intermediate cases. This will become increasingly important in future work where the objective will be to evaluate the variation of residual stress with depth. In that case the required quantity of calibration data is massively larger.

For optical measurements of surface deformations [4–7], the deformation at every measured point can be determined directly using Eqs. 12.4–12.14. Calibration constants for strain gauge measurements are more involved because of their difference format illustrated in Eq. 12.3. A practical strain gauge takes on a more involved form as a summation over all the grids of the strain gauge used. However, in practice, the details of the strain gauge grids tend to apply a fairly constant factor to the overall indicated strain. Thus, a useful result for the finite thickness case can be achieved by proportioning the published "thick" calibration constants by the calibration results in Figs. 12.7 and 12.8 as follows:

$$\bar{a}\,(h, W) = \bar{a}_{ASTM}\,(h, \infty) \times \frac{\tilde{a}\,(h, W)}{\tilde{a}\,(h, \infty)} \qquad \bar{b}\,(h, W) = \bar{b}_{ASTM}\,(h, \infty) \times \frac{\tilde{b}\,(h, W)}{\tilde{b}\,(h, \infty)} \tag{12.15}$$

where $\bar{a}\,(h, W/a)$ is the calibration coefficient for a hole of depth h in a specimen of thickness W, $\bar{a}_{ASTM}\,(h, \infty)$ is the corresponding value for an ASTM strain gauge rosette with hole depth h and "infinite" specimen thickness, and $\tilde{a}\,(h, W)$ and $\tilde{a}\,(h, \infty)$ are the calibration constant values determined using Eq. 12.3. The $\bar{b}$ quantities follow the same pattern. The resulting calibration constants will not be "exact", perhaps with errors up to 10 %, but they remain useful given that the discrepancy is comparable to the experimental accuracy level achievable with practical hole-drilling measurements.

The calibration constants presented in Figs. 12.7 and 12.8 are for hole drilling in residual stress fields that are uniform with depth from the surface. The measured deformation response at the surface to such hole drilling reaches a limiting value at hole depths approximately equal to the hole diameter (=2 hole radii, $h/a = 2$). Experience with hole drilling in "thick" specimens indicates that the sensitivity to specific stresses reaches down to just one hole radius, i.e., h/a [8]. Further surface deformations beyond this hole depth occur only from the geometry change of the hole, without influence of the deeper interior stresses. It is for this reason that the interpolated calibration constants are focused on the hole depth range zero to one hole radius.

12.7 Conclusions

Figures 12.7 and 12.8 demonstrate that the surface deformation response caused by hole drilling into a stressed material changes significantly with specimen thickness. The change is non-linear, and in the axisymmetric case governed by the quantity $\bar{a}$, a prominent peak in the response occurs due to local bending effects around the hole. The results show that the effect of the lower boundary becomes significant for specimens of thickness less that about three times the hole radius. At the other extreme, the analytical model described in Eqs. 12.4–12.12 realistically describes the hole-drilling response of thin plates of thickness less than one hole radius. In the intermediate thickness region, the interpolation described in Eqs. 12.13 and 12.14 closely follows the finite element solutions.

The presented results are significant because they extend the possible use of hole-drilling residual stress measurements to specimens of a wide range of thicknesses. The interpolation based on the analytical model described in Eqs. 12.4–12.12 is significant because its use allows the classical "infinite" thickness solution to be extended to the finite thickness case in a compact way, without having to store the specific results of numerous intermediate thickness cases. The interpolation scheme gives calibration constant values typically within 5 %, occasionally within 10 %. These are reasonable results given that the experimental error of typical hole-drilling measurements commonly exceeds 10 %.

The calibration constants presented here are for hole drilling in residual stress fields that are uniform with depth from the surface. Work is presently underway to extend the analysis to the "stress profiling" case, where the residual stresses vary with depth from the surface. This will be described in a future report.

Acknowledgments The authors sincerely thank the Natural Sciences and Engineering Research Council of Canada (NSERC), American Stress Technologies, Pittsburgh, PA, and the Institute for Computing, Information and Cognitive Systems (ICICS) at UBC for financially supporting this research. They also gratefully thank Prof. R.L. Taylor for his kind support with the finite element analysis.

References

1. Schajer GS, Whitehead PS (2013) Hole drilling and ring coring. In: Schajer GS (ed) Practical residual stress measurement methods, Chapter 2. Wiley, Chichester, UK
2. Vishay Measurements Group (1993) Measurement of residual stresses by the hole drilling strain gage method. Vishay Measurements Group, Raleigh, NC, Tech Note TN-503-6
3. Rendler NJ, Vigness I (1966) Hole-drilling strain gage method of measuring residual stresses. Exp Mech 6(12):577–586
4. Steinzig M, Ponslet E (2003) Residual stress measurement using the hole drilling method and laser speckle interferometry: Part I. Exp Tech 27(3):43–46
5. Nelson DV, McCrickerd JT (1986) Residual-stress determination through combined use of holographic interferometry and blind-hole drilling. Exp Mech 26(4):371–378
6. McGinnis MJ, Pessiki S, Turker H (2005) Application of three-dimensional digital image correlation to the core-drilling method. Exp Mech 45(4):359–367
7. Nelson DV, Makino A, Schmidt T (2006) Residual stress determination using hole drilling and 3D image correlation. Exp Mech 46(1):31–38
8. American Society for Testing and Materials (2013) Standard test method for determining residual stresses by the hole-drilling strain gage method. American Society for Testing and Materials, West Conshohocken, PA, Standard Test Method E837-13
9. Muskhelishvili NI (1953) Some basic problems in the mathematical theory of elasticity. Noordhoff, Groningen, Holland, pp 202–204
10. Schajer GS (1981) Application of finite element calculations to residual stress measurements. J Eng Mater Tech 103(2):157–163
11. Bueckner H (1958) The propagation of cracks and the energy of elastic deformation. Trans Am Soc Mech Eng 80:1225–1230
12. Wilson EL (1965) Structural analysis of axisymmetric solids. AIAA J 3(12):2269–2274
13. Zienkiewicz OC, Taylor RL (2005) The finite element method, 6th edn. Elsevier, Oxford
14. Schajer GS (1993) Use of displacement data to calculate strain gauge response in non-uniform strain fields. Strain 29(1):9–13
15. Abraham C (2011) Hole-drilling residual stress measurement in an intermediate thickness specimen. MASc Thesis, Dept. Mechanical Engineering, University of British Columbia, Vancouver, Canada
16. Vinson JA (2005) Plate and panel structures of isotropic composites and piezoelectric materials including sandwich construction. Springer, Norwell, MA

Chapter 13
Residual Stress Response to Peening in Metallic Glass

B. Jayakumar, M. Allahkarami, and J.C. Hanan

Abstract Normally, the residual stresses as a result of shot peening, include compressive surface stresses on the treated side and tensile stresses on the subsurface. Residual stresses from shot peening of Bulk Metallic Glasses have been beneficial for improving their plasticity in compression. Recently, significant residual stresses were observed in abrasive treated metallic glass ribbons using a process similar to shot peening. Significant thermal residual stresses have also been predicted in amorphous metals, from the rapid quenching needed to retain their structure. Early measurements used neutron diffraction to identify the thermal stresses in metallic glasses. Peening using glass beads leads to curvature as a result of surface stresses. Optical measurements show a reduction in ribbon thickness. XRD measurements on the abrasive treated metallic glass also showed shifts of the broad amorphous diffraction hump compared to untreated ribbons, indicative of the change in residual strain. The resolution of X-ray strain measurements on amorphous metals and the relaxation of thermal residual stresses were considered for validating the prediction.

Keywords Amorphous metal • Residual stress • Curvature • Modulus • Layer removal

13.1 Introduction

Metallic glasses, with their exceptional mechanical properties including Yield strengths up to 5 GPa [1] and Elastic Modulus up to 195 GPa [2] are of significant interest for structural applications. Thermal residual stresses are expected in metallic glasses due to the high cooling rates (~10^6 K/s) required in their production to retain the amorphous phase. Analytical models have been developed to predict residual stresses also in Bulk Metallic Glasses (BMGs) using an instant freezing model, supported by results from layer removal methods [3]. One of the main drawbacks of metallic glasses is their near zero ductility in tension, resulting due to shear localization [4]. In order to improve their plasticity and prevent catastrophic failure, introduction of a second phase has been widely studied in processing BMG based composites [5–9]. Residual stresses have been considered detrimental in metallic glasses for several reasons including: degradation of magnetic properties [10], interference with measurements of hardness [4], and altering fracture toughness [11]. However introducing compressive residual stress in BMGs using indentation [12] and shot peening [4] were other approaches sought for improving plasticity. Characterizing the magnitude and depth of residual stresses and understanding their impact on mechanical properties is critical. Substrate curvatures have been used in measuring residual stresses in BMGs as a result of shot peening [4] and also in soda alumina silicate glasses as a result of tempering [13]. A comparison of measured compressive residual stresses using different treatments and methods has been reported in literature [14]. As a result of plastic deformation from peening impact, metallic glasses have also shown to change their structure [14]. Bond orientation ordering and relaxation of the glassy structure have been attributed to enhanced atomic mobility [15]. Other studies have attributed structural relaxation to cyclic elastic loading [16]. Measuring density is a direct method to show a change in free volume as a result of peening, but smaller processing-induced changes have limited their use [17, 18]. Measurement and comparison of Elastic moduli have also been suggested to understand the change in free volume [19].

B. Jayakumar • M. Allahkarami • J. Hanan (✉)
Mechanical and Aerospace Engineering, Oklahoma State University, Tulsa, OK 74106, USA
e-mail: Jay.Hanan@gmail.com

N. Sottos et al. (eds.), *Experimental and Applied Mechanics, Volume 6: Proceedings of the 2014 Annual Conference on Experimental and Applied Mechanics*, Conference Proceedings of the Society for Experimental Mechanics Series,
DOI 10.1007/978-3-319-06989-0_13,

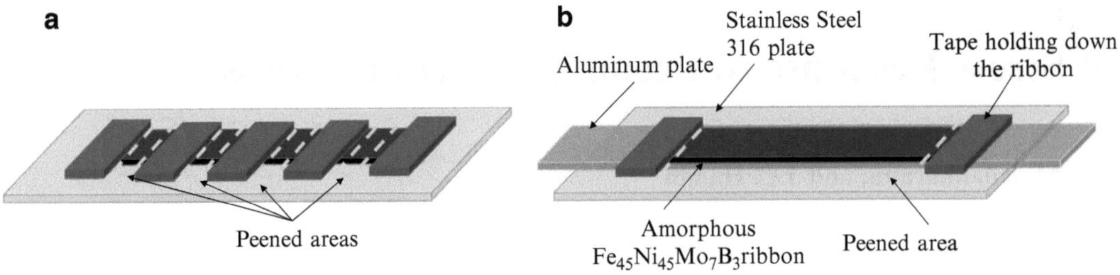

Fig. 13.1 Schematic of the Peening set-up. (**a**) Condition used for observing multiple curvatures, (**b**) Condition used to evaluate effect of peening on mechanical properties using tension testing

This work investigates the response of amorphous ribbons to shot peening using glass particles. XRD studies were carried out on treated and untreated ribbons and the change in peak positions were estimated. Elastic strains in the ribbon were calculated using peak shifts. Change in curvatures and thicknesses were measured. Several peening pressures were investigated and the change in mechanical properties including elastic modulus and strength were measured before and after peening.

13.2 Materials and Methods

Amorphous $Fe_{45}Ni_{45}Mo_7B_3$ ribbons of 29 μm thickness and 50.8 mm width were used in this study. Owing to surface chilling in manufacture of the ribbon, one side of the ribbon had a reflective surface and the other a matte surface. Ribbon samples were peened on both sides by holding them on plates made of 50.8 mm wide Aluminum 2024, secured on a stainless steel template. The reflective side was treated first in all of the samples followed by treatment of matte side.

Peening was performed in an enclosure[1] equipped with a reclaimer and a dust bag. Glass beads of the type BT-8 were used as the impact media (0.15–0.2 mm in diameter). Several pressures were used for treating the ribbons from a 114 mm nozzle distance. With a significant curvature observed, the ribbon's response for a pressure of 0.48 MPa (70 psi) is of interest in this study. Untreated and peened ribbons were etched using a 3 % Nital solution to observe a change in curvature. X-ray diffraction (XRD) measurements were also conducted on ribbon samples. All were treated using the two conditions shown in Fig. 13.1a: (1) untreated, (2) on the curvature (denoted as "peened") after peening. X-ray diffraction[2] was performed with Cu-Kα radiation at tube parameters of 40 kV/40 mA. The detector distance to the center of diffraction was kept at 30 cm, which covers approximately the area of 20° in 2θ and 20° in χ with 0.02° resolution. A motorized five axis (X, Y, Z (translation), χ (tilt), φ (rotation)) stage was used to move the measurement spot to the instrument center within 12.5 μm position accuracy. Sample positioning was controlled by a video-laser positioning system before each exposure to ensure diffraction patterns come from the impact treated regions of the ribbon.

Samples for tension testing were prepared from peened samples of 50.4 mm width. A total of four samples were prepared for bulk tension tests, with each sample 12.5 mm wide. Tension tests were performed using a hydraulic driven Instron UTM. ASTM E345—93(2008) "Standard Test Methods of Tension Testing of Metallic Foil" was followed. Aluminum tabs were glued to the ribbon test specimen ends to avoid slipping from the grips. Applied strain was computed using a Laser Extensometer[3] along with a data-logger suitable for recording compliance free strain measurements. Micrographs on the ribbon surface and thickness measurements were taken using a field emission scanning electron microscope (FESEM)[4] equipped with a back scatter detector and EDAX system. The mass of the ribbon to 0.01 g accuracy was measured before and after peening.

[1] Zero INEX 3048R.

[2] Bruker D8 Discover XRD[2] micro-diffractometer equipped with the General Area Diffraction Detection System (GADDS) and Hi-Star 2D area detector.

[3] Model LE-05.

[4] Hitachi S4800-high resolution.

13.3 Results

Upon removing boundary constraints from the template fixture, a curvature was observed after peening of the amorphous ribbons shown in Fig. 13.1. Similar to earlier work [20], a change in surface texture was also observed. Ribbon samples treated under a 0.48 MPa pressure and a 114 mm distance, were analyzed further. Samples for XRD measurements were prepared using conditions shown in Fig. 13.1a, and mechanical tension testing was prepared using conditions shown in Fig. 13.1b. 1D data from XRD measurements for untreated and peened conditions are presented in Fig. 13.2. There were no signs of crystallization observed when comparing results from the two samples. In comparing the amorphous diffraction patterns of each of these conditions, a change in nearest neighbor spacing was expected. To estimate the center of the amorphous peak, a Gaussian function was fit to the humps found at the following 2θ ranges: 12–24°, 30–39° and 39–55°. The peak shift was higher for lower angles as compared to the higher 2θ angles. Corresponding displacement errors were calculated.

Peak shifts for curved and stretched conditions were measured by comparing to the peak at the untreated condition. XRD elastic strains were calculated using Bragg's equation given by

$$\lambda = 2d \sin \theta \tag{13.1}$$

Where λ is the wavelength of the Cu-Kα X-ray source (1.5 Å), d is the average lattice spacing and θ is the diffraction angle (Fig. 13.2).

A change in atomic spacing was calculated from the change in Gaussian fit. Computed strains and error associated with the peak shift are presented in Table 13.1.

The Young's modulus measured from tension testing of the treated ribbons was compared to untreated ribbons. Results including surface areas of the peened ribbons were also measured (Fig. 13.3).

Due to a change in ribbon curvature across the width, treated samples had a tendency to warp. Some samples, failed in shear near the grips, with failure initiating from to the micro-cracks seen on ribbon edges (Fig. 13.4).

The mass of the samples was measured before and after treatment and showed no difference with a 0.1 mg accuracy. Thickness measurements show distorted thickness on the ribbon edges and an average reduction of 17 % (Fig. 13.5).

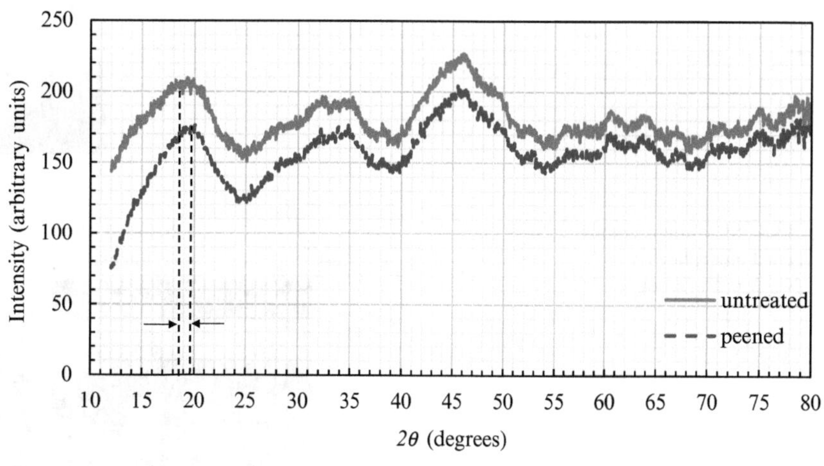

Fig. 13.2 XRD peak pattern for untreated and peened samples. The error in displacement (Δ2θ) was calculated using a 12 μm accuracy at 0.00008 rad compared to (Δ2θ) from strain calculated at 0.01 rad

Table 13.1 Strains calculated from XRD peak shift after peening

Peak angle	% Compressive strain	Peak fit error %	Δ2θ from 12 μm displacement (%)
18	3.42	0.15	0.0082
33	0.83	0.16	0.0079
45	0.12	0.1	0.0076

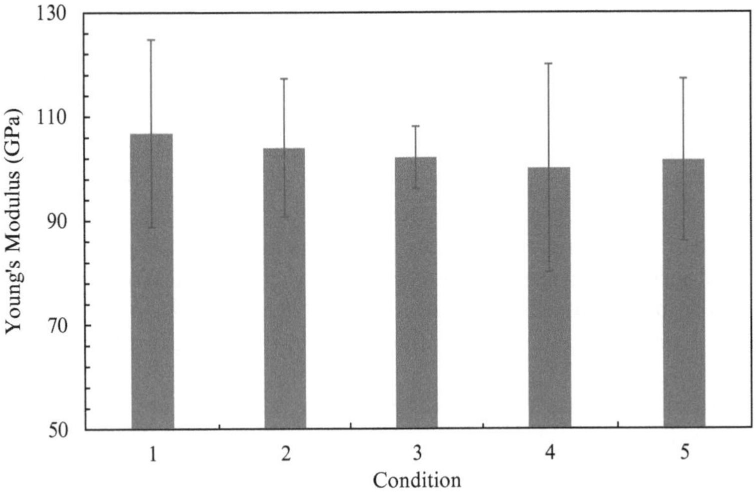

Condition	
1	Untreated
2	90 degree bend created by plastic deformation.
3	Abrasive treated – 328 mm^2, Figure 1 (a)
4	Abrasive treated – 656 mm^2, Figure 1 (a)
5	Abrasive treated – 640 mm^2, Figure 1 (b)

Fig. 13.3 Comparison of Young's modulus measured from tension testing under different conditions

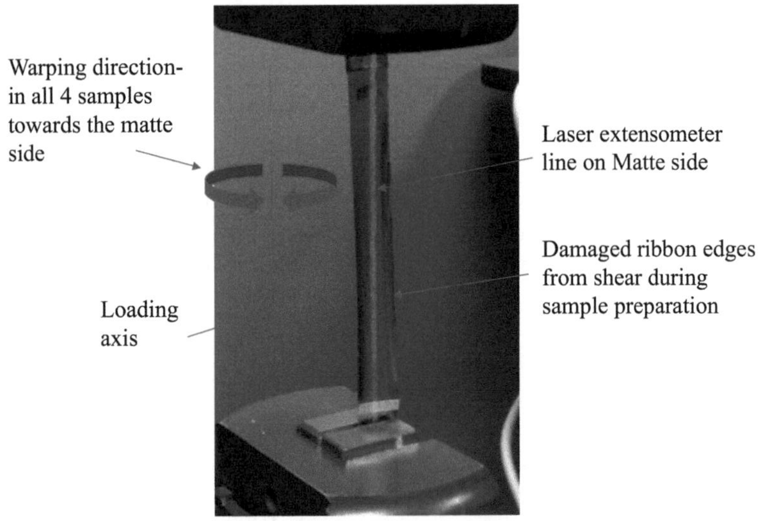

Fig. 13.4 Picture of peened ribbon under tension

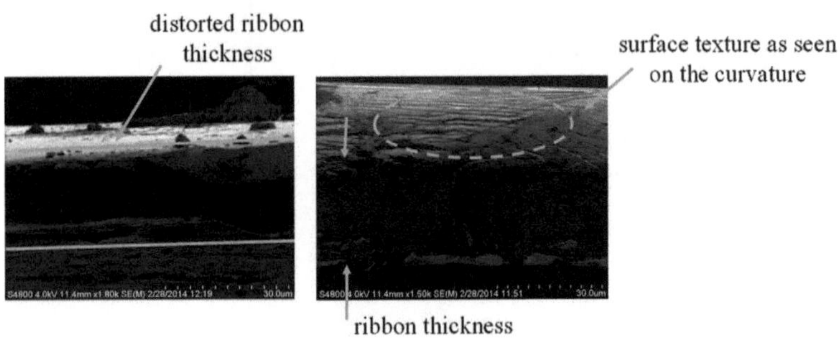

Fig. 13.5 SEM micrographs of peened amorphous $Fe_{45}Ni_{45}Mo_7B_3$ ribbon

13.4 Discussion

A change in ribbon curvature as a result of peening is indicative of a change in the residual stress state of the ribbon. Peak shifts measured from XRD show a change in average atomic spacing resulting from treatment. XRD elastic strains at low angles measured from peak shift was at least 5× more than the peak fit error, confirming the presence of residual stresses from peening. Using a stage with a 12.5 μm position accuracy, the calculated displacement errors ($\Delta 2\theta$) were at least 2 orders of magnitude less than the measured values. Corresponding peak fit errors are presented in Table 13.1.

Local shear transformation requires free volume to allow atomic mobility [21]. Deformation is associated with a change in bond arrangement. Peak shifts indicate a change in average atomic spacing. A drop in diffracted intensity was also observed. This can be attributed to the change in surface roughness caused by peening and also obvious in the SEM images. Other work [21] using XRD on amorphous metals shows bonding rearrangements are possible in both elastic and plastic deformation where bond exchange initiates deformation, while the total number of bonds remains un-changed. Bond orientation anisotropy has been reported as a result of local bond deformation measured in a mechanical creep study of metallic glasses [22]. Atoms prefer to move when there is free volume, therefore a change in bond length and a change in free-volume can be related to density and elastic modulus, both of which are material properties. The in-plane elastic modulus measured from bulk tension testing of treated and untreated amorphous $Fe_{45}Ni_{45}Mo_7B_3$ ribbons did not show a significant difference while, peak shifts from XRD studies show out of plane strains up to 3 %. A reduction in ribbon thickness, measured from scanning electron micrographs is also seen on the curvature as a result of processing. Most of the change in atomic arrangement could be in the out of plane direction, perpendicular to peening.

Chemical etching of untreated ribbons did not show a significant change in curvature, suggesting the treatment is causing the residual stress change and not relieving an existing residual stress profile. However, early observations from etching of peened ribbons show a reduction in curvature. Thus, etching also confirms peening induces a stress profile across the ribbon which can be relaxed by removing the peened layer. Ongoing studies will further quantify the change in thickness and curvature as a result of chemical etching and look for anisotropy in the change of average bond lengths.

13.5 Conclusions

Residual stresses were observed in amorphous $Fe_{45}Ni_{45}Mo_7B_3$ ribbons as a result of peening leading to curvature. The amorphous phase of the peened ribbon remained, as evident from XRD. XRD peak shifts were observed by fitting a Gaussian function to the broad amorphous humps. Changes in atomic spacing and corresponding changes in strains were calculated. A difference in peak intensities was also observed, likely related to the surface roughness from peening. Scanning electron micrographs show distorted ribbon edges and an average thickness reduction of 17 % after peening. The elastic modulus measured from bulk tension testing did not show a significant change from peening.

Acknowledgements The Oklahoma Center for Advancement and Technology (OCAST) and the National Science Foundation (NSF) (Award No: 1214985) partially funded this work at MetCel LLC and the Helmerich Research Center through grants to the OSU Foundation.

References

1. Chen M (2008) Mechanical behavior of metallic glasses: microscopic understanding of strength and ductility. Ann Rev Mater Res 38(1):445–469
2. Johnson WL, Samwer K (2005) A universal criterion for plastic yielding of metallic glasses with a (T/T-g)(2/3) temperature dependence. Phys Rev Lett 95(19):195501
3. Aydiner CC, Üstündag E, Hanan JC (2001) Thermal-tempering analysis of bulk metallic glass plates using an instant-freezing model. Metall Mater Trans A 32(11):2709–2715
4. Zhang Y, Wang WH, Greer AL (2006) Making metallic glasses plastic by control of residual stress. Nat Mater 5(11):857–860
5. Chen G et al (2009) Enhanced plasticity in a Zr-based bulk metallic glass composite with in situ formed intermetallic phases. Appl Phys Lett 95(8), 081908
6. Hays CC, Kim CP, Johnson WL (2000) Microstructure controlled shear band pattern formation and enhanced plasticity of bulk metallic glasses containing in situ formed ductile phase dendrite dispersions. Phys Rev Lett 84(13):2901–2904
7. Sun GY, Chen G, Chen GL (2011) Design, synthesis, and characterization of bulk metallic glass composite with enhanced plasticity. J Mater Sci 46(15):5216–5220
8. He G et al (2003) Novel Ti-base nanostructure-dendrite composite with enhanced plasticity. Nat Mater 2(1):33–37
9. Sarac B, Schroers J (2013) Designing tensile ductility in metallic glasses. Nat Commun 4:2158
10. Liu YJ, Chang ITH, Lees MR (2001) Influences of oxide phases on the coercivity of mechanically alloyed multicomponent Fe-based amorphous alloys. Scr Mater 44(12):2729–2734
11. Ritchie RO, Schroeder V, Gilbert CJ (2000) Fracture, fatigue and environmentally-assisted failure of a Zr-based bulk amorphous metal. Intermetallics 8(5–6):469–475
12. Liu FX, Gao YF, Liaw PK (2008) Rate-dependent deformation behavior of Zr-based metallic-glass coatings examined by nanoindentation. Metall Mater Trans A 39(8):1862–1867
13. Sglavo VM, Bonafini M, Prezzi A (2005) Procedure for residual stress profile determination by curvature measurements. Mech Mater 37(8):887–898

14. Méar FO et al (2008) Residual-stress distribution in shot-peened metallic-glass plate. Philos Mag Lett 88(11):757–766
15. Méar FO et al (2008) Structural relaxation in a heavily cold-worked metallic glass. Scr Mater 59(12):1243–1246
16. Concustell A et al (2009) Structural relaxation and rejuvenation in a metallic glass induced by shot-peening. Philos Mag Lett 89(12):831–840
17. Wang LM et al (2000) Ultrasonic investigation of Pd39Ni10Cu30P21 bulk metallic glass upon crystallization. Appl Phys Lett 77(8):1147–1149
18. Deng D, Lu B (1983) Density change of glassy Pd77Si16.5Cu6.5 alloy during cold drawing. Scripta Metallurgica 17(4):515–518
19. Harms U, Jin O, Schwarz RB (2003) Effects of plastic deformation on the elastic modulus and density of bulk amorphous Pd40Ni10Cu30P20. J Non Cryst Solids 317(1–2):200–205
20. Jayakumar B, Hanan J (2014) In: Rossi M et al (eds) Curvature from residual stress in rapidly quenched amorphous metals using abrasive layer removal, in residual stress, thermomechanics & infrared imaging, hybrid techniques and inverse problems, Vol 8. Springer, New York, NY, pp 87–92
21. Miller M, Liaw P (2008) In: Miller M, Liaw P (eds) Bulk metallic glasses. Springer, New York, NY
22. Suzuki Y, Haimovich J, Egami T (1987) Bond-orientational anisotropy in metallic glasses observed by x-ray diffraction. Phys Rev B 35(5):2162–2168

Chapter 14
Residual Stress Modeling and Measurement in Aluminum Wrought Alloys

Bowang Xiao, Qigui Wang, Cherng-Chi Chang, and Josie E. Rewald

Abstract X-ray diffraction and hole-drilling methods are applied to measure the residual stresses in a turbo charge compress wheel made of aluminum wrought alloys for finite element model validation. Aluminum wrought alloys are usually subjected to heat treatment which includes quenching after solution treatment to improve aging responses and mechanical properties. Rapid quenching can lead to high residual stress and severe distortion which significantly affect dimension stability, functionality and particularly performance of the product. A finite element based approach was developed by coupling a nodal-based transient heat transfer algorithm with material thermo-viscoplastic constitutive model, to model residual stress and distortion during heat treatment for robust product design and durability assurance. The comparison shows that hole-drilling residual stress measurements provide more accurate and reliable results than X-ray diffraction for this particular part and material. A good agreement between residual stress measurement and FEA prediction has demonstrated that the integrated residual stress model is robust in predicting residual stresses and optimizing heat treatment of aluminum wrought alloys.

Keywords Residual stress measurement • Aluminum wrought alloys • Quenching • Simulation • Hole-drilling • X-ray diffraction

14.1 Introduction

Aluminum wrought alloys are widely used in aerospace and automotive industry to reduce weight and improve fuel efficiency because of their high strength to weight ratio. Many aluminum wrought alloys are heat-treated through solution, quenching and ageing to improve mechanical properties like tensile strengths and hardness. However, high cooling rates in water quenching can produce significant amount of residual stresses [1–3] that can result in severe distortion [4], even cracking. Tensile residual stress produced during water quenching is also detrimental to fatigue properties of the material [2, 5]. Therefore, product geometry design and heat treatment processing must be optimized so that the mechanical properties can be improved and meanwhile residual stress and distortion are minimized.

The numerical modeling of quenching process is full of challenges, due to the complexity of part geometry and boundary conditions and in particular the variation of material behaviors under different thermal and stress conditions [6, 7]. In this article, the numerical modeling of quenching, and simplified machining processes of aluminum wrought alloys is demonstrated. A turbo charge compress wheel made of aluminum wrought alloy is quenched in water and residual stress measurements are made by strain gauge rosette hole-drilling method and X-ray diffraction method. The measured residual stresses are then used to evaluate the numerical modeling. It is also very interesting to note that the advantages and disadvantages of different residual stress measurement technologies when applied to different materials and components.

B. Xiao (✉) • Q. Wang • C.-C. Chang • J.E. Rewald
GM Global Powertrain Headquarter, 823 Joslyn Ave, Pontiac, MI 48326, USA
e-mail: bowangxiao@gmail.com

N. Sottos et al. (eds.), *Experimental and Applied Mechanics, Volume 6: Proceedings of the 2014 Annual Conference on Experimental and Applied Mechanics*, Conference Proceedings of the Society for Experimental Mechanics Series,
DOI 10.1007/978-3-319-06989-0_14,

14.2 Residual Stress Modeling of Quenching

Residual stress modeling of quenching includes two parts: thermal analysis that predicts temperature distribution in the component at any given time and structural analysis that predicts residual stresses due to temperature changes governed by the material behavior under the geometry constraints.

When the hot component is merged into cool liquid, heat is transferred from the high-temperature region to the low-temperature region via convection, radiation and conduction. During quenching, heat is mainly transferred from aluminum wrought alloy component to the surrounding quench media by convection as expressed in Eq. 14.1. Heat flux and HTC data vary from location to location and play very important roles in quenching results. Temperature-time profile of the quenched part at different time stages can be predicted by FEA tools with experimentally calibrated HTC curves [8].

$$\dot{Q} = h \cdot A \cdot \left(T_{surface} - T_{water}\right) \quad (14.1)$$

where

$\dot{Q}$ = heat rate, J/s

h = heat transfer coefficient, w/m^{2o}C

A = surface area, m^2

$T_{surface}$ = work piece surface temperature, °C

T_{water} = water temperature, °C

In the followed structural simulation, the temperature–time profile of the quenched work piece from the thermal simulation is read in. The finite element code then calculates local (node to node) displacements based on the temperature distribution and the relevant materials properties. During quenching, the part shrinks because of the temperature drop. The non-uniform thermal shrinkage of the part is constrained by the geometric structure and material properties that vary with temperatures and strain rates [9–11]. Thus an accurate representation of the material behavior at different temperatures and strain rates is extremely important to the simulation results. To accurately predict the residual stress and distortion during quenching, the temperature- and strain rate-dependent material constitutive behavior needs to be modeled. Following Newman et al, a material constitutive model called Mechanical Threshold Stress (MTS), was adapted in this work for water quenching process [9].

Figure 14.1 shows the mesh model for the extrusion billet and the compressor wheel, and the locations where the wheel is machined from the billet. Figure 14.2 shows the predicted residual stress distributions in the extrusion billet, and the stress

Fig. 14.1 A rectangular resistance strain rosette [12]

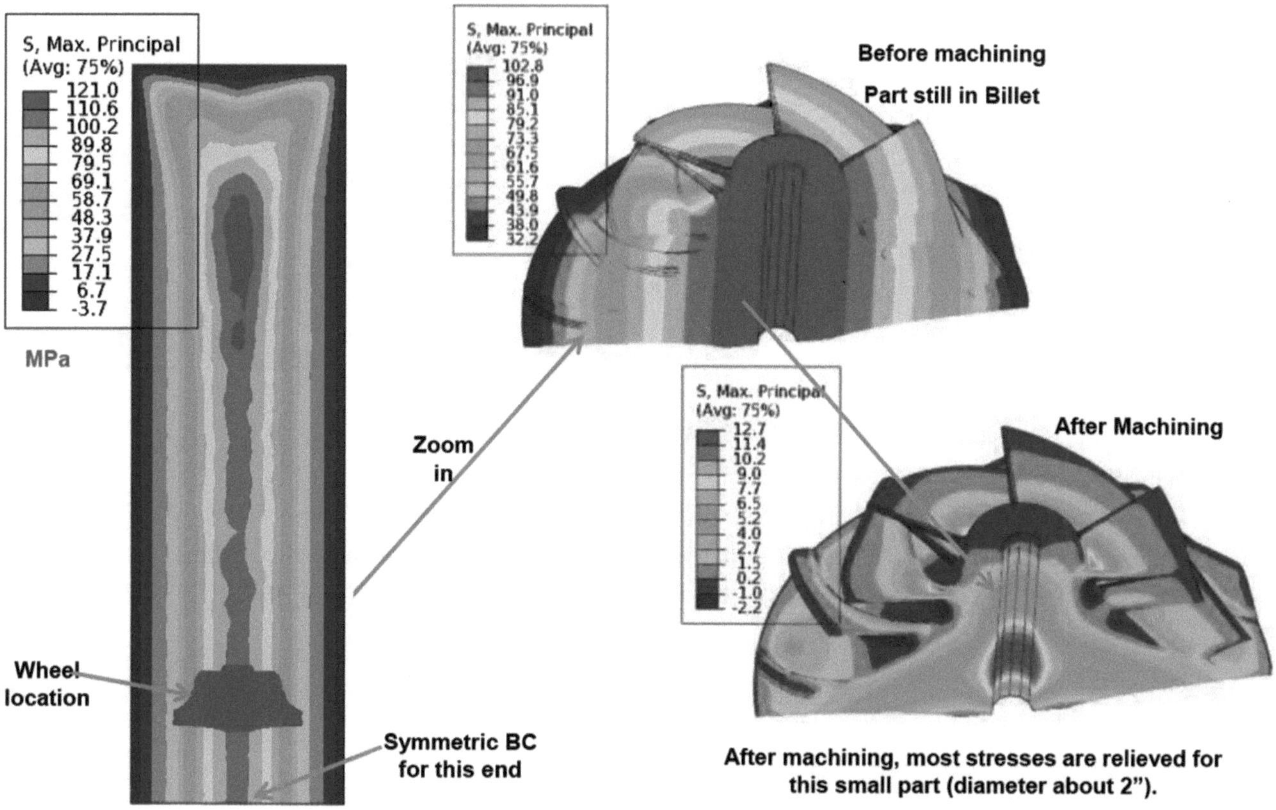

Fig. 14.2 Residual stress evolution in quenching and machining processes

distribution after the compressor wheel is machined. It is clearly shown that a big amount of residual stresses are generated in quenching process, but majority of the stresses are then relieved in the machining process because some materials are machined off.

14.3 Residual Stress Measurement

14.3.1 Strain Gauge Rosette Hole-Drilling Residual Stress Measurement

A strain gauge rosette as shown in Fig. 14.3 is mounted onto the surface of the sample and a small hole is drilled at the center of the 3 strain gauges. During the hole drilling, strains are relieved due to the release of residual stresses. The relieved strains are measured and used to back-calculate residual stresses in that small area. However, no exact solution of residual stresses in terms of relieved strains for a blind hole drilled into a plane stress field is available from the theory of elasticity [12–14]. Fortunately, it had been demonstrated that this case closely parallels the through-hole condition in the general nature of the stress distribution [12, 13]. The general expression for the relieved radial strains due to a plane biaxial residual stress state is Eq. 14.2 [12]:

$$\begin{aligned} \varepsilon_1 &= A\left(\sigma_x + \sigma_y\right) + B\left(\sigma_x - \sigma_y\right)\cos 2\alpha \\ \varepsilon_2 &= A\left(\sigma_x + \sigma_y\right) + B\left(\sigma_x - \sigma_y\right)\cos 2\beta \\ \varepsilon_3 &= A\left(\sigma_x + \sigma_y\right) + B\left(\sigma_x - \sigma_y\right)\cos 2\gamma \end{aligned} \tag{14.2}$$

where

$\varepsilon_{1,2,3}$ = measured strain relieved from strain gauge 1, 2 and 3, respectively
σ_x, σ_y = stress in x and y direction, respectively

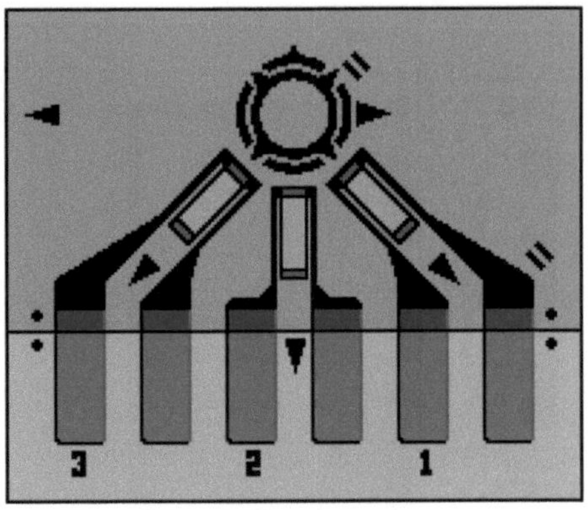

Fig. 14.3 A rectangular resistance strain rosette [12]

A, B = Calibration Coefficients
$\alpha\beta\gamma$ = angle measured counterclockwise from the x direction to the axis of the strain gauge 1, 2 and 3, respectively

For a rectangular strain gauge rosette like as shown in Fig. 14.3, the three strain gages measure the three strains along the 3 gage directions during hole-drilling, where $\alpha = 0^o$, $\beta = 45^o$, $\gamma = 90^o$. The principal stresses and their direction are solved and shown in Eq. 14.3:

$$\begin{aligned} \sigma_{\max} &= \frac{\varepsilon_1 + \varepsilon_3}{4A} - \frac{1}{4B}\sqrt{(\varepsilon_3 - \varepsilon_1)^2 + (\varepsilon_1 + \varepsilon_3 - 2\varepsilon_2)^2} \\ \sigma_{\min} &= \frac{\varepsilon_1 + \varepsilon_3}{4A} + \frac{1}{4B}\sqrt{(\varepsilon_3 - \varepsilon_1)^2 + (\varepsilon_1 + \varepsilon_3 - 2\varepsilon_2)^2} \\ \tan 2\alpha &= \frac{\varepsilon_1 - 2\varepsilon_2 + \varepsilon_3}{\varepsilon_1 - \varepsilon_3} \end{aligned} \tag{14.3}$$

Since the coefficients A and B for blind hole-drilling cannot be calculated directly from theoretical considerations, they are usually obtained by numerical procedures such as finite-element analysis [13]. Some tables of the coefficients defined in Eq. 4 were published [15, 16]. It is suggested that the coefficients A and B can be interpolated or extrapolated from the published nondimensionless coefficients [15, 16]. Errors can be introduced in this procedure due to interpolation and extrapolation. More accurate residual stresses can be calculated if the errors of interpolation can be avoided, e.g. determine the calibration coefficients directly from experiments or FEA for a specific measurement [17].

$$a = \frac{2E}{1+\nu}A \qquad b = 2E \cdot B \tag{14.4}$$

Strain gauge rosette hole-drilling residual stress measurements are made at 3 locations as shown in Fig. 14.4. Figure 14.5 shows the maximum and minimum principal stresses at these 3 locations [18]. It is clearly that the residual stress on the wheel surface is very small because the wheel fan thickness is only 0.5 mm thick.

14.3.2 X-ray Diffraction Residual Stress Measurement

X-ray and neutron diffraction methods attract a lot of attentions because they enable a nondestructive measurement of stresses. They are very useful when estimating the fatigue life of mechanical components. Compared with conventional techniques, X-ray and neutron diffraction methods enable local measurements and real-time analysis of stress [19].

When a monochromatic X-ray beam irradiates a solid material, the beam is scattered by the atoms composing the material, as shown in Fig. 14.6 [19]. Atoms are packed regularly intro a three-dimensional periodic lattice for a perfect crystalline material. When the X-ray incidences at a certain angle that meets the condition expressed in Eq. 14.5, the intensities of scattered waves sum up into a constructive interference and the diffraction pattern can be observed [19–21].

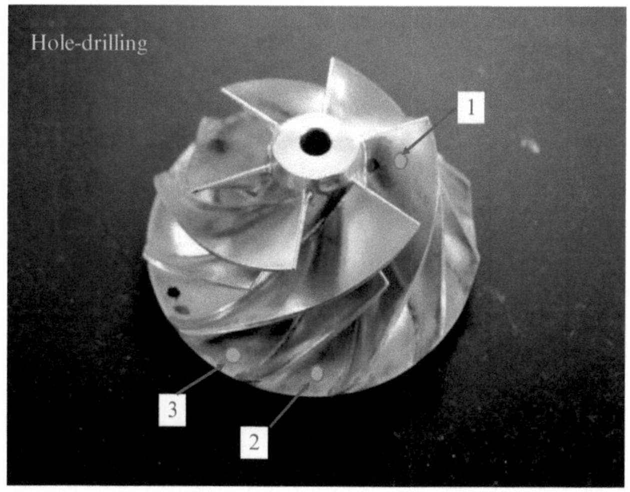

Fig. 14.4 Strain gauge rosette hole-drilling measurement on the wheel

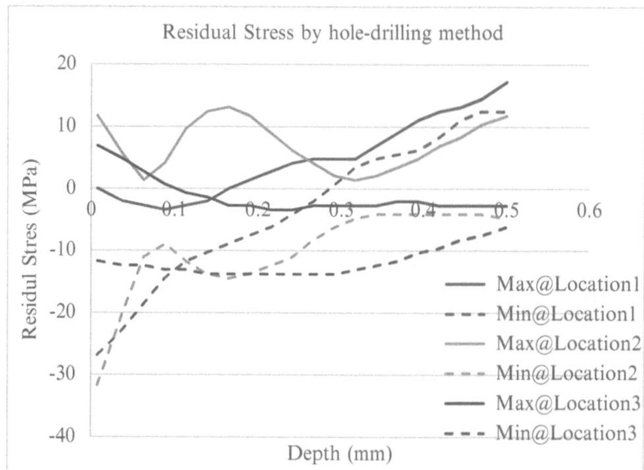

Fig. 14.5 Residual stress measurement results by hole-drilling

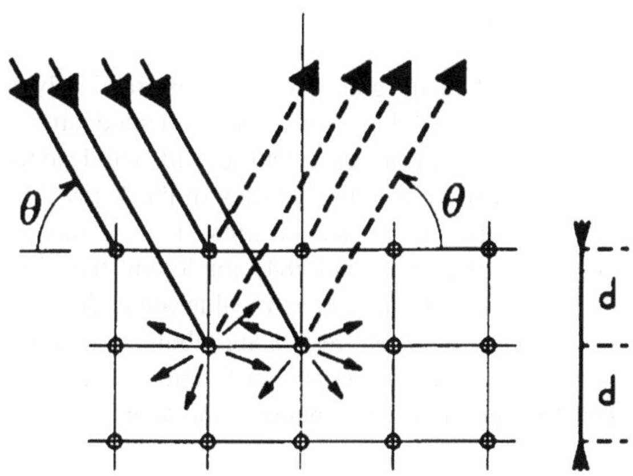

Fig. 14.6 Bragg's law for X-ray and neutron diffraction [19–21]

$$2d \sin \theta = n\lambda \tag{14.5}$$

where d is the spacing between the planes in the atomic lattice, n is an integer, λ is the wavelength of the X-rays, and θ is the angle between the incident X-ray and the scattering planes [19–21].

When an X-ray beam irradiates the surface of a crystalline material, the beam is scattered constructively only if the lattice planes orient and fulfill the Bragg's law. There will be always some grains meet the diffraction condition, if the material is

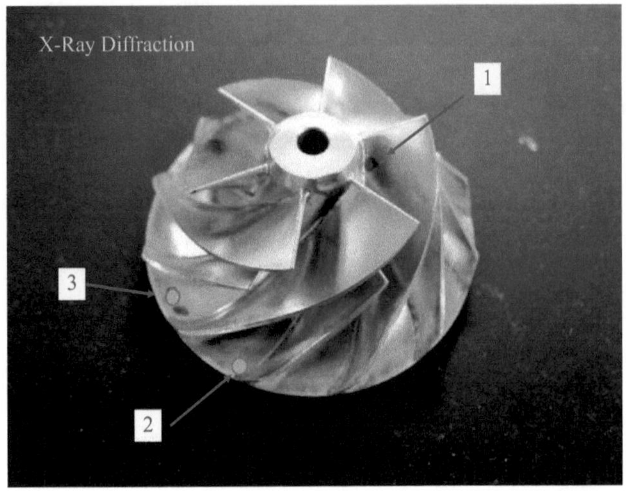

Fig. 14.7 X-ray measurement on the wheel

Table 14.1 Residual stress measurement results by X-ray diffraction method

Location	Maximun (MPa)	Minimum (MPa)	PH (°)
1	−43	−137	−13.1
2	−116	−161	−49.8
3	−80	−103	85.7

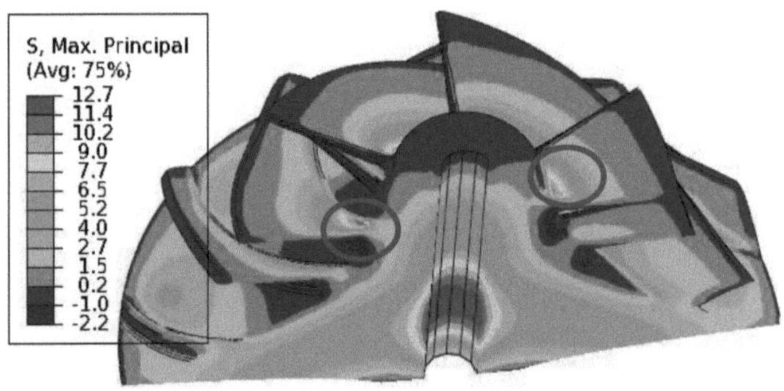

Fig. 14.8 Max. principal residual stress on the turbo wheel by numerical prediction

composed of many grains (crystallites) randomly oriented. If the grain size is big, fewer grains will be irradiated and therefore there will be fewer suitably oriented lattice planes fulfilling Bragg's Law. In this case, the local crystal defect such as dislocations, vacancies, stacking faults will lead to a local fluctuation of the lattice spacing, resulting in a peak broadening [19]. This will reduce the accuracy of the X-ray diffraction methods.

X-ray diffraction residual stress measurements were performed using a two-angle sine-squared-psi technique, in accordance with SAE HS-784, employing the diffraction of chromium K-alpha radiation from the (311) planes of the FCC structure of the A2618-T6 aluminum. X-ray diffraction residual stress measurements were made at the surface as shown in Fig. 14.7 and results are tabulated in Table 14.1. Measurements below the surface were attempted but were not successful because no suitable diffraction peak could be detected in the subsurface material. This is likely due to large grain size. Measurements were made in the hoop, 45°, and axial directions at 1, 2, and 3 as shown in Fig. 14.7 [22].

14.4 Residual Stress Comparison and Discussion

The residual stresses are predicted by finite element modeling and are shown in Figs. 14.8 and 14.9. It is shown that although considerable residual stresses are generated in the billet after water quenching, but majority of them are relieved after the machining process because the turbo wheel fan thickness is so small. Figures 14.8 and 14.9 show that principal residual stresses at most of the wheel locations are in the range from −15 MPa to 10 MPa, which is agreeable to strain

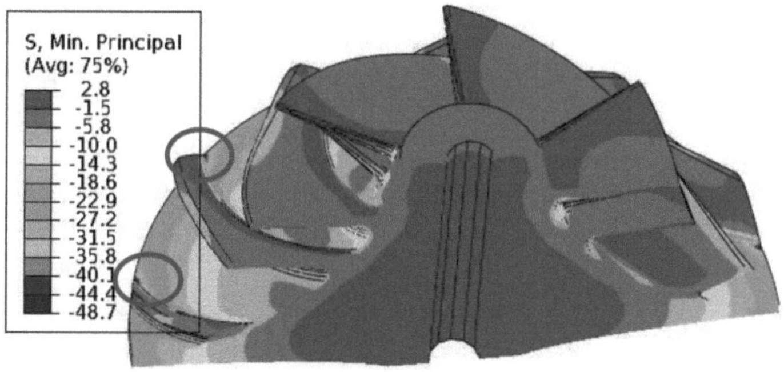

Fig. 14.9 Min. principal residual stress on the turbo wheel by numerical prediction

gauge rosette hole-drilling measurements. On the other hand, X-ray over measured the residual stress on the surface due to its limitation. For example, the large grain size in this aluminum wrought alloys material narrows the range for suitable diffraction peak.

14.5 Conclusions

To understand and optimize heat treatment and stretching processes, a computational model for residual stress analysis has been used in aluminum wrought alloy. In addition, residual stress measurements by strain gauge rosette hole-drilling method and X-ray diffraction method are made to validate the numerical prediction. The residual stress comparison shows that finite element prediction is in good agreement with strain gauge rosette hole-drilling measurements and the X-ray diffraction method over measures the residual stress, probably due to its limitation on this large grain size wrought aluminum material.

It is also noted that a significant amount of residual stress is relieved during machining, particularly for small components with simple geometry and thin walls.

References

1. Li K, Xiao B, Wang Q (2009) Residual stresses in as-quenched aluminum castings. SAE Int J Mater Manuf 1(1):725–731
2. Li P, Maijer DM, Lindley TC (2007) Simulating the residual stress in an A356 automotive wheel and its impact on fatigue life. Metall Mater Trans B 38(4):505–515
3. Xiao B, Li K, Wang Q, Rong Y (2011) Numerical simulation and experimental validation of residual stresses in water-quenched aluminum alloy castings. J Mater Eng Performance 20(9):1648–1657. doi:10.1007/s11665-011-9866-7
4. Xiao B, Wang G, Rong Y (2011) Hardenability and distortion control in high pressure hydrogen quenching. Int J Manuf Res 6(1):1–14
5. Lee YL, Pan J, Hathaway R (2005) Fatigue testing and analysis: theory and practice. Elsevier, Butterworth-Heinemann, p 402
6. Wang Q, Xiao B, Chang C-C, Paluch D (2013) Modeling of residual stresses in aluminum castings, AFS Transactions, 13-1339, pp 1–7
7. Wang Q, Chang C-C, Paluch D, Zhang G (2012) Methods of predicting residual stresses and distortion in quenched aluminum castings, US Patent 8214182 B2, 3 Jul 2012
8. Xiao B, Wang Q, Jadhav P, Li K (2010) An experimental study of heat transfer in aluminum alloy casting during water quenching. J Mater Process Technol 210(14):2023–2028
9. Newman ML, Robinson BJ, Sehitoglu H (2003) Deformation, Residual Stress, and Constitutive Relations for Quenched W319 Aluminum. Metall Mater Trans A 34(7):1483–1491
10. Sehitoglu H, Smith T, Qing X (2000) Stress–strain response of a cast 319-T6 aluminum under thermomechanical loading. Metall Mater Trans A 31(1):139–151
11. Estey CM, Cockcroft SL, Maijer DM (2004) Constitutive behaviour of A356 during the quenching operation. Mater Sci Eng A 383(2):245–251
12. Vishay Company Technology (1997) Measurement of residual stresses by the hole-drilling strain gage method, Strain Measurement Technology, http://www.vishay.com/brands/measurements_group/guide/guide.htm
13. Rendler, NJ, Vigness I (1966) Hole-drilling strain-gage method of measuring residual stresses, Proc., SESA XXIII, No. 2: 577–586
14. VishayMicro-Measurements (2009) Tech Note TN-503: measurement of residual stresses by the hole-drilling strain gage method, pp 16, http://www.intertechnology.com/Vishay/pdfs/TechNotes_TechTips/TN-503.pdf
15. Schajer GS (1988) Measurement of non-uniform residual stresses using the hole drilling method, part I—stress calculation procedures. ASME J Eng Mater Technol 110(4):318–342
16. Schajer GS (1988) Measurement of non-uniform residual stresses using the hole drilling method, Part II—practical application of the integral method. ASME J Eng Mater Technol 110(4):344–349
17. Xiao B, Li K, Rong Y (2011) Automatic determination and evaluation of residual stress calibration coefficients for hole-drilling strain gage integral method. Strain 47(s1):e525–e534. doi:10.1111/j.1475-1305.2009.00650.x

18. Lachtrupp TP (2013) Hole-drilling determination of the principal residual stresses in one A2618-T6 aluminum turbine wheel, a report to GM Powertrain by Lambda Research, Inc., Report #: REPORT: 277–18095, Jul 2013
19. Lu J, James M (1996) Handbook of measurement of residual stresses. Fairmont, Lilburn, GA, p 237
20. Bragg WL (1912) The diffraction of short electromagnetic waves by a crystal. Proc Camb Philos Soc 17:43–57
21. Wikimedia Foundation Inc. (2009) Bragg's Law—Wikipedia, the Free Encyclopedia.Mht, Http://en.Wikipedia.org/wiki/Bragg's_law#-cite_note-0. 15 Dec 2009
22. Lachtrupp TP (2013) X-ray diffraction determination of the surface residual stresses in one A2618-T6 aluminum turbine wheel, a report to GM Powertrain by Lambda Research, Inc., Report #: 277–17978, May 2013

Chapter 15
Notch Fatigue Behaviour of Shot Peened High-Strength Aluminium Alloys: Role of the Residual Stress Field Ahead of the Notch Root

M. Benedetti, V. Fontanari, M. Allahkarami, and J.C. Hanan

Abstract The present paper is aimed at investigating the effect of shot peening on the fatigue behavior of Al-7075-T651 samples carrying different types of notches. The Wöhler S–N curves were determined by pulsating bending for the different experimental conditions. A different improvement of the fatigue strength was found, i.e. a different effectiveness of the treatment for different notch geometries: the more critical notches received the larger beneficial effect by shot peening treatment in terms of reduction of the notch fatigue sensitivity and increment of the notch fatigue strength. The fatigue improvements with respect to the unpeened condition were discussed accounting for the residual stress effects. The residual stress field ahead of the notch root was evaluated by means of a numerical technique, making use of XRD measurements on the plain peened specimens. The numerical predictions were then compared with the results of XRD measurements conducted in the vicinity of the notch root of the peened notched samples.

Keywords Shot peening • Notch fatigue • Al-7075-T651 • Residual stresses

15.1 Introduction

Aluminium alloys are an attractive class of materials for aircraft and automotive industry because of their high specific static strength. In aerospace, aluminium alloys face ever stiffer competition from composites. In the automotive context, more and more engine parts are being made from them. Usually, high static mechanical properties are induced in aluminium alloys by dispersion hardening through solution and ageing heat treatments. However, aluminium alloys exhibit poor plain fatigue resistance [1] and high notch fatigue sensitivity [2]. Accordingly, stress raisers, like holes, fillets and grooves, always present in machine parts, are particularly detrimental to the fatigue response of these alloys, thus limiting their use in highly stressed mechanical components characterized by complex shapes. For this reason, aluminium alloys are frequently subjected to shot peening, particularly effective in incrementing the plain and notch fatigue strength of steels and light alloys. Shot peening mainly results in three fatigue related modifications of the surface layers: roughness, residual stresses and work hardening. Clearly, the surface roughening after shot peening is detrimental to the fatigue resistance due to the stress concentration exerted by the surface dimples. In the literature, it is commonly accepted that the improvement of fatigue strength is mainly induced by the introduction of compressive residual stresses in the surface region, responsible for both retarded fatigue crack initiation and lower small crack growth rates [3, 4]. The role of work hardening on the fatigue response is essentially indirect, since it strongly affects the stability of residual stresses by preventing them from relaxing due to accumulation of plastic deformation [5].

The authors recently studied the effect of three types of shot peening treatments on the plain and notch high-cycle fatigue response of the Al-7075-T651 alloy under reverse bending loading conditions ($R = -1$) [6–8]. In particular, it was shown that:

M. Benedetti (✉) • V. Fontanari
Department of Industrial Engineering, University of Trento, via Sommarive 9, 38123 Trento, Italy
e-mail: matteo.benedetti@ing.unitn.it

M. Allahkarami • J.C. Hanan
Department of Mechanical and Aerospace Engineering, Oklahoma State University, 700 N. Greenwood Ave., Tulsa, OK 74106, USA
e-mail: Jay.Hanan@okstate.edu

N. Sottos et al. (eds.), *Experimental and Applied Mechanics, Volume 6: Proceedings of the 2014 Annual Conference on Experimental and Applied Mechanics*, Conference Proceedings of the Society for Experimental Mechanics Series,
DOI 10.1007/978-3-319-06989-0_15,

1. Shot peening is an effective method to improve the fatigue endurance of Al-alloys even in the presence of geometrical discontinuities. The fatigue improvement is more pronounced with increasing stress concentration.
2. The residual stress field is supposed to concentrate in the proximity of the notch in a similar manner as stresses due to external loading. Since the experimental determination of residual stresses near geometrical discontinuities is very difficult, the authors proposed to numerically reconstruct the residual stress field on the basis of far-field residual stress measurements [7, 8]. For this purpose, the residual stress field created by the surface treatments is introduced into a Finite Element (FE) model by means of a fictitious temperature distribution varying along the sample's thickness and responsible for a hydrostatic initial strain distribution. This is supposed to depend only on the peening treatment and not on the sample geometry, so the same temperature distribution was applied to the notched components.
3. The major cause of residual stress modification during the fatigue life is plastic flow occurring when the superposition of residual and external stresses exceeds the material's cyclic yield strength. This can be predicted by a numerical model, once the cyclic stress strain response of the shot peened layers is known.
4. The fatigue strength can be accurately predicted by a multiaxial fatigue criterion incorporating stabilized residual stresses as mean stresses and combined with a line method based on the critical distance theory to account for the notch sensitivity.

The current state-of-the-art-knowledge lacks for investigations proving (1) whether the beneficial effect of shot peening is still present in the very-high cycle fatigue regime, (2) how residual stresses are distributed in the vicinity of notches and affect the corresponding fatigue behaviour. Therefore, the present work is aimed at investigating the very-high cycle fatigue regime of shot peened Al-7075-T651 alloy and the residual stress field in the vicinity of notches. For this purpose, pulsating bending fatigue tests ($R = 0.05$) have been carried out on smooth and notched samples exploring fatigue lives comprised between 10^5 and 10^8 cycles. The residual stress field ahead of the notch root was mapped along the notch bisector using the micro X-ray diffraction technique. These measurements are of paramount importance for validating the numerical approach proposed in [8] to reconstruct the residual stress field in the vicinity of the notch root of the peened notched samples.

15.2 Materials and Experimental Procedures

The experimentation has been performed on the aluminium alloy Al-7075-T651, widely used for aeronautical applications, supplied in the form of 4 mm thick rolled plate. The monotonic tensile properties of this alloy can be found in Ref. [6].

The fatigue tests have been carried out on plane specimens whose geometry, according to the standard ISO 3928, is illustrated in Fig. 15.1. The microstructure has been tested with the stress axis parallel to the L-direction. The first type of samples (Fig. 15.1a), termed smooth samples, presents a fillet radius large enough to make any notch fatigue effects negligible. The notched specimens, whose geometry is illustrated in Fig. 15.1b, were manufactured by Electric Discharge Machining (EDM). Particular attention was paid in accurately machining the two edge V-notches (with aperture angle of 90°), which significantly affect the stress distribution. The notch root fillet radius was set to 2 mm ("blunt" notched samples), 0.5 mm ("sharp" notched samples) and 0.15 mm ("very sharp" notched samples), so that the theoretical stress concentration factor is 1.52, 2.37 and 3.76, respectively.

Part of the fatigue specimens was subjected to a controlled shot peening treatment, whose process parameters are summarized in Table 15.1. Details about the air-blast machine used for shot peening can be found in Ref. [6]. The treatment employs small ceramic beads leading to a gentle and superficial effect, which allow for higher fatigue performance as compared with larger shots [9]. Moreover, the usage of ceramic beads instead of steel shots prevents undesired galvanic effects.

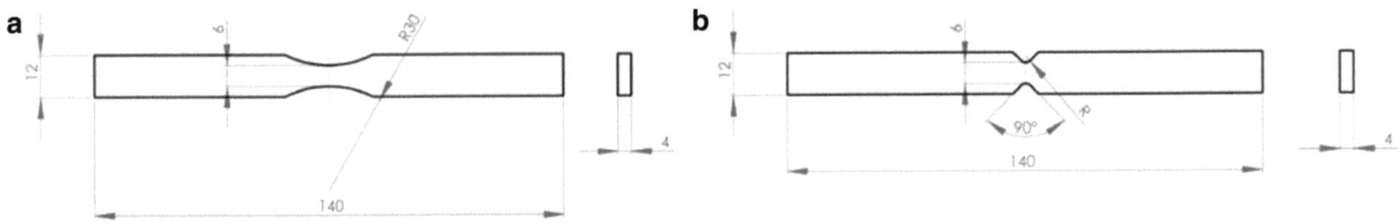

Fig. 15.1 Geometry of specimens used in this study: (**a**) smooth and (**b**) notched bending fatigue specimens. All dimensions are given in mm. Two types of notched samples were used; the notch root fillet radius R is 2 mm, 0.5 mm and 0.15 mm in the blunt-notched, sharp-notched and very sharp-notched samples, respectively

Table 15.1 Shot peening parameters

Treatment	Material	Bead size (μm)	Bead hardness (HV_1)	Bead composition	Almen intensity	Angle of impingement	Coverage (%)
B120	Ceramic	60–120	700	ZrO_2 67 % SiO_2 31 %	4.5 N	90°	100

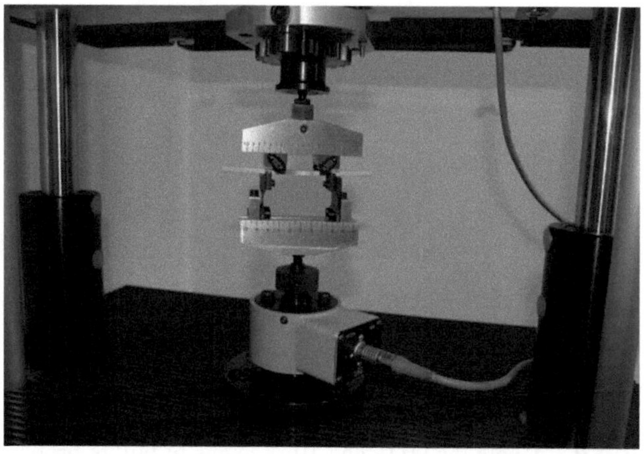

Fig. 15.2 Four-point bending fatigue test apparatus used in the present study

Table 15.2 Principal results of fatigue tests

		Wöhler curve				Gain (%)	
Sample geometry	Condition	Equation	k	σ_{P50} (MPa)	$T\sigma$	@5 × 10^5 cycles	@5 × 10^7 cycles
Smooth	As-received	15.1b	1.60 × 10^6	127	1.24	42	0
	Peened B120	15.1a	12	540	1.18		
Blunt notched (R2)	As-received	15.1b	1.25 × 10^6	86	1.14	70	24
	Peened B120	15.1a	13	400	1.13		
Sharp notched (R0.5)	As-received	15.1b	2.41 × 10^6	58	1.17	68	64
	Peened B120	15.1a	45	140	1.08		
Very sharp notched (R0.15)	As-received	15.1b	4.91 × 10^6	38	1.13	44	69
	Peened B120	15.1b	2.41 × 10^6	64	1.07		

k, σ_{P50}: fitting coefficients of the SN curve according to Eqs. 15.1; $T\sigma$: results scatter; gain: increment of the fatigue resistance of the shot peened condition as compared with the as-received material at a given fatigue life

Pulsating (R = 0.05) load-controlled 4-point bending fatigue tests were carried out in air, at room temperature, and at a nominal frequency of 110 Hz using a resonant testing machine Rumul Mikrotron 20 kN equipped with a 1 kN load cell (Fig. 15.2). Different stress levels corresponding to fatigue lives in the range between nearly 10^5 and 10^8 cycles have been considered. Tests were terminated at 10^8 cycles when no fracture occurred.

Two types of fatigue curves were obtained, i.e. without and with knee located around 10^6 fatigue cycles. The former fatigue curves, corresponding to 50 % of failure probability, were represented by a S–N curve in the form of:

$$\sigma_{P50} = \sigma_{0P50} \times N_f^{-\frac{1}{k}} \tag{15.1a}$$

The latter fatigue curves, corresponding to 50 % of failure probability, were represented by a S–N curve in the form of:

$$\sigma_{P50} = \sigma_{0P50} + \frac{k}{N_f} \tag{15.1b}$$

It should be noted that the meaning of the coefficients of Eqs. 15.1a and 15.1b is different, but they have been taken the same for a concise presentation in Table 15.2. In both cases, the uncertainty range was assumed to be constant and approximated by its centroid value. As a representative value of the scatter, the following expression was used:

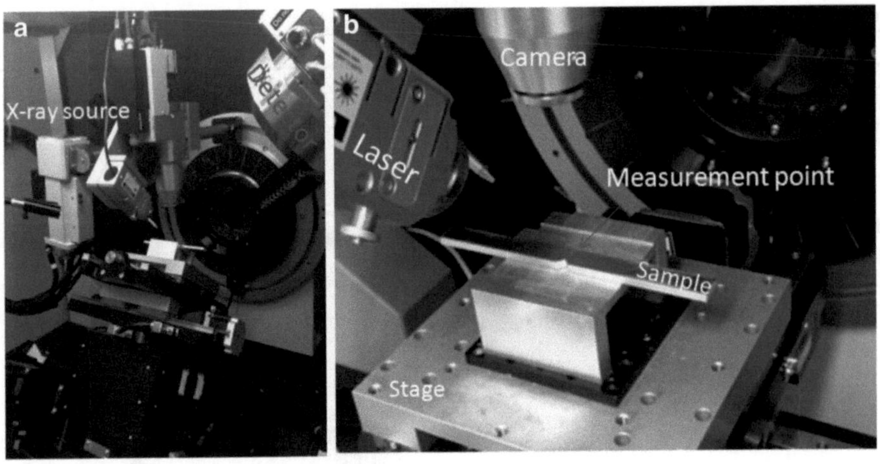

Fig. 15.3 Photograph of the XRD2 micro-diffractometer used to map the residual stress field in the notched specimens

$$T\sigma = 1 : \sigma_{P90}/\sigma_{P10} \tag{15.2}$$

P_{90}, P_{10} denote the 90 % and 10 % levels of failure probability, respectively.

The notch sensitivity factor q at a given number of cycles to failure N_f was computed as:

$$q(N_f) = \frac{K_f(N_f) - 1}{K_t - 1}; \quad K_f(N_f) = \frac{\sigma_{P50,smooth}(N_f)}{\sigma_{P50,notched}(N_f)} \tag{15.3}$$

The modifications of the surface layers of the shot peened samples were investigated through microhardness, surface roughness and residual stress measurements. The results of microhardness and surface roughness measurements are presented in the companion paper [10].

The analysis of the residual stress field induced by the peening treatment was performed both on plain and notched samples.

XRD measurements were carried out on plain samples to investigate the in-depth residual stress profile and its evolution throughout the fatigue life. For this purpose, an AST X-Stress 3000 X-Ray diffractometer was used. Measurements were made with Cr Kα radiation in the longitudinal direction in the gage region. The analysis zone was limited by a collimator of 1 mm^2 in area. The classical sin$^2\psi$ method was applied for stress evaluation with the use of 9 diffraction angles (2θ) scanned between −45° and +45° for each stress value. The ⟨311⟩ diffracting planes were chosen (1) in order to obtain high angle measurements (2θ angle 139.0°) with higher strain sensitivity, and (2) because they do not accumulate significant intergranular stresses and hence exhibit similar behavior as that of the bulk. Calibration of the system was checked by collecting a diffraction pattern from a standard polycrystalline Al powder prior to conducting the experiment. The in-depth measurements were conducted step-by-step using an electro-polishing device by removing a very thin layer of material in a region localized (2 mm × 2 mm) at the gauge section of the specimens. Both initial and stabilized residual stress fields were measured. For this purpose, measurements were performed on tested smooth specimens after failure in a region far enough from the fracture surface (about 2 mm) so that the material rupture was supposed not to have altered the residual stress field [6]. Measurements were carried out on both specimen sides, subjected to tensile and compressive bending stresses, respectively.

XRD measurements were carried out on notched specimens in order to map the residual stress field in the vicinity of the notch root. For this purpose, a Bruker's D8 Discover XRD2 micro-diffractometer equipped with General Area Diffraction Detection System (GADDS) and Hi-Star 2D area detector was employed. Tube parameters of 40 kV/40 mA using Cu-Kα radiation at a detector distance of 30 cm which covers approximately the area of 20° in 2θ and 20° in γ with 0.02° resolution was used. For line scans, a motorized five axis (X, Y, Z (translation), ψ (tilt), φ (rotation)) stage was used for positioning which can move the measurement spot to the instrument center within 12.5 μm position accuracy and 5 μm repeatability. A photograph of experimental set up is shown in Fig. 15.3. The analysis zone was limited using an 800 μm pinhole collimator. Exposure time per frame was 120 s. The ⟨311⟩ diffracting planes were selected in order to include material isotropy and the possibility of obtaining high angle measurements with higher stress sensitivity. For the given angular resolution of 0.01°, a residual stress with magnitude of ±1.04 MPa can be measured. The surface residual stress profile along the notch bisector was mapped through 10 measurements made on points spaced with 50 μm pitch starting from the notch tip. Using a sequential list, the system automatically brings each predetermined point to the diffractometer centre and performs

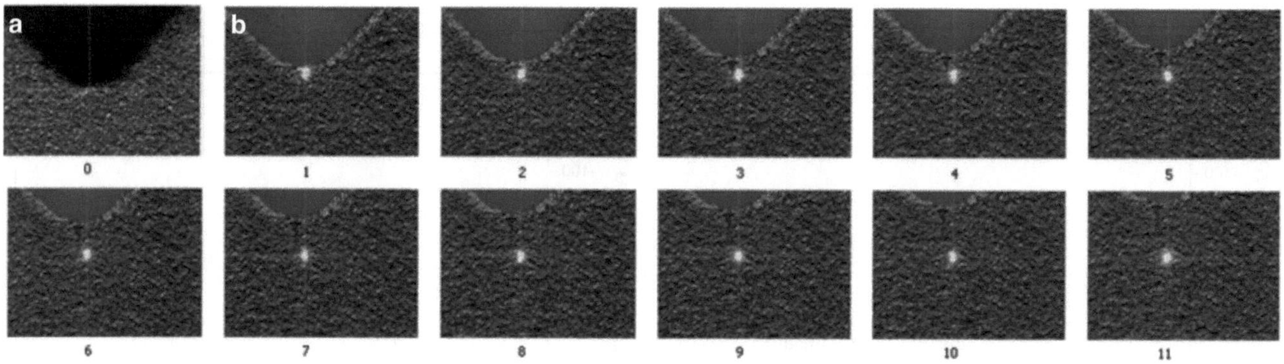

Fig. 15.4 (**a**) Top view from optical microscopy of the notched fatigue sample, superimposed with a target points. (**b**) Camera cross hair and laser point coincide at X-ray measurement points

final height adjustments automatically in the out-of-plane direction with an auto video-laser positioning system before each exposure. Initial assigned target points and actual measurement points are shown in Fig. 15.4. Two dimensional frame data were imported to Bruker's LEPTOS software for residual stress evaluation. Each frame was divided to sub regions and locally integrated along the y direction to obtain 1D diffraction pattern for each sub region. Correction on polarization, background, $K\alpha_2$ was performed. The peak positions were determined by fitting Pearson VII type functions. Due to the large number of points available for a diffraction peak, the Pearson VII function is statistically reliable.

15.3 Results and Discussion

15.3.1 Residual Stress Field

XRD measurements were carried out on plain fatigue samples in order to characterize the remote residual stress field prior to and after fatigue testing. The obtained stress profiles are illustrated in Fig. 15.5. Two measurements were performed on the untested specimens in order to account for the variability in the residual stress field. It can be noted that the sub-superficial compressive residual stress peak is located nearly 15 μm below the surface and that the depth of the surface layer subjected to compressive residual stresses is equal to about 50 μm. The repeatability of both residual stress measurement and peening treatment is very good within a depth of about 15 μm, where the fatigue response is mostly dictated [6].

In addition, Figs. 15.5a and 15.7b illustrate the evolution of the residual stress field at two loading levels on the specimen surface subjected to compressive and tensile bending stresses, respectively. It can be noted that some relaxation of the residual stress field occurred on the outer layer of the specimen side subjected to compressive bending stresses, the more pronounced the higher the load levels. Conversely, significantly lower relaxation occurred on the specimen side subjected to tensile bending stresses.

The initial far-field residual stress profile shown in Fig. 15.5 was used to numerically reconstruct the residual stress field in the neighbourhood of the notch root of the notched specimens. For this purpose, the same procedure presented in [6, 7] was followed. Specifically, the residual stress field created by the surface treatments on the plain specimens has been introduced into the FE model by means of a fictitious temperature distribution varying along the specimen's thickness and responsible for a hydrostatic initial strain distribution (ISD), according to the approach proposed by Beghini and Bertini [11]. Then, it has been assumed that the ISD, and therefore the fictitious temperature field, (1) varies with normal distance from the shot peened surfaces only, (2) is independent of specimen geometry (either plain or notched), (3) depends solely on the shot peening treatment. Hence, the same fictitious temperature field, found for the plain specimens subjected to the same shot peening treatments, has been applied to the notched ones.

The FE model, depicted in Fig. 15.6, has been set up with the ANSYS 13 commercial code using eight nodes isoparametric brick elements. In order to take advantage of the symmetry, only a quarter of the specimen has been modelled. The mesh in the surface region is particularly refined with the purpose of better reproducing stress and strain gradients in the hardened layer due to the peening treatment. The minimum thickness of the elements in the surface region is 12 μm.

Figure 15.7 illustrates the comparison between numerically predicted and experimentally measured residual stress distribution along the *y* coordinate aligned with the notch bisector and centred on the notch apex (see Fig. 15.6). It can be

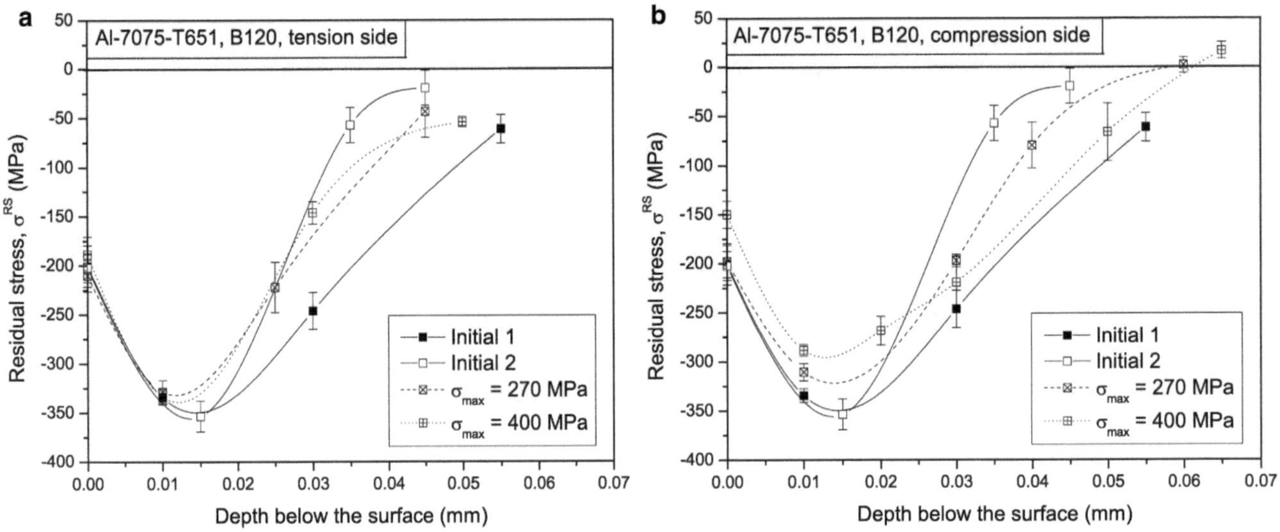

Fig. 15.5 Evolution of the residual stress profile during fatigue life in samples subjected to B120 treatment. (**a**) Tensioned, (**b**) compressed side. Specimen tested at 270 (400) maximum stress MPa failed after 44.6 × 10^6 (530,000) cycles

Fig. 15.6 Overview of the FE model of one quarter of the blunt-notched fatigue specimens illustrated in Fig. 15.1

Fig. 15.7 Comparison between experimentally measured and numerically predicted residual stress distribution along the notch bisector

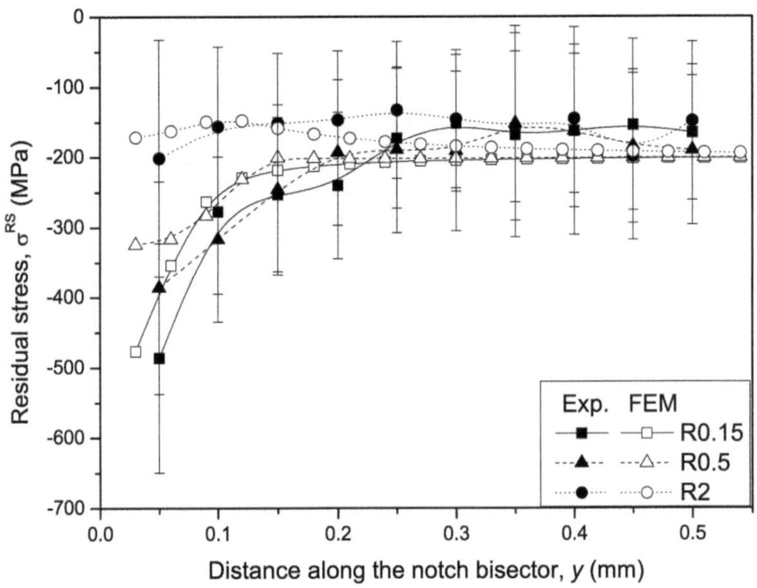

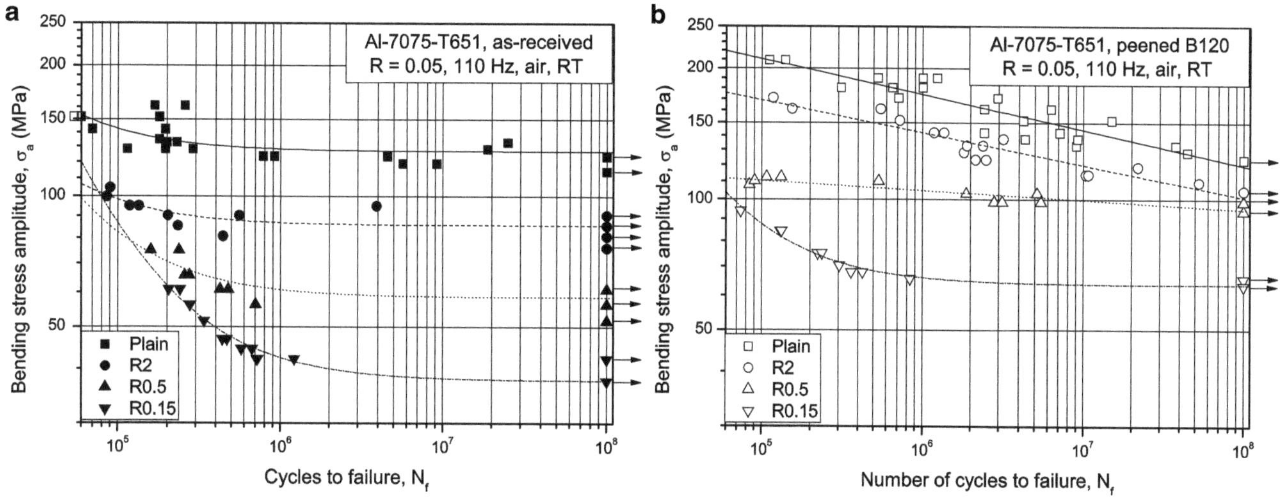

Fig. 15.8 Pulsating bending fatigue curves of the plain and notched specimens: (**a**) as-received, (**b**) peened specimens. Run-out tests are marked by arrows

noted that the blunt notch does not significantly affect the residual stress field, which in turn remains approximately constant along the bisector. Conversely, the two sharp notches result in a remarkable residual stress concentration, whose entity scales with the notch severity. The region interested by the residual stress concentration spans about 0.2 mm from the notch root. A satisfactory agreement between numerical predictions and experimental measures can be observed, especially in terms of the extension of the notch affected region. Nevertheless, the numerical simulations are affected by a systematic underestimation of the residual stresses close to the notch root on the order of about 30 %. This might be imputed to the simplifying hypothesis that the residual stress field is created by an ISD which depends only on the depth below the frontal faces of the specimens. Actually, the specimens are impacted also on the lateral faces, hence, near the notch root, additional plastic deformation due to shot peening is expected and this can further contribute to the residual stress field. Further investigation is foreseen in the near future to shed light on this aspect as well as on the stability of such intense residual stresses throughout the fatigue life.

15.3.2 *Fatigue Behaviour*

The results of the pulsating bending fatigue tests as well as the P50 fatigue lines are compared in Fig. 15.8a, b for unpeened and peened specimens, respectively, in the different material variants considered. The parameters representing the fatigue curves corresponding to 50 % of failure probability, according to Eqs. (1) and the results scatter, expressed by Eq. 15.2 are listed in Table 15.2.

The presence of a stress raiser in the as-received material suppresses the occurrence of failures at fatigue lives longer than 4×10^6 cycles and this phenomenon becomes more remarkable with increasing notch severity. All the fatigue curves of the as-received condition are well represented by Eq. 15.1b showing a knee around 1×10^6 cycles. On the contrary, the fatigue curves of the peened samples steadily decline with the fatigue life. They are then well represented by the linear relation in double-logarithmic scale represented by Eq. 15.1a, and show occurrence of failure throughout the explored fatigue life interval. The only exception is given by the very sharp notched samples, whose fatigue curve resembles those of the unpeened conditions.

The peening treatment was effective in prolonging the fatigue life of the material, even in the presence of stress concentration, as well as in reducing the large scatter in fatigue results displayed by the virgin material. This improvement depends on both the applied load and the notch severity. For smooth and blunt-notched specimens, it is more remarkable for load levels corresponding to shorter fatigue lives, leading however to higher values of the slope in the P50 fatigue line. Therefore, the increment in fatigue resistance due to shot peening steadily declines during fatigue life. While the plain fatigue performance at 10^8 cycles of the peened variants is comparable with that of the virgin material, an increment of about 20 % in the fatigue resistance of blunt-notched specimens is still present in this fatigue regime. On the contrary, the increment due to shot peening is retained at longer fatigue lives in the case of sharp notches and for very sharp notches the

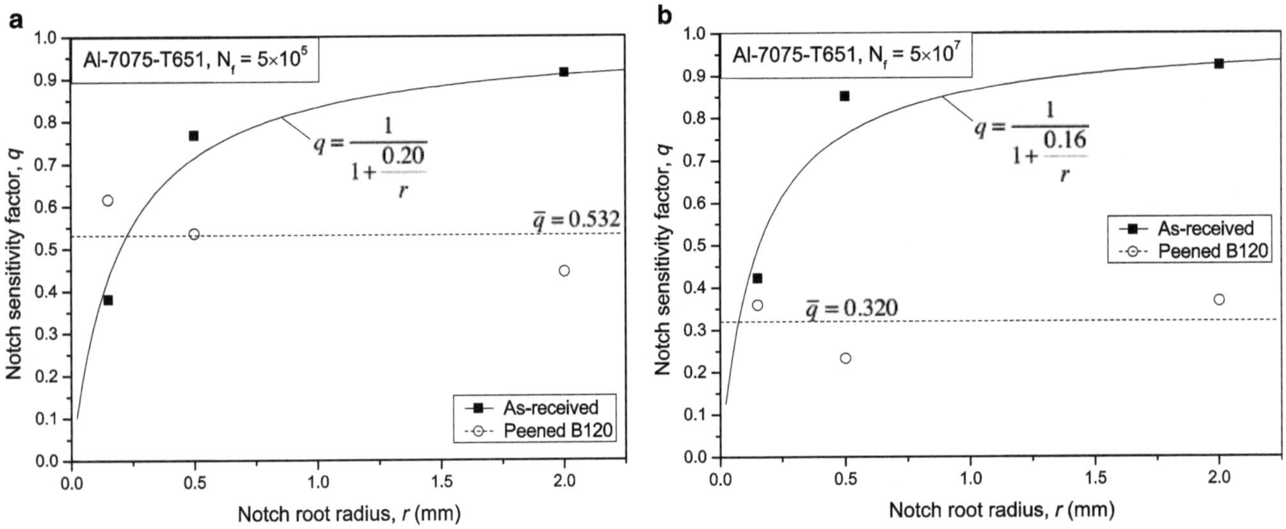

Fig. 15.9 Notch sensitivity factor dependence upon notch root radius for a fatigue live corresponding to (**a**) 5×10^5 and (**b**) 5×10^7 cycles

increment is even higher for longer than for shorter fatigue lives. The remarkable improvement in fatigue resistance due to shot peening in the sharp-notched specimens is presumably due to the high residual stress concentration at the notch tip shown in Fig. 15.7.

Figure 15.9a, b illustrate the dependence of the notch sensitivity factor q on the notch root radius r for a fatigue life of 5×10^5 and 5×10^7, respectively. A remarkable reduction in notch sensitivity due to shot peening is evident, especially for the blunt notch R2. The dependence of the factor q upon the notch root radius for the unpeened samples is well represented by the Peterson equation:

$$q = \frac{1}{1 + \frac{a}{r}} \tag{15.4}$$

where the material characteristic length a ranges between 0.16 and 0.20 mm. Conversely, the peened variants exhibit a factor q less sensitive to the notch root radius, with an average value comprised between 0.53 and 0.32 at the lowest and highest fatigue life, respectively. Notably, for the very sharp-notched specimens, the q factor of the unpeened samples is lower at shorter and higher at longer fatigue life with respect to the peened samples. This confirms that the peening treatment on the most severe notches is more effective in the very high cycle fatigue regime.

15.4 Conclusions

The plain and notch fatigue strength of shot peened aluminium alloy Al-7075-T651 has been experimentally investigated exploring fatigue lives comprised between 10^5 and 10^8 cycles. The fatigue improvements with respect to the unpeened condition were discussed accounting for the residual stress effects. The residual stress field ahead of the notch root was evaluated by means of a numerical technique, making use of XRD measurements on the plain peened specimens. The numerical predictions were then compared with the results of XRD measurements conducted in the vicinity of the notch root of the peened notched samples. The following conclusions can be drawn:

1. Shot peening is an effective method to improve the fatigue endurance of Al-alloys even in the presence of geometrical discontinuities. Notch sensitivity is greatly reduced by shot peening.
2. For smooth and blunt-notched specimens, the beneficial effect of shot peening diminishes with increasing fatigue cycles. Conversely, the increment due to shot peening is retained at longer fatigue lives in the case of sharp notches and for very sharp notches the increment is even higher for longer than for shorter fatigue lives.
3. Residual stress relaxation in smooth samples occurs mainly when compressive bending stresses are superimposed to the surface compressive residual stress field.

4. The sharp notches induce a remarkable residual stress concentration, whose entity scales with the notch severity. The residual stress distribution in the vicinity of the notch root was fairly well reconstructed on the basis of measurements made on the plain samples. Nevertheless, a systematic underestimation of the residual stresses close to the notch root was found, presumably because the contribution of shot impacts on the lateral surfaces was neglected.

References

1. ASM handbook, Properties and selection: nonferrous alloys and special-purpose materials, Vol. 2, 10th edition, Materials Park, Ohio: American society for metals, 1991.
2. Bian JC, Tokaji K, Ogawa T (1994) Notch sensitivity of aluminium-lithium alloys in fatigue. J Soc Mat Sci Japan 43:840–846
3. Guagliano M, Vergani L (2004) An approach for prediction of fatigue strength of shot peened components. Eng Fract Mech 71:501–512
4. Wagner L (1999) Mechanical surface treatments on titanium, aluminum and magnesium alloy. Mater Sci Eng A263:210–216
5. McClung RC (2007) A literature survey on the stability and significance of residual stresses during fatigue. Fatigue Fracture Eng Mater Struct 30:173–205
6. Benedetti M, Fontanari V, Scardi P, Ricardo CLA, Bandini M (2009) Reverse bending fatigue of shot peened 7075-T651 aluminium alloy: the role of residual stress relaxation. Int J Fatigue 31:1225–1236
7. Benedetti M, Fontanari V, Monelli BD (2010) Numerical simulation of residual stress relaxation in shot peened high-strength aluminium alloys under reverse bending fatigue. ASME J Eng Mater Technol 132:011012-1-9
8. Benedetti M, Fontanari V, Santus C, Bandini M (2010) Notch fatigue behaviour of shot peened high-strength aluminium alloys: Experiments and predictions using a critical distance method. Int J Fatigue 32:1600–1611
9. Oguri K (2011) Fatigue life enhancement of aluminum alloy for aircraft by Fine Particle Shot Peening (FPSP). J Mat Proc Technol 211:1395–1399
10. Benedetti M, Fontanari V, Bandini M (2014) Very high-cycle fatigue resistance of shot peened high-strength aluminium alloys: role of surface morphology. Proceedings of the SEM 2014 annual conference & exposition on experimental and applied mechanics, Greenville, SC, USA, 2–5 June 2014
11. Beghini M, Bertini L (1990) Residual stress modelling by experimental measurements and finite elements analysis. J Strain Anal 25(2):103–108

Chapter 16
Residual Stress of Individual Aluminum Grains from Three Dimensional X-Ray Diffraction

M. Allahkarami, B. Jayakumar, and J.C. Hanan

Abstract Residual stress measurement by interpreting diffraction rings is well developed. The ring is produced by many grains contributing individual diffraction spots, each from their local stress environment. Local stresses limit the bulk strain resolution of X-ray stress methods. The local stresses can be so large, traditional methods fail to produce meaningful results. However, if the local stress environment could be understood, it provides additional value for measuring and predicting material behavior. Experimental determination of the internal stress for one grain from a ring has several challenges. In some special cases, these have been overcome using synchrotron radiation. Here, using a Bruker D8 laboratory X-ray diffractometer and a 2D Hi-star detector, a method of sensing and analyzing X-ray diffraction cones in three dimensions was introduced. After a certain sample detector distance, individual grains can be resolved in spots on the ring. The method requires collecting a sequence of 2D frames at increasing sample to detector distances. The entire three-dimensional X-ray diffraction pattern (XRD^3) could be used to determine the average 2θ ring position. This allows new types of strain measurements. Other applications for tracking spots from grains are in development.

Keywords Residual stress • 3D X-ray diffraction (XRD^3) • Diffraction cone • 2D diffraction frame

16.1 Introduction

X-ray diffraction is a well known nondestructive and powerful technique to determine the stress state of crystalline materials. The X-ray based methods allow studding both residual stresses and the phase of diffracting materials [1, 2]. The fastest way to collect a diffraction signal is to use high-flux synchrotron sources [3, 4]. The data collection time is fast enough to record changes in the diffraction pattern of a material under dynamic load [5]. Although synchrotron sources are the primary choice for accurate residual stress measurement with high speed, they are not as accessible as conventional laboratory diffractometers or even portable systems [6]. Of course, laboratory point detector systems have a intensity less than the latest synchrotron that make them impractical for some measurements that would require long exposures time. A linear position sensitive detector is about 100 times faster than a point detector, and 2D area detectors are 10^4 times faster than point detectors [7, 8]. Reasonable speed, large dynamic range, and simultaneous collection of 2D data are some advantages of 2D detectors for residual stress measurement. Bob He et al., measured residual stress on the inside surface of a small spring with both conventional and 2D based methods and showed different methods agree very well [7]. Recently, diffraction patterns from 2D detectors have been used for several unique applications such as phase mapping on flat samples and even of complex curved geometries [9, 10]. In this paper we introduce a method of detecting and analyzing partial X-ray diffraction cones in three dimensions using a Bruker D8 laboratory X-ray diffractometer and a 2D Hi-star detector. The method is based on collecting a set of 2D frames at increasing sample to detector distances. The local stresses from a single grain were observed.

M. Allahkarami • B. Jayakumar • J.C. Hanan (✉)
Mechanical and Aerospace Engineering, Oklahoma State University, Tulsa, OK, USA
e-mail: jay.hanan@gmail.com

N. Sottos et al. (eds.), *Experimental and Applied Mechanics, Volume 6: Proceedings of the 2014 Annual Conference on Experimental and Applied Mechanics*, Conference Proceedings of the Society for Experimental Mechanics Series,
DOI 10.1007/978-3-319-06989-0_16,

16.2 Material and Experimental Setup

A rectangular 2 × 5 mm aluminum bar was bent to create a U-shape sample shown in Fig. 16.1. The aluminum alloy was 6061 which is one of the common alloys for general purpose use. The aluminum sample was bent to create a gradient of residual stress. The inside region is expected to be under compressive residual stress, while the outside is under tensile.

All XRD frames were collected with Cu Kα radiation at tube parameters of 40 kV/40 mA using a Bruker D8 Discover XRD2 micro-diffractometer equipped with the general area diffraction detection system (GADDS) and a Hi-Star 2D area detector. At each set of measurement, 31 diffraction patterns collected starting with a 30 cm sample to detector distance (SDD) to 15 cm at 0.5 cm steps. The detector distance changed manually at each step. All parameters except the SDD were kept constant between steps. A schematic of the experimental setup with the detector at two different positions is shown in Figs. 16.1 and 16.2a. Figure 16.2b shows a photograph of experimental setup. The exposure time was 60 s per frame using an 800 μm diameter beam. The X-ray beam incident angle θ_1 and detector angle θ_2 were 40°.

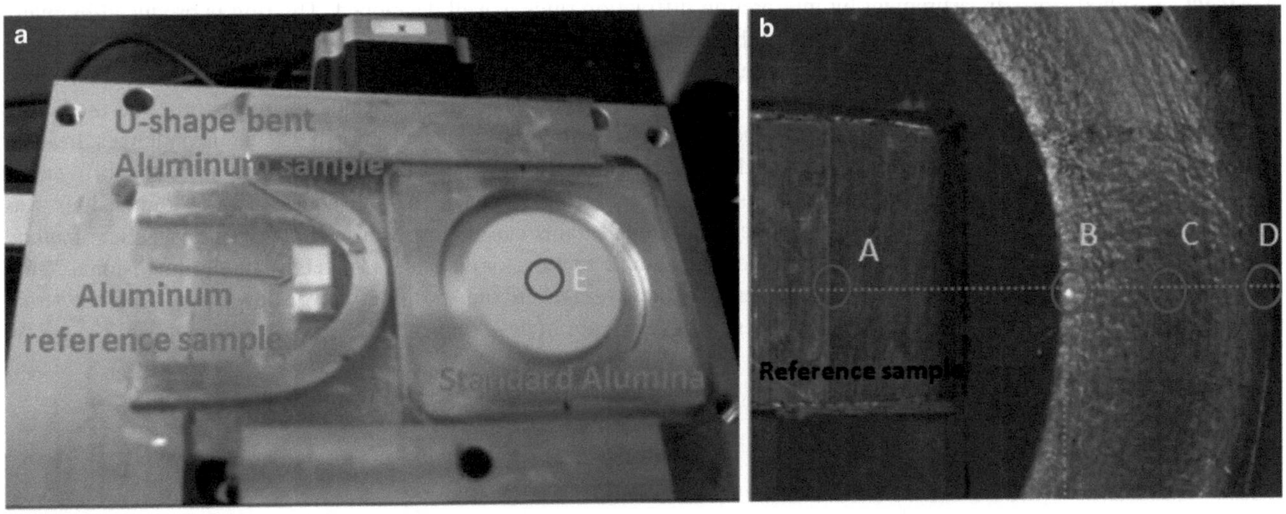

Fig. 16.1 (**a**) Samples on measurement stage (**b**) Strain free reference aluminum sample and aluminum U-bar sample (bent sample)

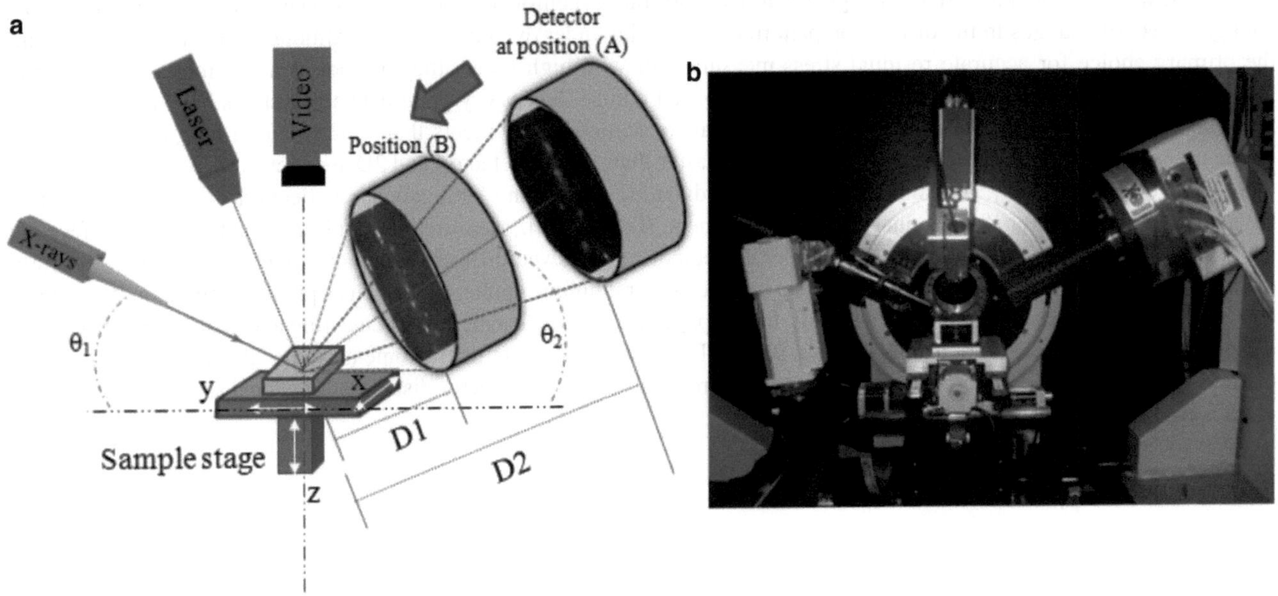

Fig. 16.2 (**a**) Schematic of setup showing of monochromatic X-ray beam and detector to sample distance at two near and far positions (**b**) Photograph of experimental setup

16.3 Microstructures of Aluminum

In order to observe microstructure sections of the sample were cut using a band-saw. Samples were mounted in epoxy, sanded and polished to a mirror finish and etched using a 0.5 % HF solution. Elongated grains in the tensile side, smaller grains in the compressive side and similar grains in the neutral and untreated surfaces were observed (Fig. 16.3).

16.4 Results and Discussion

16.4.1 Calibration

Before collecting data with the D8 Discover with GADDS detector, calibration is required. The routine way is to calibrate the detector at each distance. This first method was used by Allahkarami et al. for three-dimensional X-ray diffraction detection and visualization [11]. The calibration includes determining the precise sample to detector distance (SDD) and beam center which involves flood field correction and spatial correction [12]. The disadvantage of this method is that normally it takes 1 h of skilled user time to complete the calibration at each detector location. The other way around this is to collect a frame on a Corundum standard sample and use it as a displacement reference for calibrating the frames from the aluminum sample. This is similar to mixing silicon powder with unknown powder as internal reference peak. Translation oscillations or angular rotation was used to bring more crystallites into the diffraction condition while collecting the diffraction from the standard sample. Figure 16.4 shows a mosaic pattern of frames collected along different detector distances from the NIST 1976 a standard Corundum sample. The frame collected at the furthest distance (30 cm) has the smallest number of rings and the frame collected at the closest distance (30 cm) has the largest number of rings. It was observed that the background intensity decreases by increasing the SDD. Images with 1,024 $\times$ 1,024 pixel size with ".TIF" format was created from raw ".gfrm" 2D data files. All 31 frames were imported[1] as a sequence of images as shown in Fig. 16.5. The visualization software provides

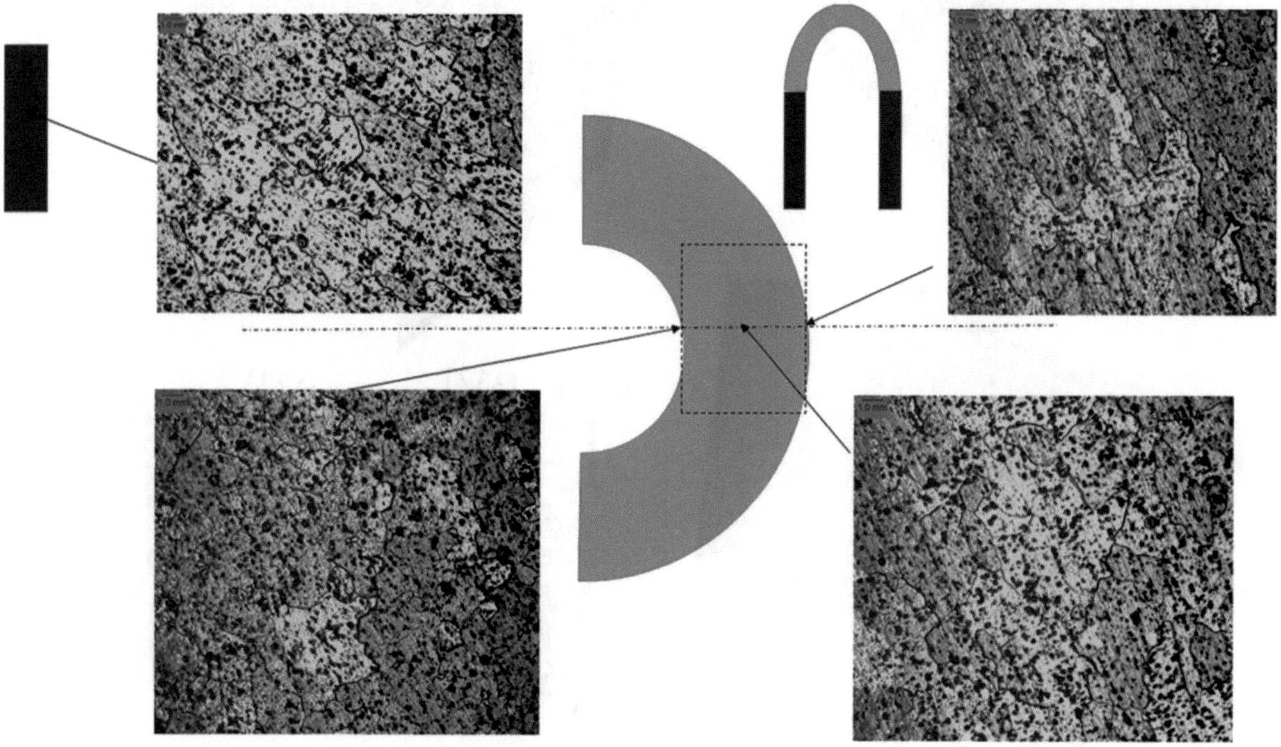

Fig. 16.3 Microstructures of Aluminum showing grain boundaries under optical microscope

[1] Into Amira 3.1.1, a three dimensional data analysis and visualization software.

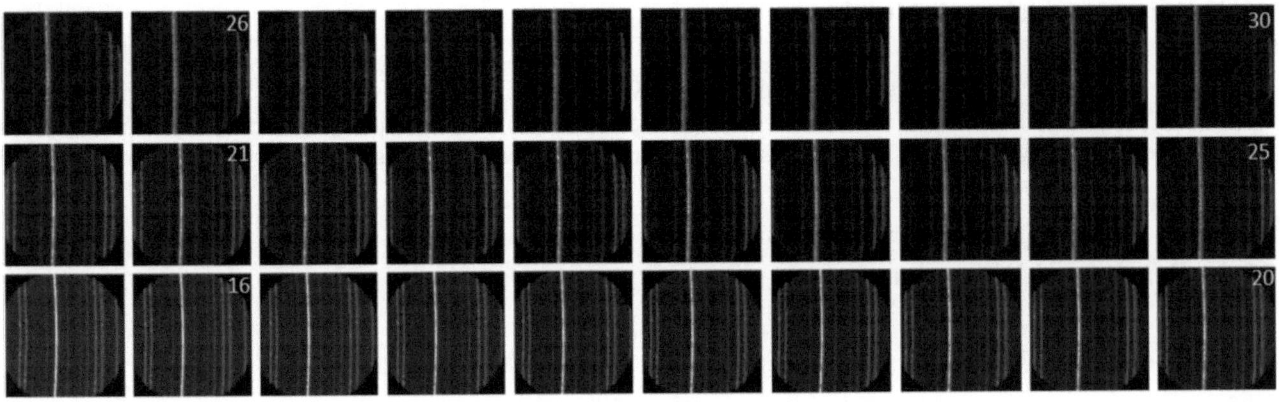

Fig. 16.4 Mosaic pattern of 2D XRD frames with different detector distance starting from 30 to 15 cm with 5 mm steps. The number on right top corner of the frame represents the detector distance

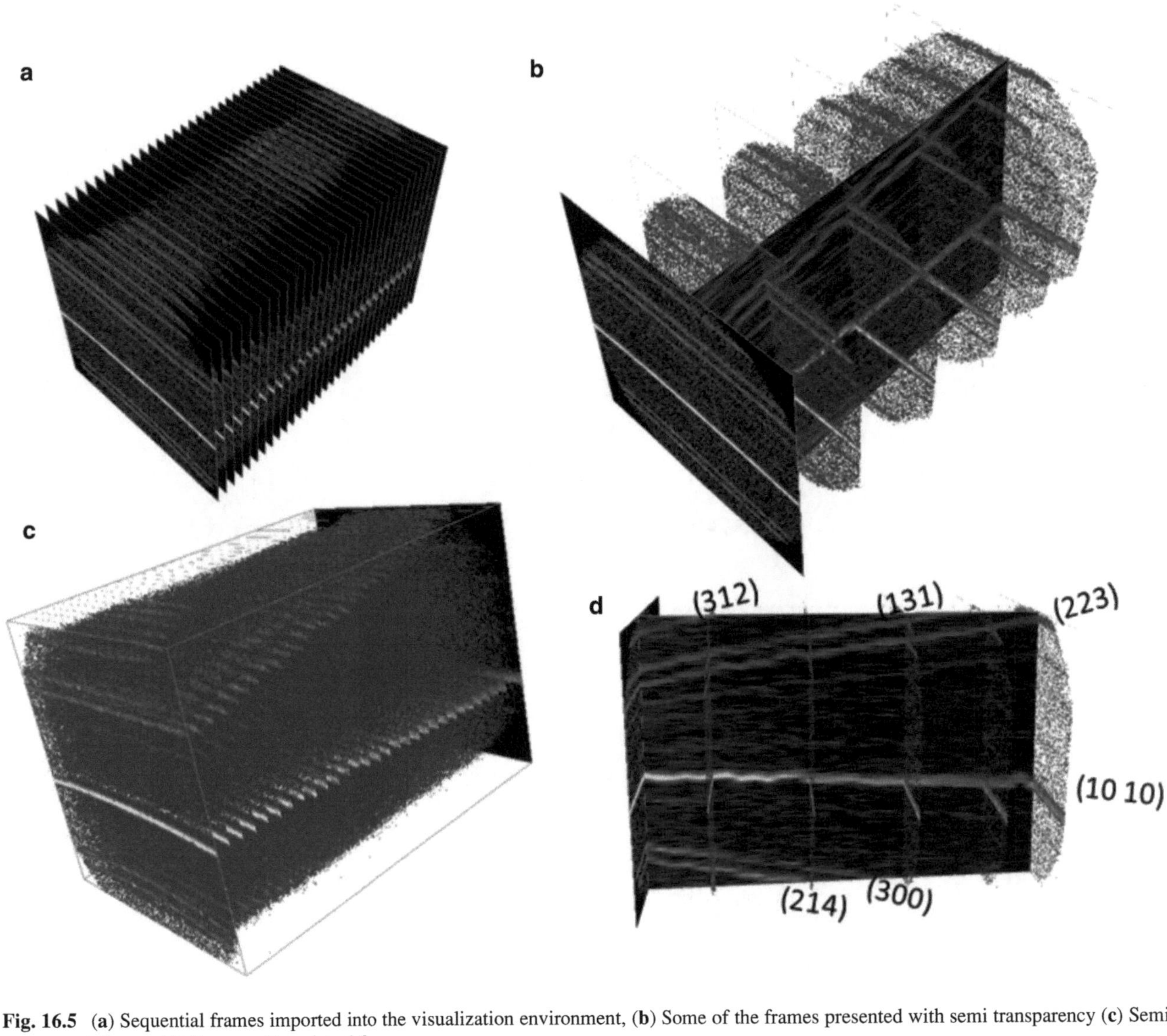

Fig. 16.5 (a) Sequential frames imported into the visualization environment, (b) Some of the frames presented with semi transparency (c) Semi transparent frames (d) Cross section virtual frame

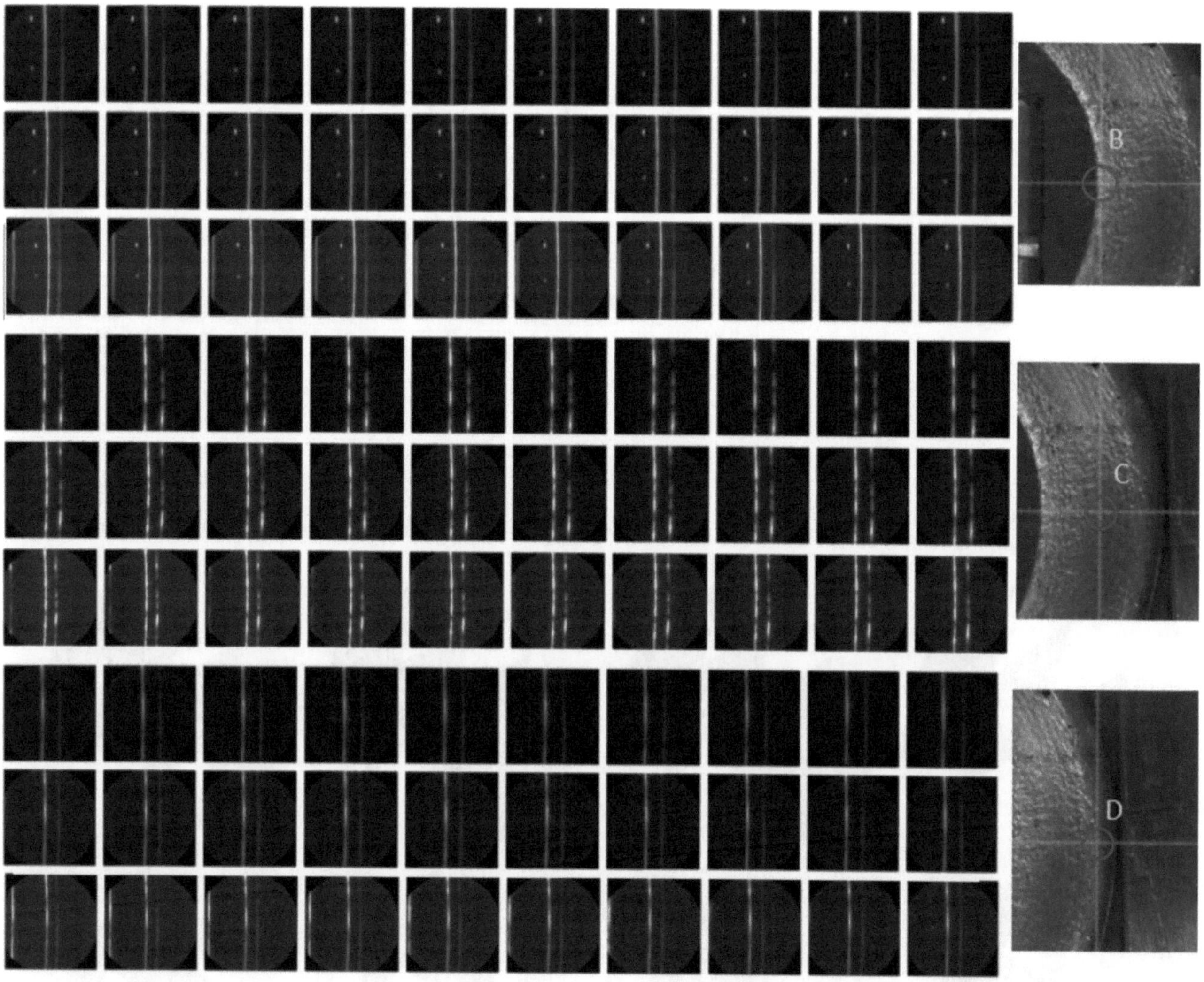

Fig. 16.6 Mosaic patterns of frames collected along different detector distances from inside "B", center "C", and outside "D" locations across the width of the bent aluminum bar

arbitrary cross section through the data set. Orthogonal cross sections are shown in Fig. 16.5a, b from two diffract perspectives. The *hkl* miller indexes are labeled on the virtual cross section. The 046-1212 ICDD Powder diffraction file was used for peak matching.

16.4.2 3D Diffraction from Bent Aluminum

Figure 16.6 shows mosaic patterns of frames collected along different detector distances from three locations across the width of the bent aluminum bar. Figure 16.7 shows 1D dimensional diffraction produced by integrating the frames collected at 30 cm detector distance for locations A, B, and C. As it can be seen, diffraction peaks are shifted due to compressive or tensile residual stress. With a similar process explained for calibration above, all 31 frames were imported as a sequence of images. Only four frames are presented for better visualization in Fig. 16.8.

16.4.3 Stress from a Single Grain

As it can be observed from Fig. 16.9, individual grains are far away from average reference 2θ values. A grain contributing to the (311) diffraction cone was chosen to estimate the stress from a single grain. For the grain observed in the in Fig. 16.9

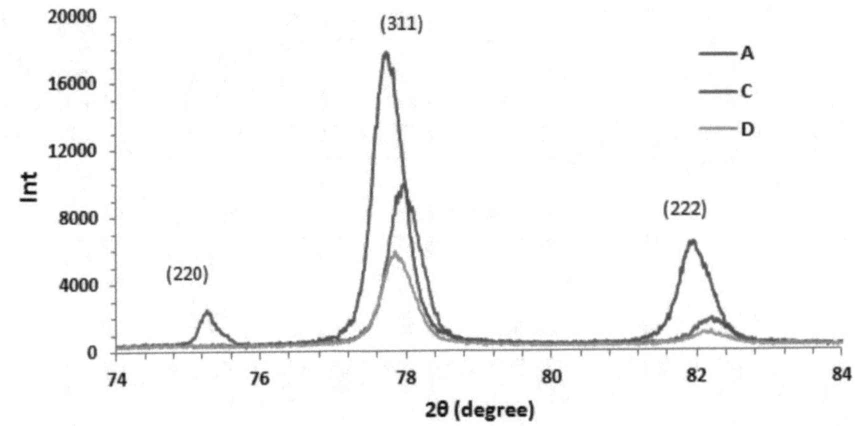

Fig. 16.7 1D X-ray diffraction patterns of selected representative frames for tensile and compressive residual stresses

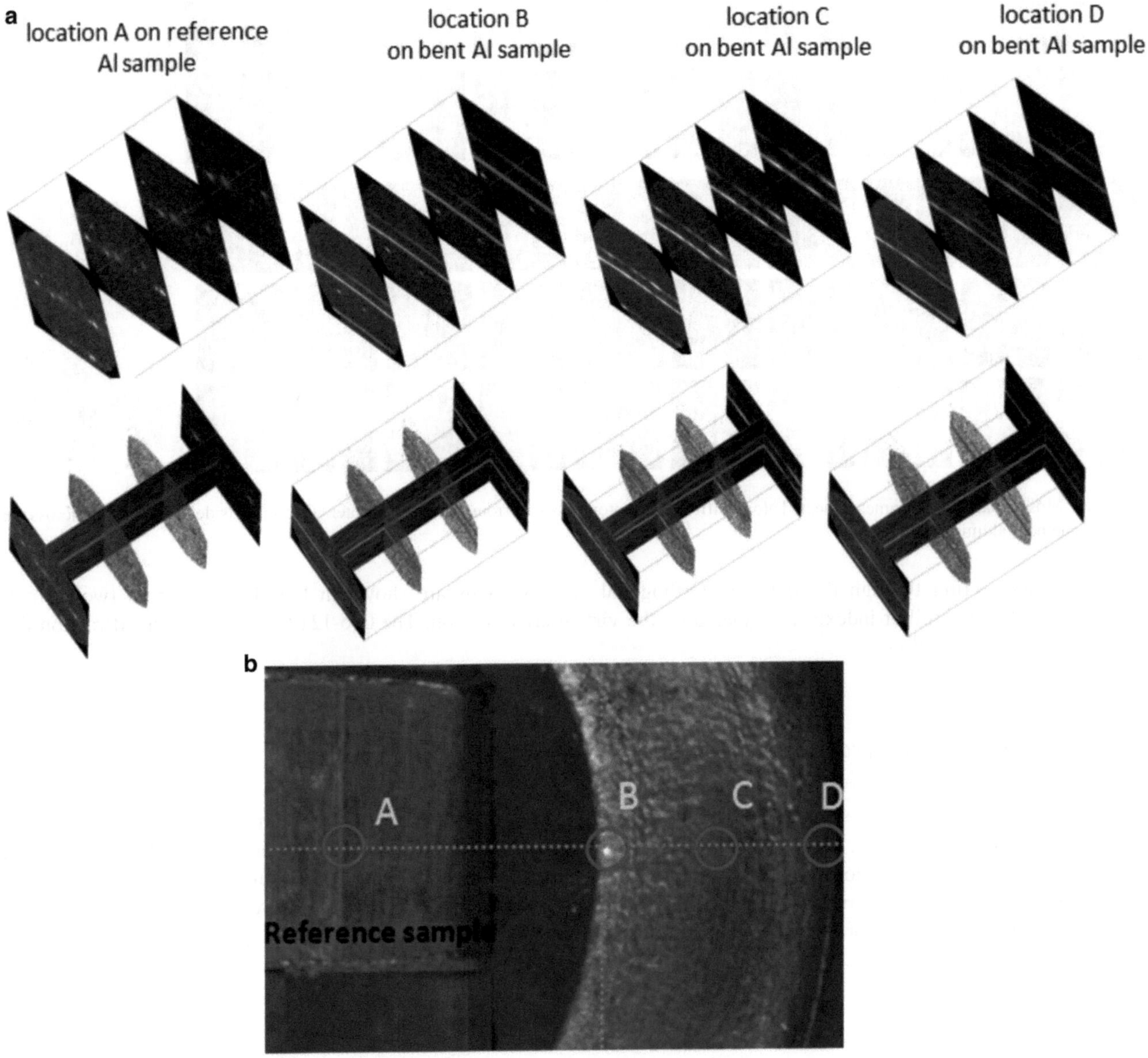

Fig. 16.8 (**a**) 3D stack of 2D diffraction frames (**b**) Measurement locations are labeled on the aluminum strain free reference sample and bent samples

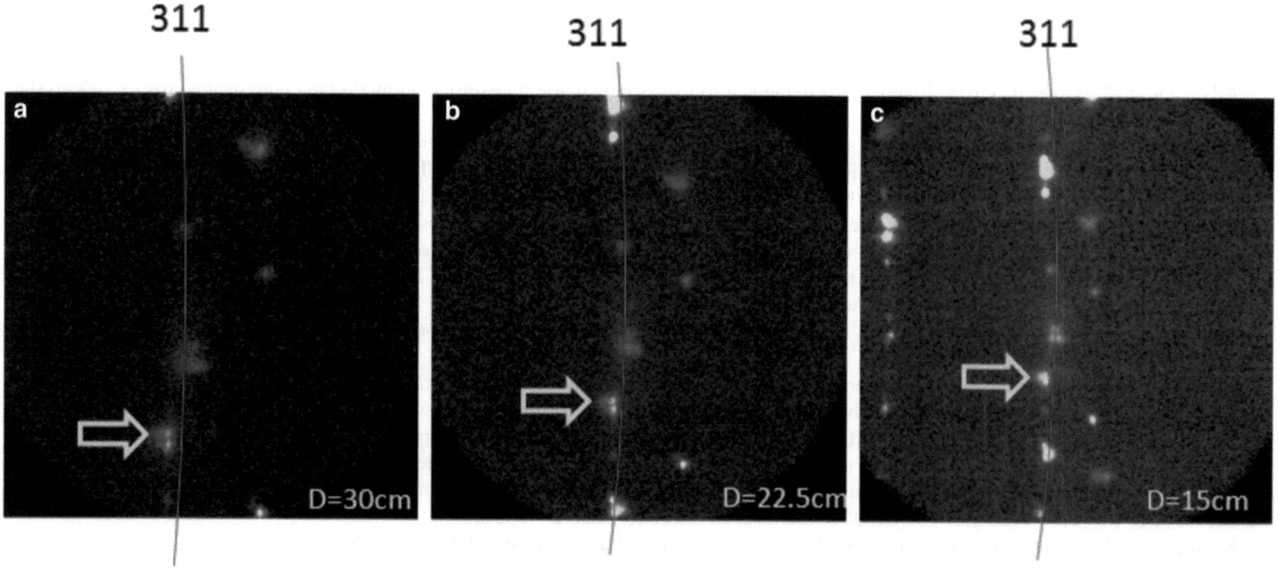

Fig. 16.9 (**a**) 2D frame at 30 sample-detector distance (**b**) 2D frame at 22.5 sample-detector distance (**c**) 2D frame at 15 cm sample-detector distance

marked by an arrow, the measured strain based on the 2θ value was 8.8×10^{-3}, which corresponds to a stress of 616 MPa. When material grains are too large or X-ray diffraction beam size is small, the diffraction signal appears as spotty arc on a 2D frame instead of a uniform ring as shown in Fig. 16.4. Stress free spots should center at the same place as a powder diffraction ring. In the case of a bent aluminum bar, it can be seen that they spots are distorted due the elastic change in interplanar distance related to the residual stresses. Another source for deviation of diffraction ring position is related to errors due to displacement of grains out of the center of diffraction.

The contribution of grain displacement out of the focused beam point in to the diffraction angle change is given by [13]:

$$\Delta(2\theta) = \frac{2\delta \cos \theta}{R} \tag{16.1}$$

The R is diffractometer radius and δ sample displacement from the true center. The 2θ is the ideal peak location.

A trace of the individual grain was calculated for its depth from the surface to determine the amount of error contributed to the residual stress measurement from the single grain that was used for stress calculation. The center of the diffraction spot on each frame was found. Using least squares refinement, a relation between the coordinates of the center of the spots and detector distance was determined. The depth then would be the intersection of the fitted line with the detector at zero distance. The depth for the grain was found to be 6.5 μm, that can cause $\Delta 2\theta$ of 9.3×10^{-6} with a 30 cm diffractometer radius, which corresponds to 1.218×10^{-7} strain.

16.5 Conclusion

New experimental techniques in X-ray science are achievable by detecting the complete or partial XRD cone in 3D space.

Using a Bruker D8 laboratory diffractometer, a new approach for characterizing materials using XRD has been investigated.

With a fixed experimental set-up at different sample detector distances, sequences of frames were recorded for 3D diffraction visualization. As a typical application, residual stress from a single grain of bent aluminum was measured, and the trace of the individual grain was used to calculate its depth from the surface. The method is new and many more data sets are required for development of the method.

References

1. Allahkarami M, Hanan JC (2011) Mapping the tetragonal to monoclinic phase transformation in zirconia core dental crowns". Dent Mater 27(12):1279–1284
2. Allahkarami M, Hanan JC (2012) Residual stress delaying phase transformation in Y-TZP bio-restorations". Phase Transit 85(1–2):169–178
3. Reimers W, Broda M, Brusch G, Dantz D, Liss K-D, Pyzalla A, Schmackers T, Tschentscher T (1998) Residual stress; high energy synchrotron diffraction; energy dispersive diffraction. J Nondestruct Eval 17(3):129–140
4. Bale HA, Tamura N, Hanan JC (2010) Cyclic impact fatigue and macroscopic failure considering grain-to-grain residual stress in ceramic dental restorations. SEM 2010 annual conference and exposition on experimental and applied mechanics
5. Hanan JC, Üstündag E, Beyerlein IJ, Swift GA, Almer JD, Lienert U, Haeffner DR (2003) Microscale damage evolution and stress redistribution in Ti–SiC fiber composites. Acta Mater 51:4239–4250
6. Reimers W, Pyzalla A, Broda M, Brusch G, Dantz D, Schmackers T (1999) The use of high-energy synchrotron diffraction for residual stress analyses". J Mater Sci Lett 18:581–583
7. He B, Preckwinkel U, Smith KL (2003) Comparison between conventional and two-dimensional XRD. Adv X-ray Anal 46:37–42
8. He B, Preckwinkel U, Smith KL (2000) Advantages of using 2D detectors for residual stress measurements". Adv X-ray Anal 42:429–438
9. Allahkarami M, Hanan JC (2011) X-ray diffraction mapping on a curved surface. J Appl Crystallogr 44:1211–1216
10. Allahkarami M, Hanan JC (2014) Residual stress and quantitative phase mapping on complex geometries. Powder Diffract, 29(2), 176–185
11. Allahkarami M, Hanan JC (2014) Three-dimensional X-ray diffraction detection and visualization. Meas Sci Technol 25(3)055604 pp. 1–10
12. Bruker AXS Inc. 2005 M86-EXX007 GADDS User Manual ch 12 (Madison, WI, USA)
13. Krawitz AD (2001) Introduction to diffraction in materials, science, and engineering. John Wiley, New York

Chapter 17
Incremental Ring Core by Optical Methods: Preliminary Results

Antonio Baldi and Filippo Bertolino

Abstract The ring core method is a well-known technique for residual stress measuring. It consists of milling a circular ring around the point of interest and measuring the surface deformations of the core. The method is more sensitive than hole drilling, but its sensitivity decreases with depth to become null when ring depth is equal to one third of the diameter. To overcome this problem, in literature an incremental version of the technique has been proposed consisting of removing the core, re-installing the strain gauge rosette and re-performing the measurement. Although the idea is interesting, its practical implementation is quite difficult, in particular re-installing the rosette is almost impossible when depth becomes significant, thus the incremental measurement is never performed.

In this work we propose to replace the strain gauge rosette with an optical technique. In this way the incremental approach becomes viable, even though, depending on the optical technique used, some practical problems have to be addressed.

Keywords Residual stress • Ring core • In-plane sensitive speckle interferometry • Reverse method

17.1 The Ring Core Method for Residual Stress Measurement

Residual stresses are self-balanced stress fields which exist inside a component even when external loads are null. They are named "residual" because they are usually a side effect (a residue) of the technological treatments performed during component machining.

Since they exist inside the component, they add to stresses induced by external loads, thus they may be a significant technical problem. Even though engineers sometimes voluntarily induce residual stress fields (usually compressive) in components to enhance their strength (consider for example the shot pinning technique, used to enhanced fatigue strength, or the press and shrink fits), nevertheless the residual stress level must be known to be able to estimate the safety margin of mechanical components.

Since residual stress fields are self-balanced, measuring them is quite difficult. You have either to rely on their influence on some stress-sensitive physical quantity (e.g. the effective elastic parameter of X-ray diffraction [1]) or you have to break their self-balanced state, usually by removing some material, to indirectly measure the residual stress level from the resulting displacements/strains. The latter is by far the approach most used and several different techniques have been proposed. In the hole drilling approach [2] a small hole is drilled on the specimen and residual stress values are computed from surface strains around the hole, whereas in the ring core approach a circular slit is milled and the residual stress level is estimated from the surface deformations of the core.

Even though the two approaches appear to be specular (in the former, some material is removed from the center, and outside strains are monitored, in the latter, material is removed from the outside and the inside is monitored) they are

A. Baldi (✉) • F. Bertolino
Dipartimento di Ingegneria Meccanica, Chimica e dei Materiali, Università degli Studi di Cagliari,
Via Marengo, 2, 09123 Cagliari, Italy
e-mail: antonio.baldi@dimcm.unica.it

N. Sottos et al. (eds.), *Experimental and Applied Mechanics, Volume 6: Proceedings of the 2014 Annual Conference on Experimental and Applied Mechanics*, Conference Proceedings of the Society for Experimental Mechanics Series,
DOI 10.1007/978-3-319-06989-0_17, © The Society for Experimental Mechanics, Inc. 2015

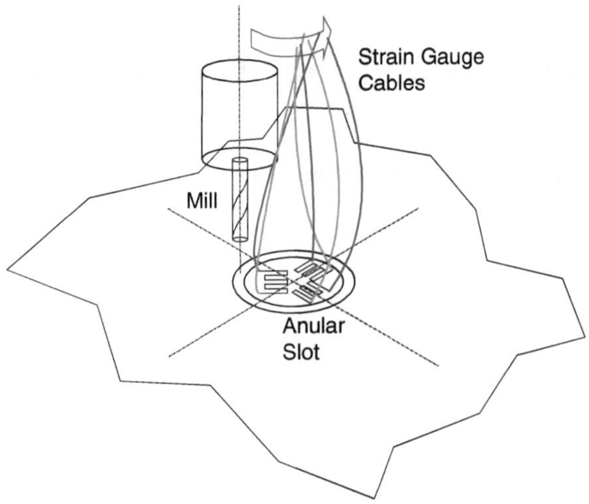

Fig. 17.1 Ring core technique. Note how milling the slot may be complex due to cables of strain-gauge rosette

not equivalent both from the practical and theoretical viewpoints. Indeed, in the ring-core approach the central region is significantly less constrained than the outer region in the hole-drilling technique, thus the former grants higher sensitivity; however, the larger attitude to release stresses significantly reduces the ability of the method to measure sub-surface fields; moreover in its standard implementation the technique is significantly more complex from the practical viewpoint owing to the presence of the cables of the strain gauge rosette which must pass through the axle of the milling machine (Fig. 17.1).

Both ring-core and hole-drilling techniques are usually implemented using strain-gauge rosettes; however their use is not implied by the measuring principle and alternative techniques can be used to inspect displacements/strains around the hole (inside the core). In particular, interferometric optical methods have been successfully used in combination with hole drilling by several authors [3–6]; combining optical techniques with the ring-core method is in principle not more difficult, but the only example of application is, to our knowledge, the optical strain rosette approach of Li and Ren [7, 8].

In this work we will focus on the application of in-plane speckle interferometry to the ring-core method. In particular, we will first show that performing the measurement in the ring-core case is no more complex than hole drilling, then we will try to extend the method, showing that an incremental approach is possible, thus allowing measurement of residual stress at a greater depth from the surface.

17.2 Incremental Ring Core

As already noted, from the practical viewpoint the ring core method is somewhat more complex to implement, owing to the location of the rosette and its cables. Moreover, the slit is significantly more invasive than a small hole, thus this technique is less used than hole drilling. However, when the displacement field is measured using an optical method, most of the difficulties disappear because there is no rosette and no cable which may be damaged during milling. Thus, the only significant difference between this technique and hole-drilling is the somewhat larger magnification required for the imaging optics.[1] However ring-core has an unique feature: in fact, once the maximum depth is reached (at about 1/3 of rosette diameter), the measurement can be restarted. The idea is quite simple (Fig. 17.2): it suffices to remove the core, install a new rosette and perform a new measurement [9].

Even though this idea appears to be a significant advantage with respect to hole-drilling, actually it is not so. Indeed, applying the proposed methodology is very difficult because it requires installing a new rosette at the bottom of a small, deep hole and then re-connecting at least six cables. Consequently the incremental ring-core method does not appear to be used.

[1] Actually the required magnification is not significantly larger. Indeed, in the ring-core method the active area is the core, which should fill the imaged area, whereas in hole-drilling no theoretical limitation exists, but displacements decay rapidly with distance from the axis of the hole, thus imaging more than 5–6 diameters is practically useless.

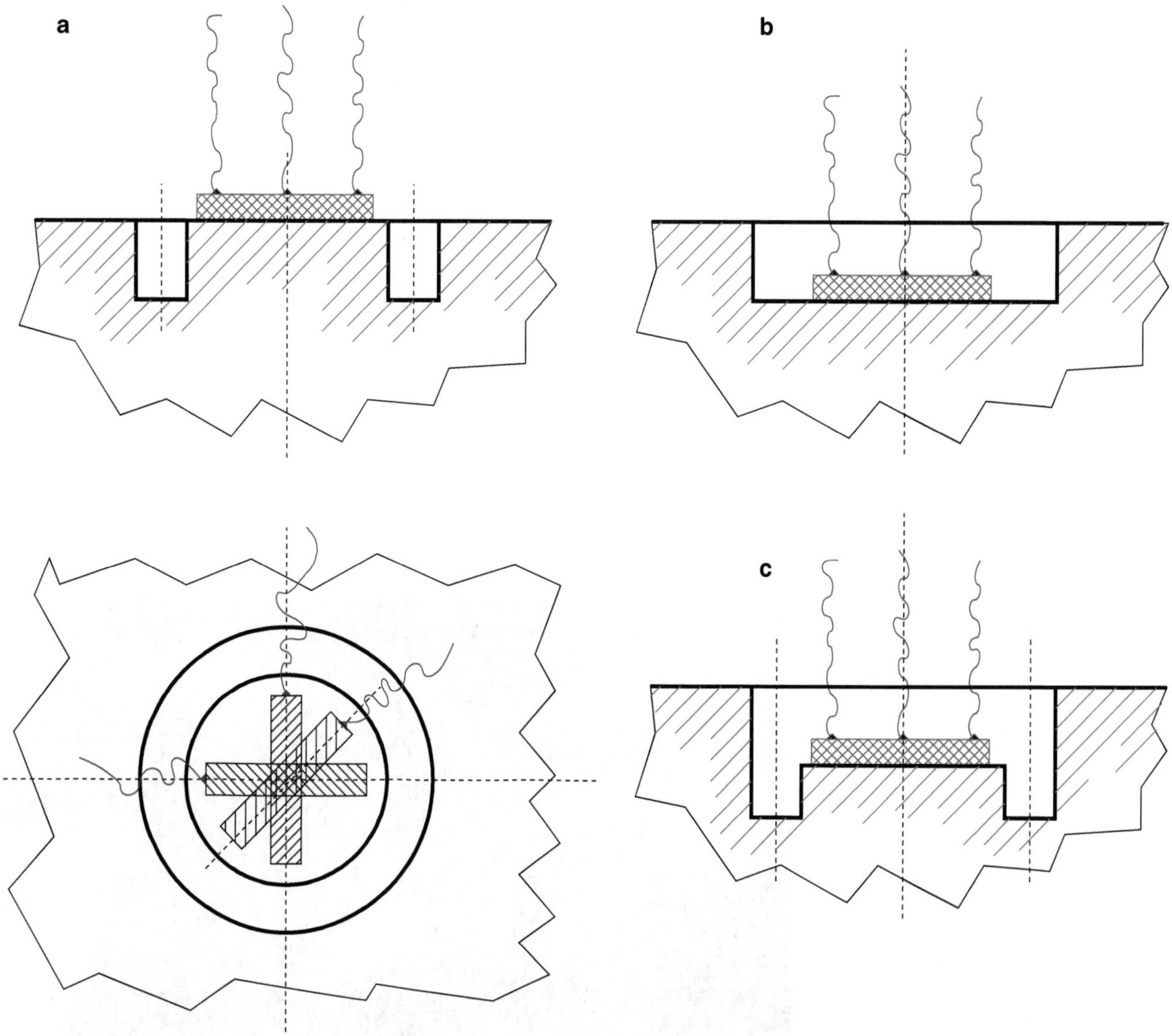

Fig. 17.2 Incremental ring core. After milling first ring (**a**), the core is removed (**b**) and a new ring is milled (**c**)

At first glance, using optical methods appears to solve this problem: in particular speckle based methods[2] require little or no treatment of the surface, thus all problems related to the strain gauge rosette[3] disappear. However, this is not the case. Looking at Fig. 17.3, it is apparent that as the hole becomes deeper, shadows start to cover a significant part of the imaged area. Given that the shadowed areas are illuminated by the symmetrical beam, they may be difficult to detect because they appear darker than the central region, but not black, when observed through the imaging system. Moreover, due to the coherent illumination, a speckle field will appear, but as the shadowed area is lightened by a single beam, no interference will develop. A simple geometric analysis shows that the limit condition (shown in the figure above) is obtained for a depth $h = R\ \mathrm{cotg}(\theta)$, where R is the external radius of the ring and θ is the angle between the beam and the hole axis, thus the improvement obtained is quite limited; to give a quantitative example, assuming the external diameter of the ring equal to 12 mm and an illumination angle of 30°, the limit condition is obtained for $h = 10.4$ mm i.e. only two incremental restarts are theoretically possible.

[2] Using moiré interferometry is obviously impossible. Indeed, using this technique would simply move the problem from gluing the rosette to replicating a grating at the bottom of a deep hole.

[3] Note that also the optical strain rosette presents significant problems: indeed, it requires micro-indenting the surface at the bottom of the hole; moreover the diffracted beams quite probably will be blocked by the lateral surface of the hole.

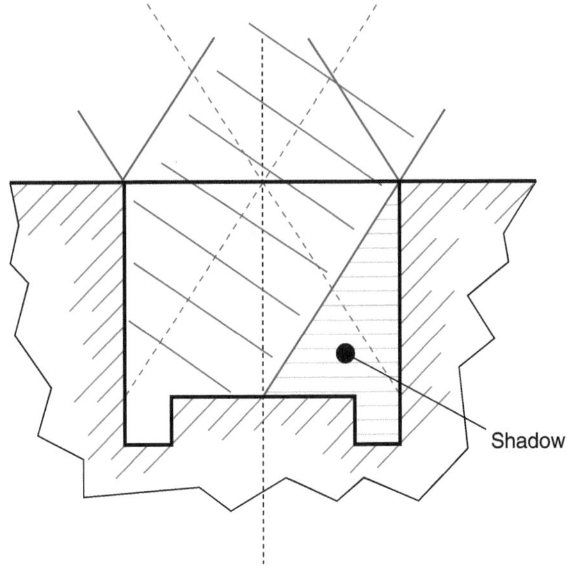

Fig. 17.3 Using in-plane speckle interferometry, the inclined illumination de-facto limits maximum depth

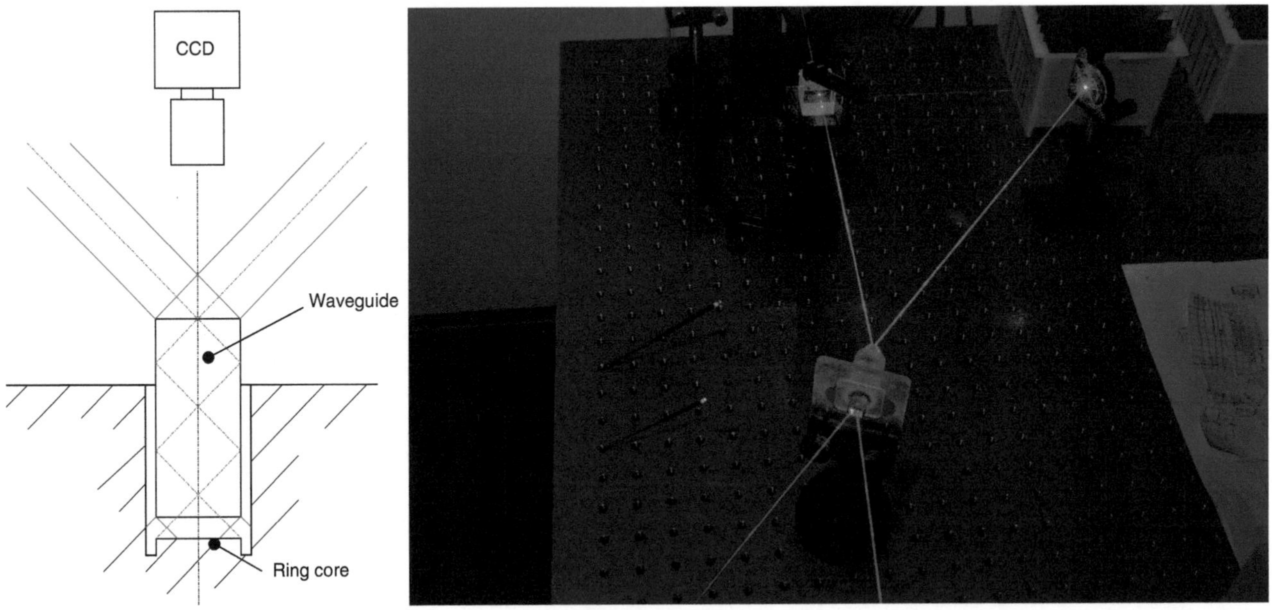

Fig. 17.4 *Left*: inserting a waveguide allows illumination of the core without any shadow. Moreover looking through the waveguide it is possible to image the core. *Right*: experimental demonstration of the waveguide. The two beams exit from the bottom of the waveguide parallel to the entry beams

The problem can be solved by using a waveguide (Fig. 17.4): instead of directly illuminating the surface of the core, a waveguide is inserted in the hole and the illumination is moved to the top of it. The two beams impinging the top surface will exist with the same angle from the bottom, thus allowing the interference to develop. Moreover, by positioning the camera on top of the waveguide, it is possible to image the surface of the core through the waveguide. Note that, depending on the gap between the waveguide and the top surface of the core, part of it may be incorrectly illuminated; however the inactive area is usually quite small and can be controlled (reduced) by adjusting the vertical position of the waveguide. Note also that the waveguide has to be removed to perform a milling increment. Thus a removable support, allowing waveguide removal and repositioning, has to be installed on the specimen.

Figure 17.4 (right) shows an experimental validation of the proposed approach: two beams enter the waveguide and exit from the other side parallel to the entry ones. The two burning incense sticks visible in the image are used to generate some smoke to make the beams visible. The green tone of the image is due to the color of the lamp used to illuminate the scene (normally used to develop red-sensitive holographic plates).

17.3 Experimental Validation

Figure 17.5 shows the experimental setup used for the initial validation of the proposed methodology. Since we have no specific facility for milling the core, we opted to use a kinematic mount, thus allowing removal of the specimen and its installation in our numerically controlled milling machine in the mechanical workbench of the department. The kinematic mount consists of a reference plate, fixed to the optical table, with 6 spheres mounted on the top (to allow use of two orthogonal sensitivity directions). The specimen is mounted on a removable support, presenting a hole, a v-shaped slit and a reference plane in the bottom, thus realizing an isostatic support. The accuracy requirements for the milling machine are significantly lower, thus we used a simple pin-based system for positioning the specimen in the working table of the CNC. We decided to image an area significantly larger than the core to be able to show also the fringe system related to the external area, i.e. a sort of combined hole-drilling/ring-core test. Note that the displacement field inside the core is similar, but not equal to the hole-drilling case: indeed, in the former displacements are proportional to the radius, whereas in the latter they are inversely proportional to the normalized radius (i.e. in the hole-drilling case displacements go to zero at infinity). Figure 17.6 shows the acquired phase field at 0.05 mm (A) and 1 mm (C) depth. Sub-images (B) and (D) correspond to the same fields (respectively B to A and D to C) after rigid body removal. To make the displacement more visible, the unwrapped phase fields have been re-wrapped after subtraction of the fitting plane.

Even though between the A and C images the specimen has been removed and repositioned 20 times, correlation is still satisfactory, thus proving the effectiveness of the experimental procedure. Note that the main problem is not the repositioning but the small oil drops released by the milling machine, which contaminate the surface of the specimen and make (locally) the comparison of the phase fields difficult before and after the incremental step.

Using the above described procedure we milled the ring down to 2.8 mm depth. Owing to the oil drops, the bottom right part of the image was almost unusable, but fringes were easily detectable in the top one. 2.8 mm correspond to almost the maximum allowable depth for a standard ring core, thus from this point on we installed the waveguide. Figure 17.7 shows the unwrapped phase field obtained at the successive step. Apart from some noise on the border, no fringe is visible and the phase field is almost flat.[4] This corresponds to the expected result because the residual stress field was induced by yielding the specimen in a four-point bending test and the expected stress at 2.8 mm is a few MPa.

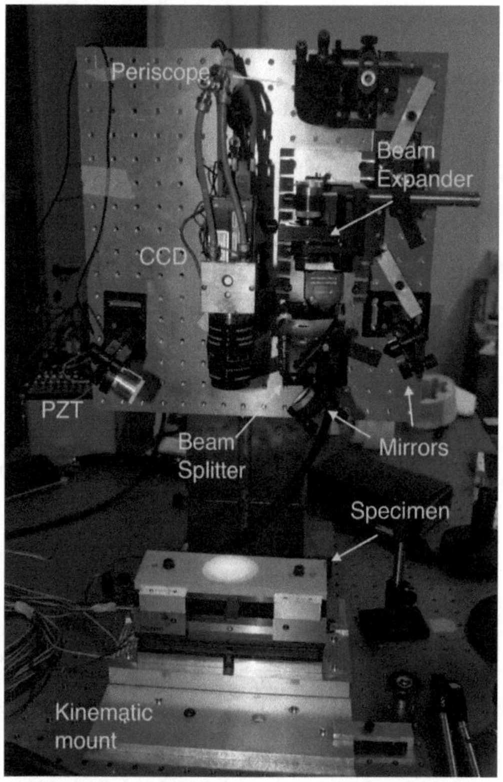

Fig. 17.5 Experimental setup. Standard in-plane configuration

[4] Note that all previous phase fields result from the subtraction of current phase from the reference one (i.e. the phase field of the original surface). Because we removed the core, we had to shift the reference field, which now corresponds to the bottom of the hole at 2.8 mm depth.

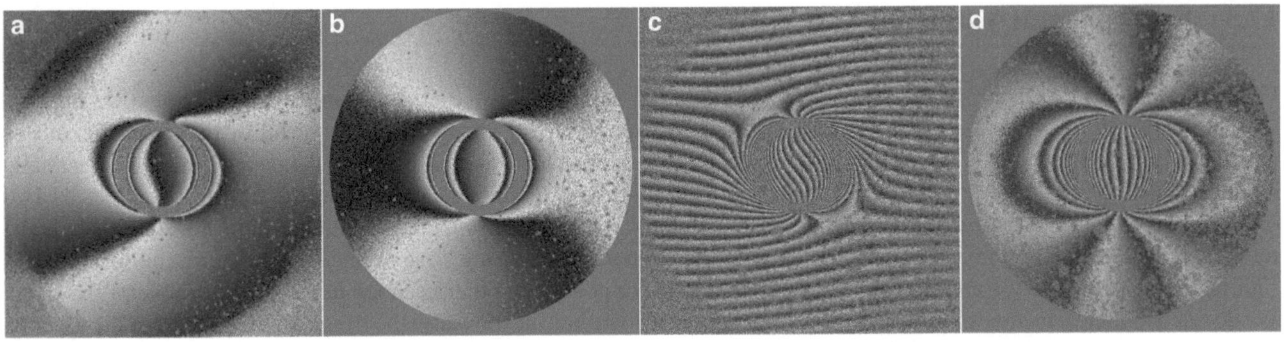

Fig. 17.6 Wrapped (and rewrapped) phase field before (**a**, **c**) and after (**b**, **d**) rigid body motion removal

Fig. 17.7 Unwrapped phase field (waveguide acquired data)

17.4 Conclusions

This work is a first step toward the assessment of the incremental ring-core method. The experimental results show that the ring-core approach can be easily combined with optical measurement methods, thus realizing a robust and relatively simple residual stress measuring procedure. The use of the optical waveguide allows measurement at a significantly greater depth, but the experimental results are not definitive because the absence of residual stress may also correspond to "inability to read" and more tests are required.

References

1. ASTM E1426 98(2009)e1 Standard test method for determining the effective elastic parameter for x-ray diffraction measurements of residual stress. Tech. rep., American Society for Testing and Materials, West Conshohocken, PA, 1998. doi:10.1520/E1426-98R09E01
2. ASTM E837-08e1 Standard test method for determining residual stresses by the hole-drilling strain-gage method. American Society for Testing and Materials, West Conshohocken, PA, 2008. doi:10.1520/E0837-08E01
3. Baldi A (2005) A new analytical approach for hole drilling residual stress analysis by full field method. J Eng Mater Technol 127(2):165–169
4. Focht G, Schiffner K (2002) Numerical processing of measured full-field dis-placement data around holes for residual stress determination. In: Mang HA, Rammerstorfer FG, Eberhrdsteiner J (eds) Fifth World Congress on Computational Mechanics, Vienna, Austria
5. Lin ST, Hsieh CT, Lee CK (2005) Full field phase-shifting holographic blind-hole techniques for in-plane residual stress detection. In: Honda T (ed) Int. Conf. on Applications of Optical Holography, Bellingham, Washington, Proceedings of SPIE, vol 2577, 226–237
6. Ponslet E, Steinzig M (2003) Residual stress measurement using the hole drilling method and laser speckle interferometry part ii: analysis technique. Exp Tech 27(4):17–2
7. Li K, Ren W (2007) Application of Minature ring-core and interferometric strain/slope rosette to determine residual stress distribution with depth. Part I: Theories. J Appl Mech 74(2):298–306
8. Ren W, Li K (2007) Application of miniature ring-core and interferometric strain/slope rosette to determine residual stress distribution with depth. Part II: experiments. J Appl Mech 74(2):307–314
9. Bender N, Hofer G, Lenz O, Stücker E (1979) Method of determining internal stresses in structural members, U.S. Patent no. 4155264

Chapter 18
Uncertainty Quantification in VFM Identification

P. Wang, F. Pierron, O.T. Thomsen, M. Rossi, and P. Lava

Abstract This article presents a methodology to optimize the design of a mechanical test to characterize all the material stiffness parameters of a PVC foam material in one single test. The two main experimental techniques used in this study are Digital Image Correlation and The Virtual Fields Method. The actual image recording process was mimicked by numerically generating a series of deformed synthetic images. Then the whole measurement procedure was simulated by processing the synthetic images with DIC and VFM routines. This procedure was used to predict the uncertainty of the measurements (systematic and random errors) by including the most significant parameters in actual experiments. By using these parameters as design variables and defining several error functions, the optimization study was performed to minimize the uncertainty of the identification and select the optimal test parameters. The simulated confidence intervals of the identified parameters were defined based on the predicted systematic and random errors. The simulated experimental results also indicated that averaging multiple images could lead to a significant reduction of the random error. The experimental validation was conducted using the optimized test parameters obtained from the numerical study. The results displayed a very good consistency with the simulation.

Keywords Virtual fields method • Composites • Arcan test • Uncertainty • Full-field measurements

18.1 Introduction

Cellular polymer closed cell foams are broadly used as core material in lightweight sandwich structures. Ideally, polymer foam core materials can be considered as homogenous isotropic materials. However, in practice most polymer foams display both heterogeneous and anisotropic material behaviour due to the density variations and directionality of foam cells developed during the manufacturing process. The orthotropic material behaviour of polymeric foams has been studied extensively in the literature [1–3]. Most of the studies rely on the use of several different testing methods including uniaxial tension, compression and shear, conducted along deformation measurements that are based on either point-wise or area-wise (averaged) measurement techniques like e.g. extensometers or strain gauges. More recent work by Zhang et al. [4] and Taher et al. [5] have characterized the cross linked PVC foam material Divinycell H100 using Digital Image Correlation (DIC). Both of these studies characterized the orthotropic properties of PVC foam, and the results obtained were in good agreement with the datasheets from the manufacturer [6]. However, a significant amount of time and effort was spent on

P. Wang
Department of Mechanical and Manufacturing Engineering, Aalborg University, Aalborg, Denmark

F. Pierron (✉)
Faculty of Engineering and the Environment, University of Southampton, Southampton, UK
e-mail: f.pierron@soton.ac.uk

O.T. Thomsen
Department of Mechanical and Manufacturing Engineering, Aalborg University, Aalborg, Denmark

Faculty of Engineering and the Environment, University of Southampton, Southampton, UK

M. Rossi • P. Lava
Università Politecnica delle Marche, Ancona, Italy

N. Sottos et al. (eds.), *Experimental and Applied Mechanics, Volume 6: Proceedings of the 2014 Annual Conference on Experimental and Applied Mechanics*, Conference Proceedings of the Society for Experimental Mechanics Series, DOI 10.1007/978-3-319-06989-0_18,

designing the different test specimen shapes needed to reach a uniform stress/strain state in the gauge area. With the development of full-field measurement techniques, several novel inverse techniques have been proposed to process the heterogeneous stress/strain fields to identify the whole set of constitutive parameters [7], among which the Virtual Fields Method (VFM) [8].

The Virtual Fields Method enables characterization of the material properties directly from full-field measurements. This method takes advantage of the heterogeneous strain fields obtained through full-field measurement techniques, such as Digital Image Correlation (DIC) [9] in order to obtain more information from a single test. Since the heterogeneity of the strain fields plays an important role in the VFM identification, the accuracy of the identification of the elastic stiffness coefficients depends heavily on the test configuration as well as on the full-field measurement parameters such as camera noise, spatial resolution and smoothing levels. Therefore, the design of the experiment becomes a non-trivial issue when using the VFM technique. A methodology to optimize the test configuration for VFM identification was firstly proposed by Pierron et al. [10]. The idea was to find an optimized specimen length and orthotropic material axis angle so as to minimize a cost function based on the sensitivity to noise of the sought material stiffness components. Recently, a refined test configuration design procedure was proposed by Rossi and Pierron [11]. The study used the grid method as the full-field technique and simulated the whole measurement and identification chain, including image forming and grid method algorithm. This study provided a significant improvement of the optimization procedure by introducing the many different types of error sources into the cost function. In particular, the effect of the spatial resolution of the full-field technique was correctly taken into account which was not the case in [10]. However, this approach was not validated experimentally. Also, it relied on the grid method (also known as sampling moiré) which is not so commonly used in the experimental mechanics community.

In a recent article, an efficient experimental methodology to identify all the material stiffness parameters of a PVC foam material in one single test using Digital Image Correlation and the Virtual Fields Method was presented [12]. The study provided an optimized material test configuration with a particular objective of improving the accuracy of the identification. Although the selected test configuration led to a significant improvement of the experimental results, significant differences were found between reference values of material parameters known from literature and/or other tests and the experimental results from this that study. It was thought that one of the reasons for this is that the conducted optimization study was based on finite element simulated strain fields which do not include the sources of error that arise from real DIC measurements. In particular, the low-pass spatial filtering effect of the DIC measurements will lead to underestimation of the strains in large strain gradients areas of the test specimen, which in turn will lead to biases on the identified stiffness components. Moreover, the low signal to noise ratio associated with the measurement of the elastic material properties of polymer foams will tend to increase the random error (scatter) of the data.

In the present study, a procedure has been developed to realistically simulate the modified Arcan test for polymer foams using Digital Image Correlation and the Virtual Fields Method. The idea is to construct deformed synthetic images using finite element (FE) displacements. From this, the reference and deformed synthetic images will be processe dusing Digital Image Correlation (DIC), and the Virtual Fields Method 5VFM) will be used subsequently to extract all the stiffness parameters. This paper aims at reporting a few preliminary result of this study.

18.2 Modified Arcan Test

In order to generate heterogeneous states of stress and strain in the foam specimen, a modified Aran fixture is used as in [12]. Thick orthotropic Divinycell H100 PVC foam panels from DIAB are considered here. The orthotropy directions are denoted '1' (longitudinal), '2' (through-the-thickness direction) and '3' (depth direction). The specimens can be cut from a thick foam panel such that that the '1' direction lies at a certain angle θ from the specimen longitudinal axis 'x' (see Fig. 18.1), resulting in a material 'off-axis' configuration. Besides, the modified Arcan fixture enables to load this specimen at a certain angle α from the 'y' direction of the specimen (see Figs. 18.1 and 18.2).

In [12], a systematic study showed that certain configurations (i.e., certain sets of values for α and θ) produced better identification results than others. However, this analysis was based the strain maps from simulated data and did not allow to derive expected uncertainties reliably. This is the objective of the present work.

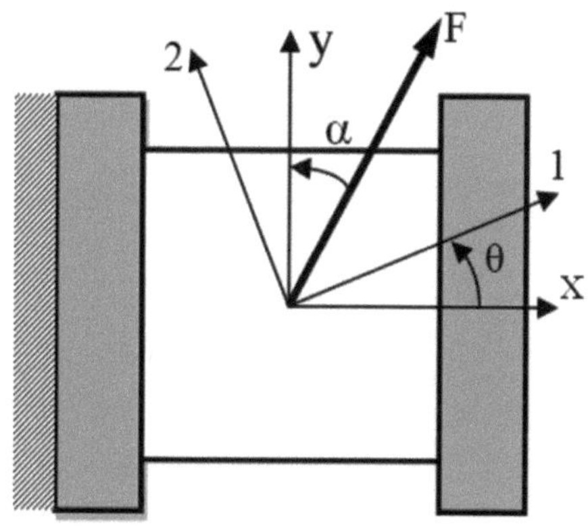

Fig. 18.1 Schematic of the modified Arcan test

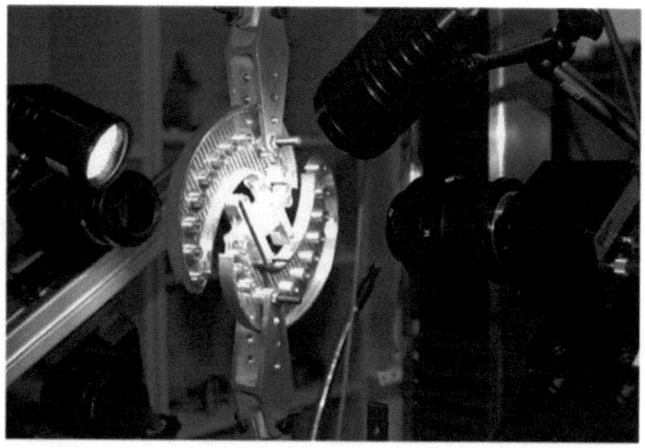

Fig. 18.2 Experimental set-up of the modified Arcan test

18.3 Identification Simulator

In order to realistically evaluate the uncertainty of the identified mechanical parameters, a simulator has been developed. It is a follow-up of the simulator presented in [11], extended to Digital Image Correlation (whereas [11] used the grid method). The details of this simulator can be found in [13]. A complete flowchart is provided in Fig. 18.3.

The optical measurement technique used here is 2D Digital Image Correlation. It was shown in [12] that this technique could produce reliable results; in particular, the effect of out-of-plane movement could be removed by the use of back to back cameras, which also enabled to average out the effect of possible through-thickness strain heterogeneities, as in [14].

The inverse technique used here to identify the full set of four orthotropic stiffness components is the Virtual Fields Method (VFM). It is beyond the scope of the current paper to detail the VFM, the reader is referred to the numerous papers on the topic [8, 10–14].

The advantage of the simulator is that the effect of the many parameters influencing the final results (DIC: speckle pattern, subset size, smoothing; VFM: number of virtual DOFs; imaging: lighting, noise level, lens distortion etc.) can be directly checked. This enables to: firstly, define optimized post-processing parameters such as subset size, step size and virtual strain gauge size; secondly, produce predictions of uncertainty levels, both systematic (bias) and random.

18.4 Experimental Results

Experiments were conducted on several configurations but only one is reported here. This configuration corresponds to a short specimen loaded off-axis (Short Off-axis Tension specimen ('SOT' specimen)). It has the potential to become a new standard test for simultaneous orthotropic stiffness identification. It corresponds to $\alpha = 25°$ and $\theta = 90°$ [15]. Figure 18.4

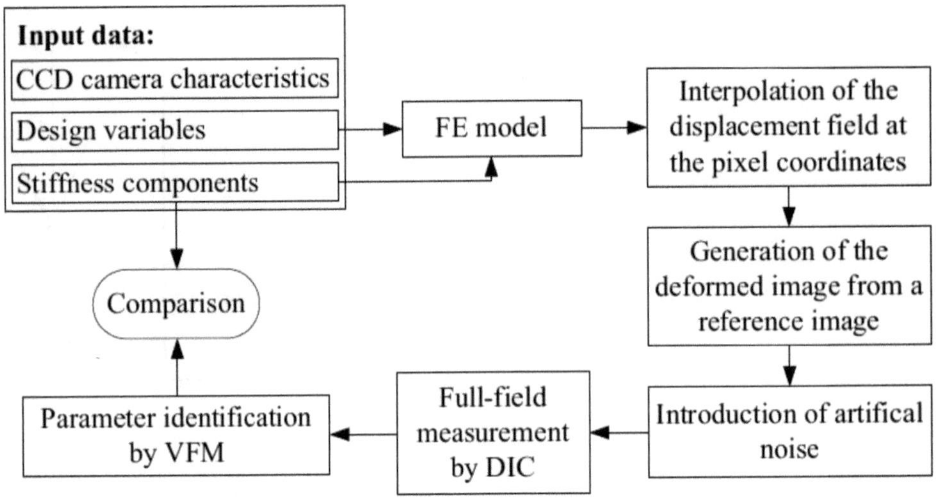

Fig. 18.3 Flowchart of the identification simulator [13]

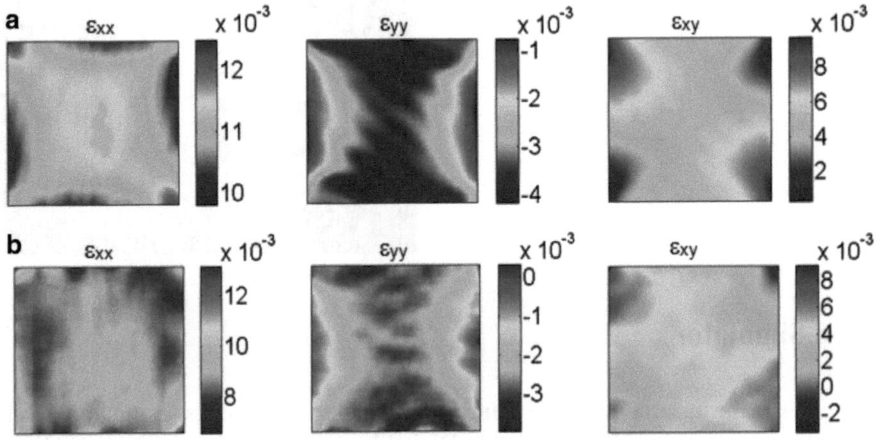

Fig. 18.4 Strain maps for the 'SOT' test. (**a**) from the simulator (**b**) from experiments [15]

Table 18.1 Means stiffness values for 20 repetitions

	Q_{11} [MPa]	Q_{22} [MPa]	Q_{12} [MPa]	Q_{66} [MPa]
Specimen 1	149.27	68.05	27.85	32.89
Specimen 2	147.63	70.84	29.64	31.05
Specimen 3	150.09	69.12	29.89	32.37
Mean	***149.00***	***69.34***	***29.13***	***32.10***
Simulator	***146.06***	***66.03***	***27.93***	***30.94***
Reference data	**143.44**	**63.41**	**26.01**	**30.12**

shows the three strain components obtained from the simulator and the experiments. The match is reasonable even though some differences can be spotted (typically between 5 and 10 % locally).

Three different specimens were tested 20 times each, in an effort to evaluate the bias, scatter between test and scatter between specimens. For each specimen, the mean and standard deviations were determined and compared to the results from the simulator and, for the mean, checked against reference values obtained from standard tests. The results are reported in Tables 18.1 and 18.2.

Table 18.2 Coefficients of variation (in %) for 20 repetitions

	Q_{11} [%]	Q_{22} [%]	Q_{12} [%]	Q_{66} [%]
Specimen 1	0.2	2.0	1.0	0.8
Specimen 2	0.3	2.0	1.0	0.7
Specimen 3	0.3	2.3	1.2	0.9
DIC Simulation	***0.2***	***1.6***	***0.8***	***0.4***

The results show consistency with expectations. First, in Table 18.1, the simulator results show a systematic overestimation of the stiffness components. This is caused by the low pass filtering effect of DIC which reduces high strain gradients. However, the effect is limited to a few percent. Using a larger spatial resolution (here, the camera was 2,048 × 2,048), for instance a 16 Mpixel camera, will reduce this bias significantly. Such cameras are now more common and their price is coming down. Experimental results tend to show good consistency with expected simulator results. The values are all systematically higher by a few percent. This slight difference is still to be studied but these first results look really promising. Examining the random error in Table 18.2, one can see that first, the coefficients of variation are very low and that they compare reasonably well with the predicted values. As can be expected, experimental values are higher as all error sources have not been implemented in the simulator (lens distortion, lighting noise etc.) This is a very important result which opens up the way to the definition of new standard tests based on full-field measurements and inverse identification.

18.5 Conclusion

This paper has presented the results from an attempt to quantify error propagation in inverse identification from full-field measurements. A modified Arcan test was used to perform heterogeneous tests on orthotropic foam block specimens. Digital Image Correlation was used to obtain strain maps and the Virtual Fields Method was employed for the inverse identification. A numerical simulator simulating the whole chain from image recording to inverse identification was developed in previous work and used here to both select optimal processing parameters (DIC subset size, smoothing, VFM Dofs etc) and predict both the random and systematic errors on the identified stiffness components. These predictions were then compared to a series of experimental results both for the mean and the coefficient of variation. These results showed high consistency. The small remaining deviations are probably caused by some other more minor sources of error like lighting noise, lens distortion etc. The simulator however provides a generic tool to enrich the data acquisition by including such effects in the future to provide even better predictions of bias and scatter.

The present work opens up the way to industrial transfer of such new generation tests with a view to establishing new test standards providing better data in a more cost efficient way. Here, the stiffness data can be obtained with only one test as opposed to the three currently necessary using traditional uniform stress approaches. There is potential to apply this new test to fibre composites in the future, including for through-thickness properties which are currently very difficult to obtain with current standard test methods.

References

1. Kanny K, Mahfuz H, Thomas T, Jeelani S (2004) Static and dynamic characterization of polymer foams under shear loads. J Compos Mater 38(8):629–639
2. Kabir ME, Saha MC, Jeelani S (2006) Tensile and fracture behavior of polymer foams. Mater Sci Eng A 429(1–2):225–235
3. Daniel I, Cho JM (2011) Characterization of anisotropic polymeric foam under static and dynamic loading. Exp Mech 51(8):1395–1403
4. Zhang S, Dulieu-Barton JM, Fruehmann R, Thomsen OT (2012) A methodology for obtaining material properties of polymeric foam at elevated temperatures. Exp Mech 52(1):3–15
5. Taher ST, Thomsen OT, Dulieu-Barton JM, Zhang S (2011) Determination of mechanical properties of PVC foam using a modified Arcan fixture. Compos Part A Appl Sci Manuf 43(10):1698–1708
6. DIAB (2011) Divinycell PVC datasheet. http://www.diabgroup.com
7. Avril S, Bonnet M, Bretelle A-S, Grédiac M, Hild F, Ienny P, Latourte F, Lemosse D, Pagano S, Pagnacco E, Pierron F (2008) Overview of identification methods of mechanical parameters based on full-field measurements. Exp Mech 48(4):381–402
8. Pierron F, Grédiac M (2012) The virtual fields method. Springer New York. ISBN 978-1-4614-1823-8
9. Sutton MA, Orteu JJ, Schreier HW (2009) Image correlation for shape, motion and deformation measurements: basic concepts, theory and applications. Springer New York

10. Pierron F, Vert G, Burguete R, Avril S, Rotinat R, Wisnom M (2007) Identification of the orthotropic elastic stiffnesses of composites with the virtual fields method: sensitivity study and experimental validation. Strain 43(3):250–259
11. Rossi M, Pierron F (2012) On the use of simulated experiments in designing tests for material characterization from full-field measurements. Int J Solids Struct 49(3–4):420–435
12. Wang P, Pierron F, Thomsen OT (2013) Identification of material parameters of PVC foams using digital image correlation and the virtual fields method. Exp Mech 53(6):1001–1015
13. Rossi M, Lava P, Pierron F, Debruyne D, Sasso M (2013) Error assessment on combining DIC and VFM to design an optimized experimental setup for material identification. Int J Solids Struct submitted.
14. Moulart R, Avril S, Pierron F (2006) Identification of the through-thickness rigidities of a thick laminated composite tube. Compos Part A Appl Sci Manuf 37(2):326–336
15. Wang P, Rossi M, Lava P, Pierron F, Thomsen OT (2014) Optimized experimental characterization of polymeric foam material using DIC and the Virtual Fields Method. (in preparation)

Chapter 19
Modal Identification of Over-Damped Structural Systems Using Extended Ibrahim Time-Domain Method

Chang-Sheng Lin and Tse-Chuan Tseng

Abstract In the previous study, the conventional Ibrahim time-domain method (ITD) using free-decay responses of structures has been extensively used in the modal-identification analysis, however, which is only applicable to identify the modal parameters of an under-damped structure. In the present paper, we propose a theoretical modification for ITD method, and extend the ITD method for modal identification of over-damped structural systems. The eigenvectors of the system matrix used in extended ITD method corresponding to the vibrating modes of a structural system in pair are sorted through the assurance index proposed in this paper based on the theory of structural dynamics. Numerical simulations confirm the validity of the proposed method for modal identification of over-damped structural systems.

Keywords Extended Ibrahim time-domain modal-identification method • Over-damped structural systems • Identification of modal parameter

19.1 Introduction

The original ITD method uses free-decay responses of a structure to identify its modal parameters. The method is effective in identifying complex modes even for closely spaced modes with distinct damping ratios. Although we can use the ITD method to identify the modal parameters of a system via an eigenvalue analysis, the ITD method applies only to problems involving response data of free vibration and sometimes its efficiency is low due to its large numerical computation. Ibrahim [1] applied the random decrement technique coupled with a time-domain parameter identification method (ITD) [2] to process ambient vibration data. Although the random decrement technique serves as an alternative method for estimating the auto-correlation and cross-correlation functions [3], it is based on an intuitive theory and does not yet have sound mathematical basis for general cases [4]. James et al. [5] developed the so-called Natural Excitation Technique (NExT) using the cross-correlation technique coupled with time-domain parameter extraction. It was shown that the cross-correlation between two response signals of a linear system with classical normal modes and subject to white-noise (stationary) inputs is of the same form as free vibration decay or impulse response. When coupled with a time-domain modal extraction scheme, this concept becomes a very powerful tool for the analysis of structures under ambient vibration. NExT has been applied to modal identification of many engineering structures, such as wind turbines [6] and a rocket during launch [7].

In the previous studies of the application of conventional ITD method, the modal identification of a system was performed by using free-decay responses of structures, however, which are only applicable for under-damped structural systems. In the present paper, we will propose a theoretical modification for ITD method, and extend the ITD method for modal identification for over-damped structural systems. Numerical simulations confirm the validity of the proposed method for modal identification of over-damped structural systems.

C.-S. Lin (✉) • T.-C. Tseng
Optics Group, Experimental Facility Division, National Synchrotron Radiation Research Center, Hsinchu, Taiwan
e-mail: lin.changsheng@nsrrc.org.tw

N. Sottos et al. (eds.), *Experimental and Applied Mechanics, Volume 6: Proceedings of the 2014 Annual Conference on Experimental and Applied Mechanics*, Conference Proceedings of the Society for Experimental Mechanics Series, DOI 10.1007/978-3-319-06989-0_19,

19.2 Original ITD Method

The conventional ITD method uses free-decay responses of a structure under test to identify its modal parameters in complex form [2]. From the measured free responses at n stations on a structure under test, each with q sampling points, we define a system matrix $\boldsymbol{S}$, which is an $n \times n$ matrix, such that

$$\boldsymbol{SX} = \boldsymbol{Y} \tag{19.1}$$

where $\boldsymbol{X}$ is an $n \times q$ matrix of measured response, and $\boldsymbol{Y}$ is an $n \times q$ matrix of time-delayed response. Generally, the number q is chosen to be larger than the number of measurement channels n. Therefore, the system matrix $\boldsymbol{S}$ can be estimated by the least-squares method.

In theory, a continuum structure has an infinite number of degrees of freedom and an infinite number of modes. In practice, we do not know in advance how many modes are required to describe the dynamical behavior of the observed structural system. However, the important modes of the system under consideration could be roughly found by examining the Fourier spectra associated with the measured response histories. The number of (real) modes m involved in the response then determines the order n of the system matrix $\boldsymbol{S}$ in Eq. 19.1, for which n is chosen to be at least twice of the number of modes of interest in order to appropriately identify the $2\,m$ complex modes. Note that n is not necessarily the same as the number of measured DOF of the system. If the number of measurement channels does not actually reach n, we may employ the technique of channel expansion [2], which uses time-delayed sampling points from the original response as new response channels, so that the total number of measurement channels can reach n. Note, however, that the identified mode shapes are composed of the components corresponding only to those physically measured response channels.

It can be shown that the natural frequencies and the damping ratios of the original under-damped vibrating system are directly related to the eigenvalues of the system matrix $\boldsymbol{S}$, and the mode shapes correspond to the eigenvectors of $\boldsymbol{S}$. Denote an eigenvalue of $\boldsymbol{S}$ and a characteristic root of the original under-damped vibrating system as $\rho_r = \beta_r + i\gamma_r$ and $s_r = \sigma_r + iv_r$, respectively. One can derive [2]

$$\begin{cases} \sigma_r = \dfrac{1}{2\Delta t}\ln\left(\beta_r^2 + \gamma_r^2\right) \\ v_r = \dfrac{1}{\Delta t}\tan^{-1}\left(\dfrac{\gamma_r}{\beta_r}\right) \end{cases} \tag{19.2}$$

from which the natural frequencies and damping ratios of the under-damped structural system can be obtained to be

$$\begin{cases} \omega_{nr} = \sqrt{\sigma_r^2 + v_r^2} \\ \xi_r = \dfrac{|\sigma_r|}{\sqrt{\sigma_r^2 + v_r^2}} \end{cases} \tag{19.3}$$

Hence, once the system matrix $\boldsymbol{S}$ is obtained via least-squares analysis from measured data, the modal parameters of the under-damped structural system can be determined by solving the eigenvalue problem associated with the system matrix $\boldsymbol{S}$.

19.3 Extended ITD Method for Over-Damped Systems

It has been shown in the above that the conventional ITD method can be used to perform the modal identification of a structure, however, which is only applicable to the under-damped structural system. Recall that for under-damped modes, ρ_r's and s_r's are complex, respectively, and occur in conjugate pairs, but for critical-damped modes, they are real, i.e. $v_r = 0$. From Eq. 19.3, when $v_r = 0$, $\xi_r = 100\ \%$ can then be derived. The preceding results therefore indicate that conventional ITD method is effective for the critical-damped structural systems at best, while could not be applied to the over-damped ones.

In the following, we will theoretically show that how to extend the ITD method to perform modal identification of over-damped structural systems. Assuming the structure is over-damped, one can derive the following equations of the relationship between ρ_r and s_r based on the concept of estimating eigenvalues of the system matrix $\boldsymbol{S}$ in the conventional ITD method

$$\begin{cases} \rho_r = e^{s_r \Delta t} \\ \rho_{r+m} = e^{s_{r+m} \Delta t} \end{cases} \tag{19.4}$$

and then deduce

$$\begin{cases} s_r = \dfrac{1}{\Delta t} \ln \rho_r \\ \\ s_{r+m} = \dfrac{1}{\Delta t} \ln \rho_{r+m} \end{cases} \tag{19.5}$$

where ρ_r's and s_r's are real, respectively. The eigenvalue problem for an over-damped structure expressed in Eq. 19.1 given $2\ m$ eigenvalues as real form

$$\begin{aligned} s_r &= -\xi_r \omega_{nr} + \omega_{nr} \sqrt{\xi_r^2 - 1} \\ s_{r+m} &= -\xi_r \omega_{nr} - \omega_{nr} \sqrt{\xi_r^2 - 1} \end{aligned} \tag{19.6}$$

where ω_{nr} and ξ_r are the natural angular frequency and damping ratio of the rth mode, respectively. Note that $s_{r+m} < s_r < 0$, $r = 1, 2, \ldots m$. From Eqs. 19.5 and 19.6, the natural frequencies and damping ratios of the over-damped structural system can be obtained to be

$$\begin{cases} \omega_{nr} = \dfrac{\sqrt{\Gamma_r^2 - 1}}{2|\Gamma_r|} (s_r + s_{r+m}) \\ \\ \xi_r = \dfrac{|\Gamma_r|}{\sqrt{\Gamma_r^2 - 1}} \end{cases} \tag{19.7}$$

where

$$\Gamma_r = \frac{s_r + s_{r+m}}{s_r - s_{r+m}}$$

and $|\Gamma_r| > 1$. However, for over-damped structural system, it may be difficult to identify the modal parameters of a system due to the eigenvalues of the system matrix $\boldsymbol{S}$ are not complex and do not occur in conjugate pairs such as the eigenvector of $\boldsymbol{S}$ corresponding to the under-damped structural system. According the theory of structural dynamics, the vibrating shapes of a structural system are orthogonal with each other, the eigenvectors of $\boldsymbol{S}$ can be therefore sorted as the vibrating modes of a structural system in pair. Define the assurance index k_{ij} between $\boldsymbol{\varphi}_i$ and $\boldsymbol{\varphi}_j$ as

$$k_{ij} = \left| \boldsymbol{\varphi}_i^* \boldsymbol{\varphi}_j \right| \tag{19.8}$$

where $\boldsymbol{\varphi}_i$ and $\boldsymbol{\varphi}_j$ represent two eigenvectors of interest, and the superscript * denotes the complex conjugate. The value of k_{ij} varies between 0 and 1. When the k_{ij} is equal to 1, the two vectors $\boldsymbol{\varphi}_i$ and $\boldsymbol{\varphi}_j$ correspond exactly the same vibrating shapes of a structural system. On the other hand, when two eigenvectors are orthogonal with each other, the k_{ij} is zero.

19.4 Numerical Simulation

To demonstrate the effectiveness of the proposed method, we consider a linear 2-DOF chain model with viscous damping. A schematic representation of this model is shown in Fig. 19.1. The mass matrix $\boldsymbol{M}$ and stiffness matrix $\boldsymbol{K}$ of the system are given as follows:

$$\boldsymbol{M} = \begin{bmatrix} 1 & 0 \\ 0 & 2 \end{bmatrix} \quad N \cdot \sec^2/m, \quad \boldsymbol{K} = \begin{bmatrix} 170 & -20 \\ -20 & 170 \end{bmatrix} \quad N/m.$$

Assume that each damping ratio is 120 %. The proposed method is employed by using the simulated free-decayed response data of the system (with a chosen sampling interval of $\Delta t = 0.01$ s). The simulated displacement responses of the system obtained using Newmark's method [8] are shown in Fig. 19.2. By examining the Fourier spectra associated with each

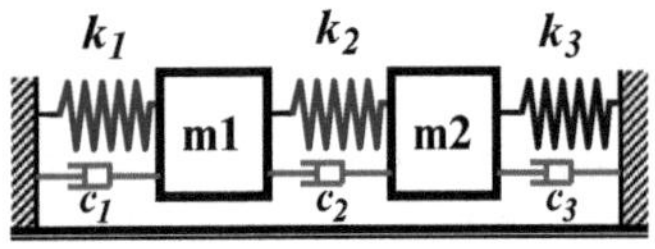

Fig. 19.1 Schematic plot of the 2-DOF chain model

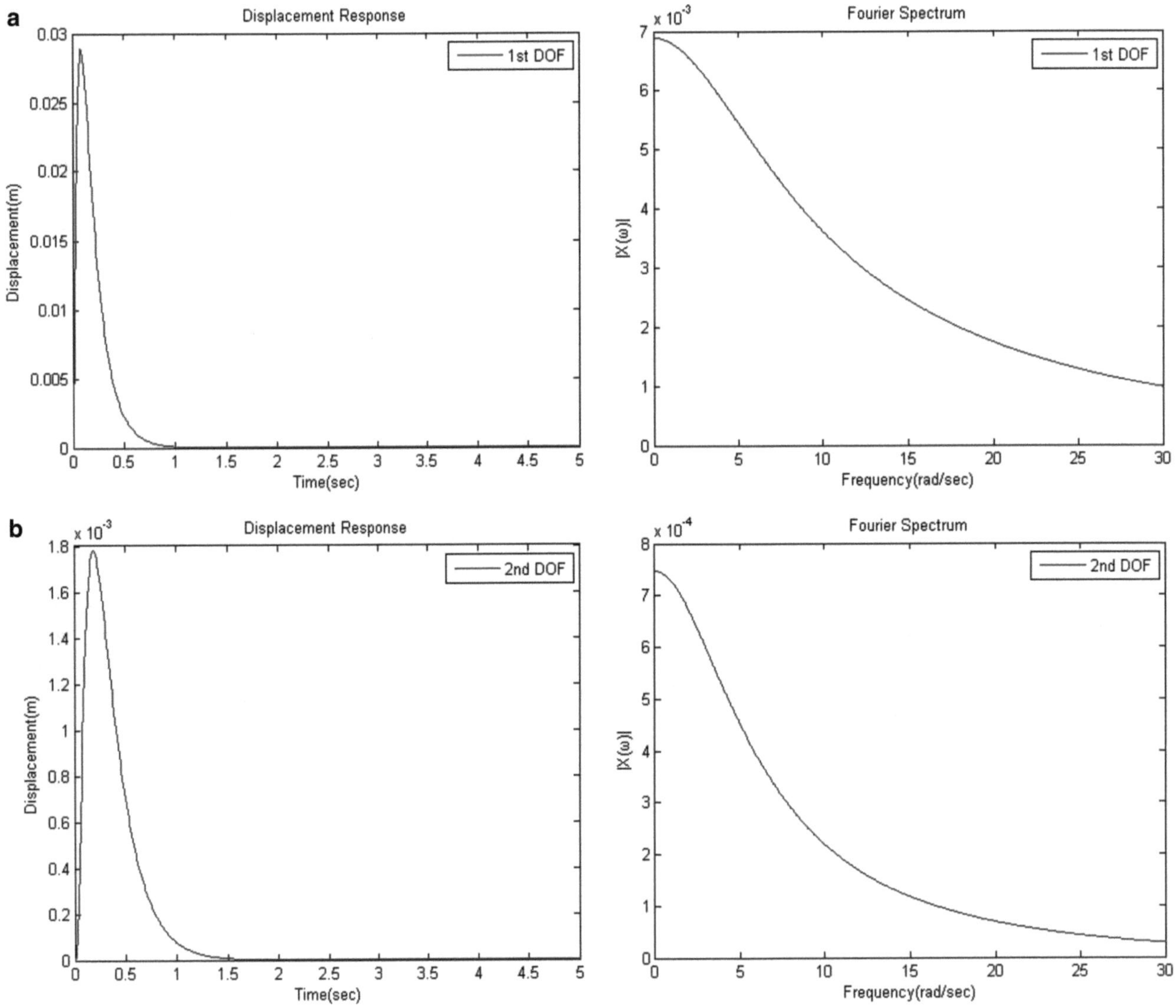

Fig. 19.2 (**a**) The free-decay response of the 2-dof over-damped system, and (**b**) the corresponding Fourier spectrum of the free-decay response associated with each of the response channel

Table 19.1 Evaluation results of assurance index k_{ij} between $\boldsymbol{\varphi}_i$ and $\boldsymbol{\varphi}_j$

i	j = 1	2	3	4
1	1	0.0593	0.0489	0.8120
2	0.0593	1	0.9326	0.4176
3	0.0489	0.9326	1	0.4372
4	0.8120	0.4176	0.4372	1

Table 19.2 Results of modal parameter identification of the 2-dof over-damped system

	Natural frequency (rad/s)			Damping ratio (%)			
Mode	Exact	ITD	Error (%)	Exact	ITD	Error (%)	MAC
1	9.09	9.24	1.65	120.00	120.78	0.65	0.99
2	13.13	12.93	1.49	120.00	119.47	0.44	0.99

of the response channel, also as shown in Fig. 19.2, in which we clearly see the serious problem of modal interference among modes of this 2-DOF system. For the original ITD method, if a system has relatively heavy damping, it becomes difficult to accurately identify the damping ratios and the mode shapes, due to the interference among modes of the system. In general, the original ITD method can yield reasonably accurate results only when modal damping ratios of the system are below 10 %.

According to the theory presented in the previous sections, we have proposed a theoretical modification for ITD method, and extended the ITD method for modal identification of over-damped structural systems. In conjunction with the proposed assurance index, we can sorted the eigenvectors of the system matrix used in extended ITD method corresponding to the vibrating modes of a structural system. Therefore, the extended ITD method could then be applied to identify modal parameters of an over-damped system. The evaluation results of assurance index k_{ij} between the eigenvectors, $\boldsymbol{\varphi}_i$ and $\boldsymbol{\varphi}_j$, of the system matrix are summarized in Table 19.1, which shows that the eigenvectors, $\boldsymbol{\varphi}_1$ and $\boldsymbol{\varphi}_4$, correspond to the vibrating modes of a structural system in one pair ($k_{14} = k_{41} = 0.8120$), as well as the other eigenvectors, $\boldsymbol{\varphi}_2$ and $\boldsymbol{\varphi}_3$, correspond to the vibrating modes of a structural system in another pair ($k_{23} = k_{32} = 0.9326$). The results of modal parameter identification are summarized in Table 19.2, which shows that the errors in natural frequencies are less than 2 % and the error in damping ratios is less than 1 %.

Furthermore, to keep track of the target modes, we utilize the MAC (Modal Assurance Criterion), that has been extensively used in the experimental modal analysis. The definition of MAC is [9]

$$MAC\left(\boldsymbol{\Phi}_{iA}, \boldsymbol{\Phi}_{jX}\right) = \frac{\left|\{\boldsymbol{\Phi}_{iA}\}^T\{\boldsymbol{\Phi}_{jX}\}^*\right|^2}{\{\boldsymbol{\Phi}_{iA}\}^T\{\boldsymbol{\Phi}_{iA}\}^*\{\boldsymbol{\Phi}_{jX}\}^T\{\boldsymbol{\Phi}_{jX}\}^*}, \tag{19.9}$$

where $\boldsymbol{\Phi}_{iA}$ and $\boldsymbol{\Phi}_{jX}$ represent two mode shape vectors of interest, and the superscript * denotes the complex conjugate. The value of MAC varies between 0 and 1. When the MAC value is equal to 1, the two vectors $\boldsymbol{\Phi}_{iA}$ and $\boldsymbol{\Phi}_{jX}$ represent exactly the same mode shape. Observing the MAC values, which signify the consistency between the identified and the theoretical mode shapes, we found all modes are identified accurately (MAC $\geq$ 0.9).

19.5 Conclusions

The conventional Ibrahim time-domain method (ITD) using free-decay responses of structures has been extensively used in the modal-identification analysis, however, which is only applicable to identify the modal parameters of an under-damped structure. In this paper, we have proposed a theoretical modification for ITD method, and extended the ITD method for modal identification of over-damped structural systems. Furthermore, it is shown that the eigenvectors of the system matrix through the extended ITD method corresponding to the vibrating modes of a structural system in pair are sorted using the propose assurance index. Numerical simulations confirm the validity of the proposed method for modal identification of over-damped structural systems.

Acknowledgements This research was supported in part of the financial funding provided by National Synchrotron Research Center (NSRRC). The first author would like to thank every member in the Optics Group of Experimental Facility Division, and Precision Mechanical Engineering Group of Instrumentation Development Division, NSRRC, for their kind assistance and friendship.

References

1. Ibrahim SR (1977) Random decrement technique for modal identification of structures. J Spacecraft Rickets 140:696–700
2. Ibrahim SR, Mikulcik EC (1977) A method for the direct identification of vibration parameters from free response. Shock Vib Bull 47(Pt. 4):183–198
3. Asmussen JC, Ibrahim SR, Brincker R (1996) Random decrement and regression analysis of bridges traffic responses. Proceedings of the 14th International Modal Analysis Conference 1, 453–458
4. Vandiver JK, Dunwoody AB, Campbell RB, Cook MF (1982) A mathematical basis for the random decrement vibration signature analysis technique. ASME J Mech Design 104:307–313
5. James GH, Carne TG, Lauffer JP (1995) The Natural Excitation Technique (NExT) for modal parameter extraction from operating structures. Modal Anal 10(4):260–277
6. James GH, Carne TG, Lauffer JP (1993) The natural excitation technique for modal parameter extraction from operating wind turbines. SAND92-1666. UC-261, Sandia National Aboratories
7. James GH, Carne TG, Edmunds RS (1994) STARS Missile—modal analysis of first-flight data using the natural excitation technique, NExT. Proceedings of the 12th international model analysis conference. Honolulu, HI, 231–238
8. Newmark NM (1959) A method of computation for structural dynamics. J Eng Mech, ASCE, 85 (EM3) 67–94
9. Allemang RL, Brown DL (1983) A correlation coefficient for modal vector analysis. Proceedings of the first international modal analysis conference, society for experimental mechanics, Bethel, CT, 110–116

Chapter 20
Structural Health Monitoring by Laser Shearography: Experimental and Numerical Investigations

Xiaoran Chen, Morteza Khaleghi, Ivo Dobrev, Weiyuan Tie, and Cosme Furlong

Abstract Non-destructive testing (NDT) is critical for many precision industries because it can provide important information about the structural health of critical components and systems. In addition, NDT can also identify situations that could potentially lead to critical failures. Specifically, NDT by optical methods have become popular because of their non-contact and non-invasive nature. Shearography is a high-resolution optical NDT method for identification and characterization of structural defects in components and has gained wide acceptance over the years; however, in practice, application of laser shearography for structural health monitoring requires loading the sample, which in some cases is not applicable or requires an experienced operator to properly perform structural testing. In this paper, a hybrid approach is proposed in which Finite Element Modeling (FEM) is used to simulate different loading conditions to obtain deformation data and in-turn, to obtain the simulated shearographic fringes. Different types of defects are embedded on the FE models and corresponding shearographic fringes are predicted. Correlating the defect and loading type to the predicted fringe pattern, in real shearographic measurements, different fringe patterns can be interpreted and classified. Also, camera calibration and image registration algorithms are used to project shearographic data onto the sample itself to locate and visualize the position of defects.

Keywords FEM • Fringe prediction • NDT • Shearography • Structural health monitoring

20.1 Introduction

This paper focuses on our recent progress in developing a non-destructive procedure based on combination of Finite Element Modeling (FEM), fringe prediction, and laser shearography to facilitate rapid defect detection for structural health monitoring. More specifically, in order to improve the process of finding defects by engineers who have limited shearography experience, a hybrid approach is proposed in which by simulating different loading conditions using FEM, corresponding shearographic fringe patterns are predicted to optimize shearography setup and achieve high sensitivity of the measurements. Additionally, a multimedia projector is used to project shearographic results onto the real surface of the object of interest. By using this approach, inspectors would have a better understanding of how to apply load to the sample by viewing the simulated deformation results of the sample. Thus, the inspector would have a better idea of how to induce the desired deformation in real testing.

X. Chen (✉) • M. Khaleghi • I. Dobrev • W. Tie • C. Furlong
Mechanical Engineering Department, Center for Holographic Studies and Laser micro-mechaTronics (CHSLT), Worcester Polytechnic Institute, Worcester, MA 01609, USA
e-mail: xchen2@wpi.edu

N. Sottos et al. (eds.), *Experimental and Applied Mechanics, Volume 6: Proceedings of the 2014 Annual Conference on Experimental and Applied Mechanics*, Conference Proceedings of the Society for Experimental Mechanics Series, DOI 10.1007/978-3-319-06989-0_20,

20.2 Principles of Laser Shearography

Principles of a digital shearography system that uses a Michelson configuration is illustrated in Fig. 20.1. In conventional Michelson interferometers, two mirrors are arranged perpendicular to each other, and therefore the reflected beams going to the CCD sensor are collinear. In a shearography system, one of the mirrors is tilted, which is called shearing mirror. As the shearing mirror is rotated, two identical, but displaced, images are recorded by the CCD camera. The two images are combined coherently, producing an interferometric speckle pattern at the CCD sensor, called shearogram [1–4].

If the surface of interest undergoes deformation, the optical path length of the incident light changes, and the optical phase difference due to this deformation is characterized by fringe-locus function, $\Omega(x, y)$. For near parallel illumination-observation conditions, out-of-plane gradient of displacement along horizontal and vertical axes are calculated by

$$\frac{\delta w}{\delta x} = \frac{\lambda\Omega}{4\pi\Delta x}, \tag{20.1}$$

$$\frac{\delta w}{\delta y} = \frac{\lambda\Omega}{4\pi\Delta y}, \tag{20.2}$$

where Ω is the fringe-locus function, λ is the laser wavelength, and Δx and Δy are the magnitude of the shear in horizontal and vertical axes, respectively [1–4].

20.3 Prediction of Shearographic Fringes by FEM and Analytical Solutions

To improve shearographic imaging in Non-Destructive Testing (NDT), the process is simulated and in-turn, optical phase maps and fringe patterns corresponding to the gradients of displacement are predicted. This allows engineers to find an optimal setup and loading conditions that provide maximum accuracy with minimal effort and inspection time.

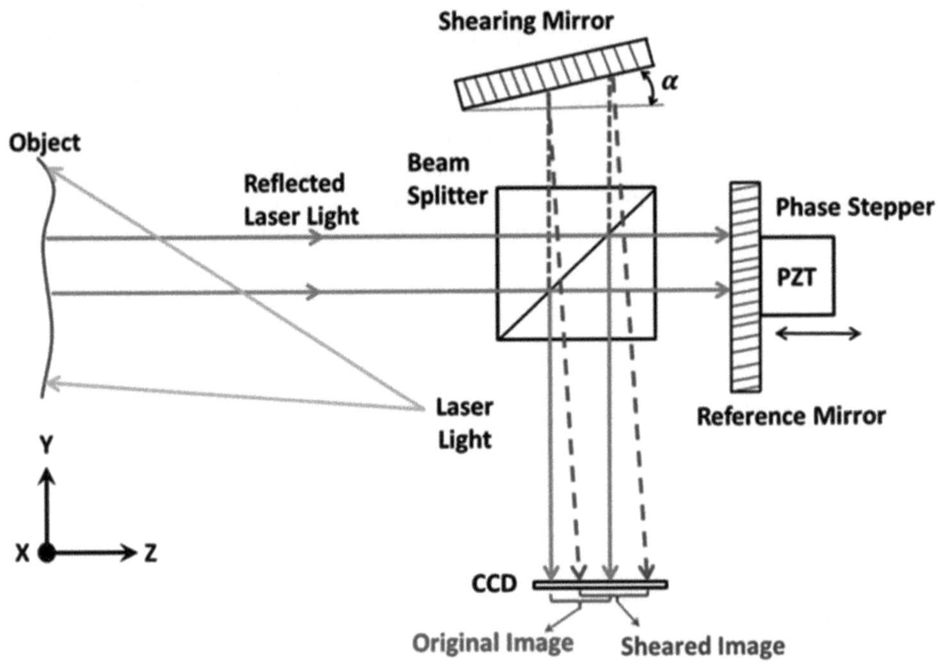

Fig. 20.1 Schematic of the optical configuration of the shearography system based on a Michelson interferometer [1–4]

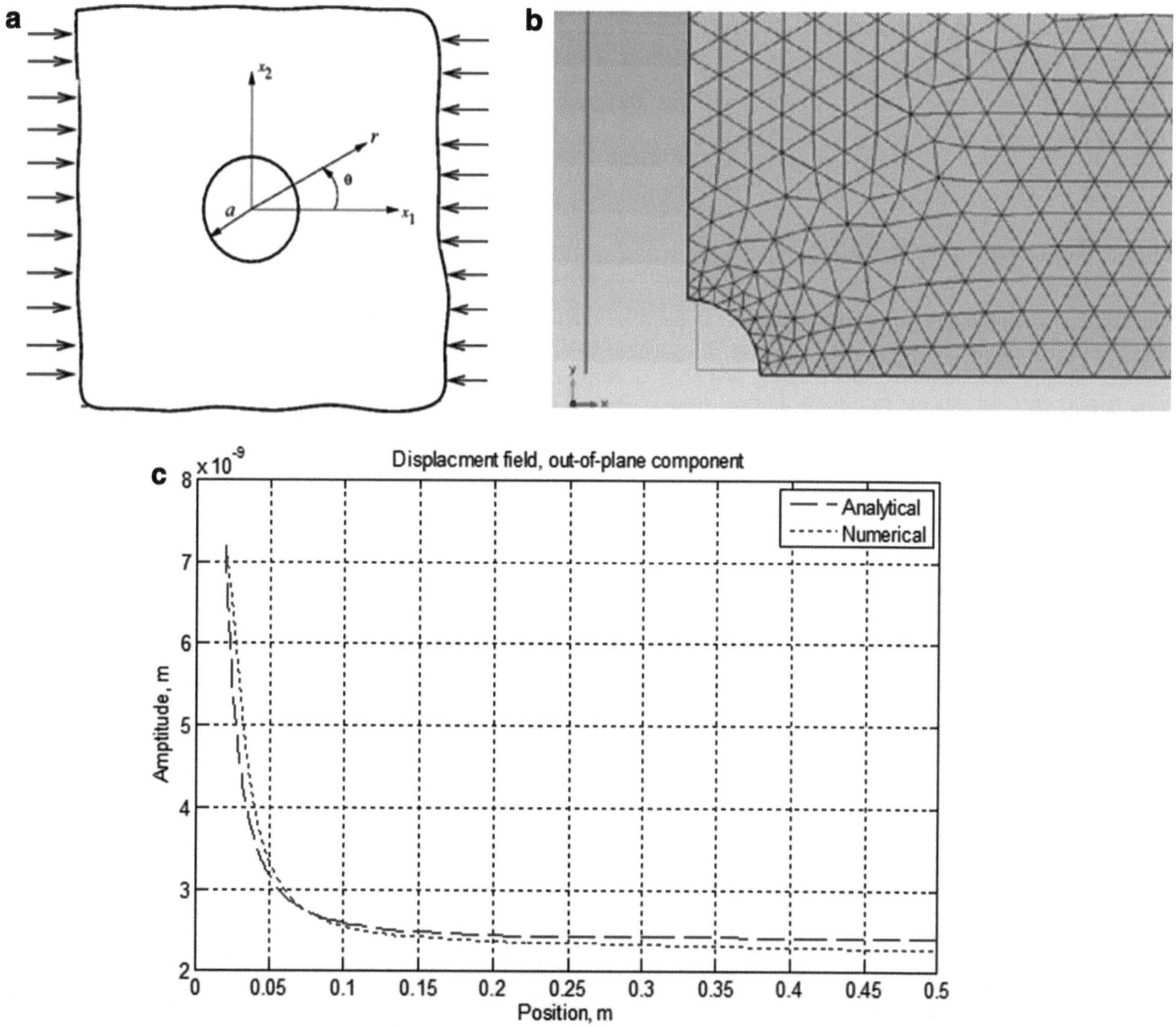

Fig. 20.2 Comparison of FEM and analytical solution for in-plane loading of a rectangular plate: (**a**) schematic of the plate with a circular hole in the center; (**b**) discretized zoomed-in view of the plate with the hole that is modelled symmetrically, and (**c**) corresponding out-of-plane displacement of a vertical cross-section, from the center of the hole to the edge of the plate, calculated by FEM and analytical solution [5–8]

20.3.1 Comparison of FEM and Analytical Solutions

To validate the accuracy of the FEM results, the analytical solution of a classical example is compared with the ones obtained by FEM. A rectangular plate with a small circular hole is subjected to an in-plane compressive load of 100 kN/m^2. The sample's material is aluminum 6061 with a Young's Modulus of 70 GPa and a Poisson ratio of 0.33. Figure 20.2a shows schematic of the problem and as shown in Fig. 20.2b the plate is discretized and considered as fully symmetric one. Figure 20.2c shows comparisons of the results of out-of-plane displacement obtained by both analytical solution and FEM [5–8].

20.3.2 Fringe Prediction

A shearography fringe prediction algorithm is developed to simulate expected fringe patterns based on experimental configurations. To improve the accuracy of the results, sensitivity vector, $\boldsymbol{K}$, is not considered to be fixed and it varies from point to point. The unwrapped phase can be calculated by

$$\Omega = \frac{\Delta\varnothing(x,y,z) - \Delta\varnothing(x+\Delta x,y,z)}{2} = \frac{\delta \boldsymbol{L}(x,y,z)}{\delta x} \cdot \frac{\pi \Delta x}{\lambda} \boldsymbol{K}(x,y,z), \tag{20.3}$$

where $\boldsymbol{L}$ is the displacement vector, $\boldsymbol{K}$ is the sensitivity vector, Δx is the shear amount, and λ is the wavelength of the laser light [1, 2].

20.4 Representative Results

Shearographic experiments are performed on a sample plate made of acrylic that has a partial hole at its center. Figure 20.3 shows an FEM model of this plate. To obtain deformation data to be used in fringe prediction, the plate undergoes first, acoustic loading and then, in-plane loading, and consequently, corresponding fringe patterns are predicted for each case. Simulations are performed in COMSOL and the dimensions of the sample are 20.32 cm, 9.525 cm, and 0.476 cm for length, width, and thickness, respectively, which are the same as in the plate used in the experiments. Diameter of the hole (defect) is 3.81 cm and its depth is 0.3175 cm.

20.4.1 Fringe Prediction Results

Acoustic loading simulations are performed using Eigen frequency analysis. Figure 20.4 shows predicted shearographic fringe patterns for the first six modes of vibration of the plate corresponding to the model shown in Fig. 20.3, constrained along the short edges. As shown in Fig. 20.4, the defect (partial hole) is either not detectable with acoustic loading or the acoustic level of excitation is high. Therefore, it is necessary to change the loading condition in order to detect the defect.

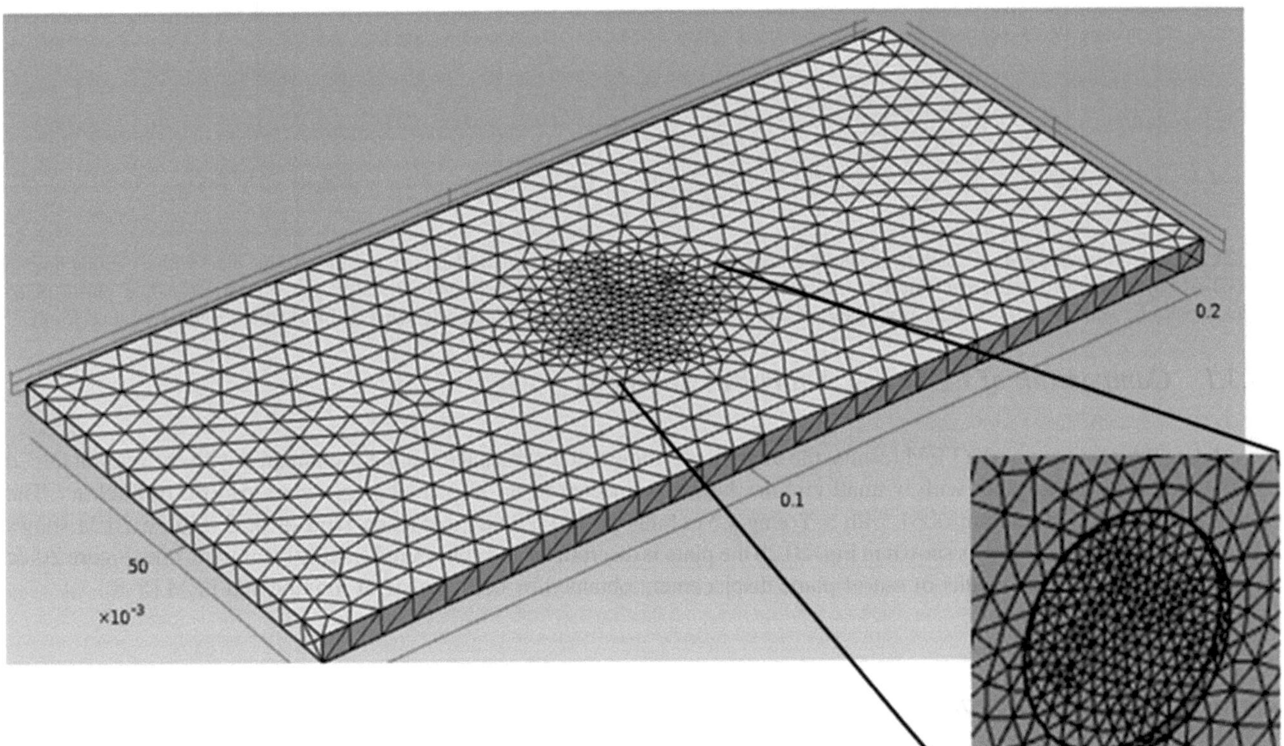

Fig. 20.3 FEM mesh of an acrylic plate containing a partial hole at its center

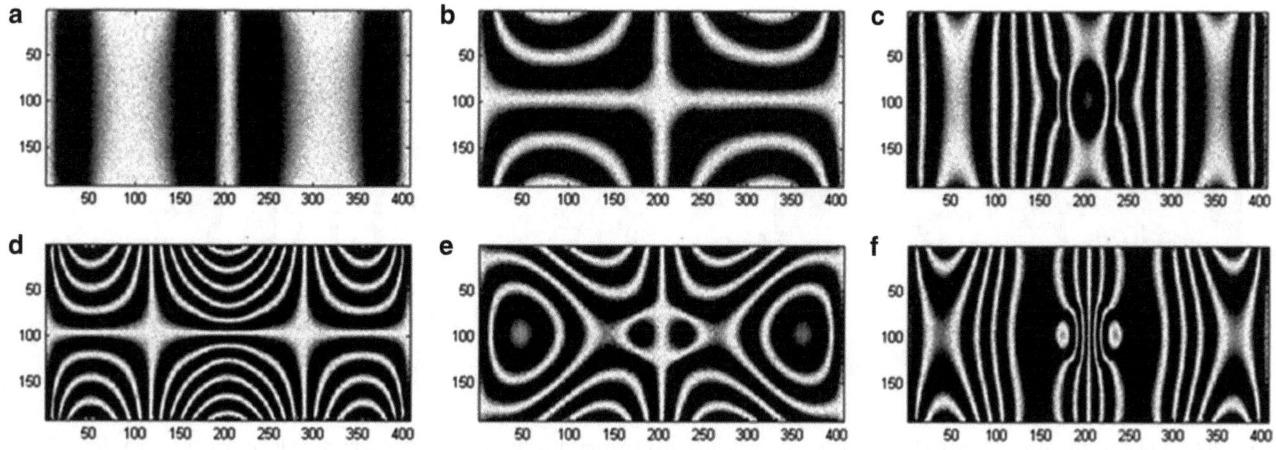

Fig. 20.4 Prediction of shearographic fringes corresponding to the first six modes of vibrations obtained from FEM analysis: (a) 205 Hz; (b) 332 Hz; (c) 545 Hz; (d) 735 Hz; (e) 1,038 Hz; and (f) 1,085 Hz, respectively

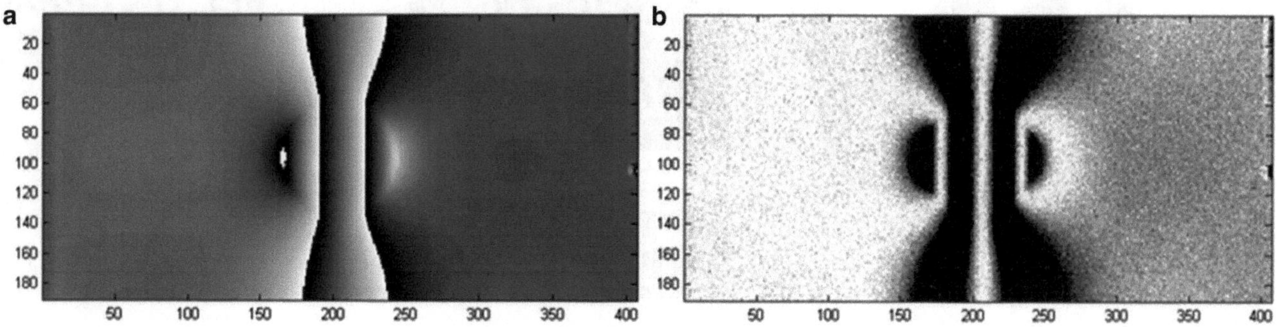

Fig. 20.5 Prediction of shearography fringes by FEM under compression loading: (a) wrapped optical phase; (b) corresponding shearography intensity fringes

Uniaxial compression loading of 220 kPa is simulated next. In this case and as shown in Fig. 20.5, the simulation results indicate that using our hybrid technique, combination of FEM and fringe prediction, inspectors can locate partial hole defect by laser shearography.

20.4.2 Experimental Results for Validating Fringe Prediction Results

The corresponding experimental result of the acrylic plate sample under compression loading is shown in Fig. 20.6. The result validates that by using appropriate loading conditions predicted by FEM, the defect can be detected with shearographic imaging.

20.4.3 Projection of Shearographic Results on the Sample

Visualization of the results generated by laser shearography can improve the identification of defects on surfaces under test [3]. As shown in Fig. 20.7, a multimedia projector has been introduced at the end of inspection process to project the shearographic results back onto the surface of the sample, so that the defect detection process becomes more evident. In this experiment, the optical head was located at a working distance of 1.8 m from the panel. A pulsed laser with pulse energy of 0.2 mJ with the wavelength of 532 nm and power of 50 mW is used for shearographic imaging.

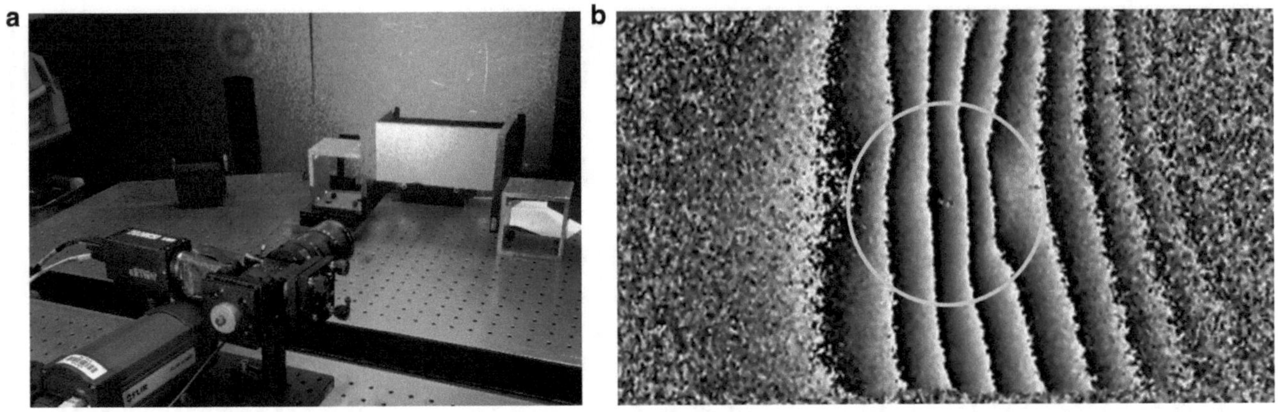

Fig. 20.6 A plate with a partial hole under shearographic imaging: (a) experiment setup; (b) wrapped optical phase containing some irregularities that corresponds to the defect

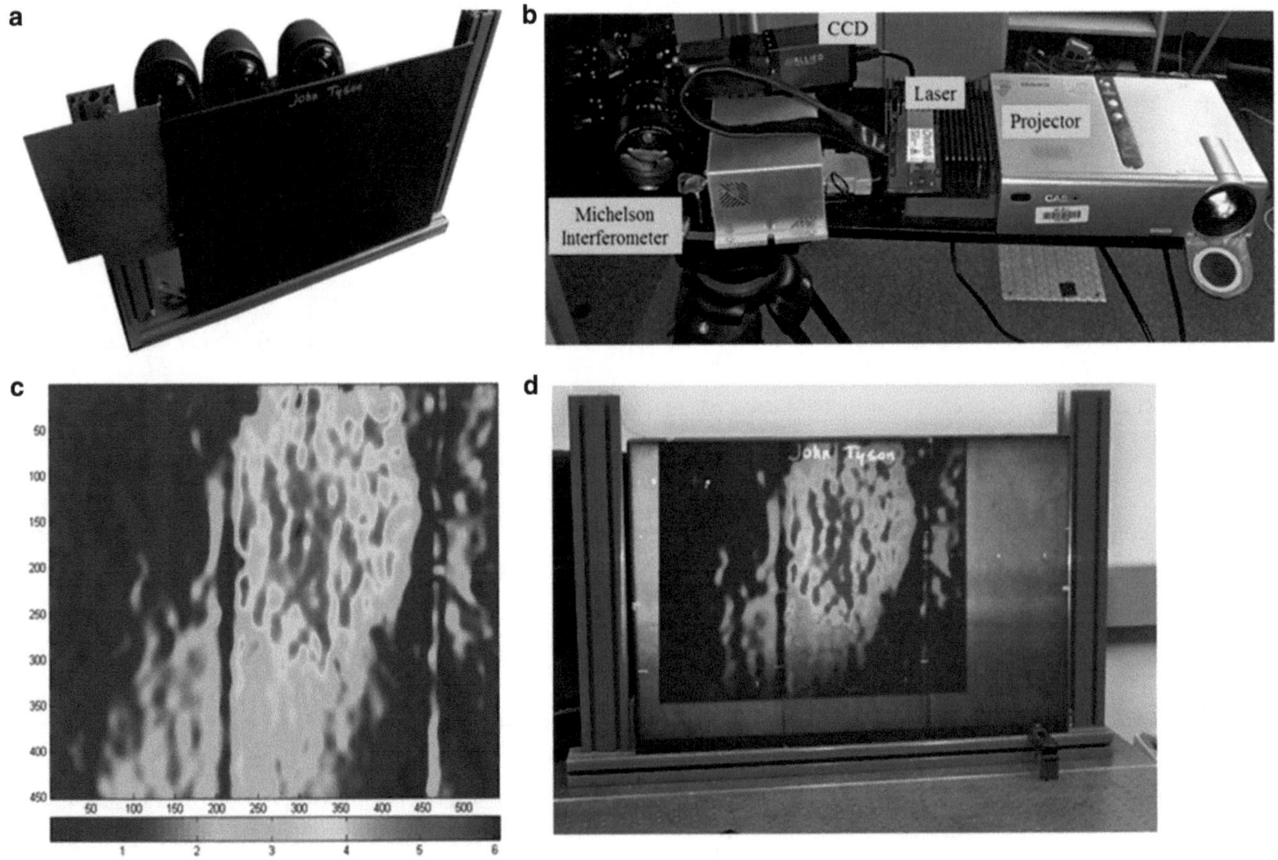

Fig. 20.7 Projection of the shearographic images using a calibrated projector: (a) sample subjected to thermal loading by incandescent sources; (b) configuration of the optical head containing laser source, CCD camera, Michelson Interferometer, and a calibrated projector; (c) the cumulative slope of the horizontal strain map which enables the identification of structural defects within the panel; and (d) back projection of shearographic results onto the sample's surface in order to localize the defect on the sample [9–11]

20.5 Conclusion and Future Work

A hybrid approach based on combination of FEM and fringe prediction is proposed to facilitate defect detection in structural health monitoring. In our approach, FEM is used to obtain simulated deformation field of the objects under different loading conditions and in turn, corresponding shearographic fringe patterns are predicted. Different types of defects are embedded on the FE models and corresponding shearographic fringes are predicted. Also, we present projection of the shearographic results onto the actual components. The results show that combinations of FEM, fringe prediction, and fringe re-projection would facilitate and increase the speed of defect detection.

Acknowledgements The authors gratefully acknowledge the support provided by NIST program at Worcester Polytechnic Institute. We would also like to thank our colleagues at the CHSLT labs and Trilion Quality Systems.

References

1. Kreis T (2005) Handbook of holographic interferometry: optical and digital methods. Wiley-VCH, New York, pp 243–266
2. Meybodi MK, Dobrev I, Klausmeyer P, Harrington E, Furlong C (2012) Investigation of thermomechanical effects of lighting conditions on canvas paintings by laser shearography. Proceedings of SPIE 8494, Interferometry XVI: Applications, 84940A; DOI:10.1117/12.958089, 2012
3. Fantin AV, Willemann DP, Albertazzi A (2012) Improvements in the marking of defects on large structures surfaces during shearographic inspection. Proceedings of SPIE Optical Engineering+ Applications. International Society for Optics and Photonics, 2012
4. Furlong C, Dobrev I, Harrington E, Hefti P, Khaleghi M (2012) Miniaturization as a key factor to the development and application of advanced metrology systems. Proceedings of SPIE, Speckle 2012, 84130T-84130T, 2012
5. Nagpal S, Jain N, Sanyal S (2011) Stress concentration and its mitigation techniques in flat plate with singularities-A critical review. Eng J 16(1):1–16
6. Connell University, ANSYS Tutorial Website: https://confluence.cornell.edu/display/SIMULATION/ANSYS+-+Plate+with+a+Hole+%28Results-Interpretation%29
7. Wikiversity website: http://en.wikiversity.org/wiki/Introduction_to_Elasticity/Plate_with_hole_in_tension
8. de Larminat PM, Wei RP (1978) Normal surface displacement around a circular hole by reflection holographic interferometry. Exp Mech 18(2):74–80
9. Harrington E, Furlong C, Rosowski JJ, Cheng JT (2011) Automatic acquisition and processing of large sets of holographic measurements in medical research. Proceedings of SPIE, Optical Measurements, Modeling, and Metrology, 5:219–228, 2011
10. Ti C, Chen X, Tyson J, Harrington E, Dobrev I, Aghazadeh BS, Furlong C (2014) High-speed shape measurements by fringe projection method: SOPRA 3D. In Advancement of Optical Methods in Experimental Mechanics, 3, 299–304
11. Khaleghi M, Dobrev I, Harrington E, Klausmeyer P, Cushman M, Furlong C (2014) Long-term effects of cyclic environmental conditions on paintings in museum exhibition by Laser Shearography. In Advancement of Optical Methods in Experimental Mechanics, 3, 283–288

Chapter 21
On Improving Thermoelastic Stress Analysis Data Near Edges of Discontinuities

W.A. Samad and R.E. Rowlands

Abstract The most serious stresses are often at the edges of geometric discontinuities, and thereby influencing the overall performance of a structure. Under adiabatic and reversible conditions, thermoelastic stress analysis (TSA) provides nondestructive full-field information of the first stress invariant in a cyclically loaded structure. However, TSA measurements at and near the edges of discontinuities, the regions of prime interest, are often unreliable due to the adverse influence of the surrounding ambient temperature as well as the movement associated with the cyclic loading. Moreover, TSA data at such locations are susceptible to nonadiabaticity because of high stress gradients, thus further supporting the need to predict stresses at edges. A method is presented here for correctly quantifying the often disregarded TSA data at the edges of a structure by making use of the linear elastic conditions of equilibrium and compatibility as well as applying the appropriate boundary conditions. The method is hybrid in the sense that experimental TSA data (excluding disregarded edge measurements) are combined with an analytical expression of the first stress invariant. The achieved improvement in thermoelastic data near discontinuities is demonstrated here for a tensile aluminum structure containing a central irregularly shaped cutout.

Keywords Hybrid • Discontinuities • Thermoelasticity • Full-Field • Edges

21.1 Introduction

The improvement in stress data at the edges of cutouts is achieved here by disregarding the measured TSA data at and close to an edge and then using the remaining TSA recorded information to predict those at the edges. The prediction of stresses at the edges is done by combining the measured TSA data away from the edge with an Airy stress function, and imposing the traction-free conditions on the edge. Assuming linear-elastic and isotropic behavior, an Airy stress function is utilized to describe the state of stress. An expression of the first stress invariant is derived in terms of the Airy coefficients, followed by a linear least squares analysis to evaluate the various Airy coefficients from the recorded TSA data. Knowing the magnitude of Airy coefficients, the individual components of stress are available throughout the structure, and most importantly, those right at the edges of cutouts. Beyond the assumptions of linear elastic and isotropic behavior, and evaluating the TSA calibration factor, the method does not require knowledge of constitutive material properties or the loading or boundary conditions. The latter can be quite advantageous since loading and boundary conditions are often not well known in real-life engineering applications. The so described TSA method is a hybrid technique as recorded temperature data is processed analytically using a stress function. In addition to improving reliable stress data at and near the edges of cutouts, the method gives the individual components of stress. This ability to provide both reliable edge data and individual stresses is demonstrated by the illustrative example of a finite aluminum tensile plate containing an irregularly shaped cutout, Fig. 21.1.

W.A. Samad (✉)
Rochester Institute of Technology, Dubai, UAE
e-mail: wascad@rit.edu

R.E. Rowlands
University of Wisconsin, Madison, WI, USA

N. Sottos et al. (eds.), *Experimental and Applied Mechanics, Volume 6: Proceedings of the 2014 Annual Conference on Experimental and Applied Mechanics*, Conference Proceedings of the Society for Experimental Mechanics Series,
DOI 10.1007/978-3-319-06989-0_21,

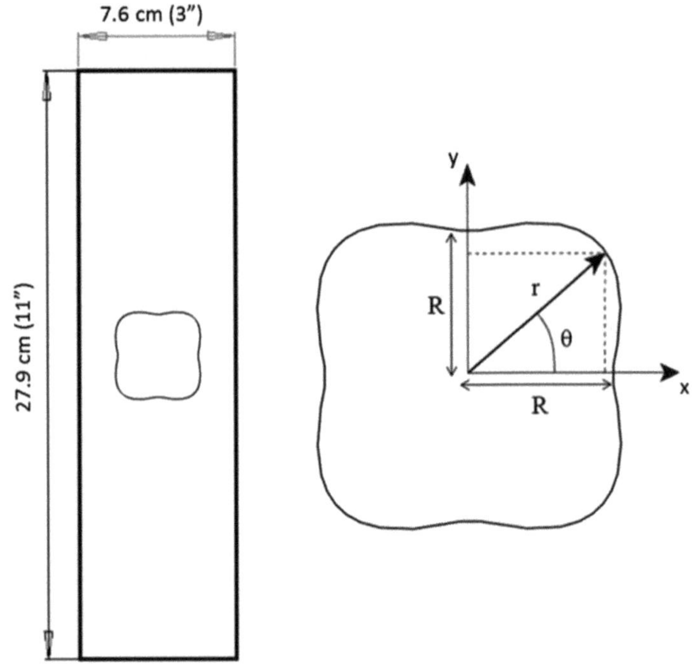

Fig. 21.1 CAD model of irregularly shaped tensile specimen

21.1.1 Thermoelastic Stress Analysis

By cyclically loading an isotropic material to achieve adiabatic and reversible conditions, the recorded load-induced thermal signal, $S*$, is proportional to the change in the sum of the normal stresses (i.e., isopachic stress) [1],

$$S* = K\Delta S = K\Delta\left[(\sigma_1 + \sigma_2) = (\sigma_r + \sigma_\theta) = \left(\sigma_x + \sigma_y\right) = \left(\sigma_\xi + \sigma_\eta\right)\right] \tag{21.1}$$

where $S*$ is the recorded TSA signal, K is the TSA calibration factor related to the relevant physical properties of the material, surface condition and TSA system parameters [1–3], S is the isopachic stress and σ_1, σ_2, σ_r, σ_θ, σ_x, σ_y, σ_η and σ_ξ are the normal stresses in the principal, polar, rectangular coordinates, and normal and tangential to the edge of the hole, respectively.

TSA recorded temperatures at and near edges are often unreliable due to the adverse influence of the surrounding ambient temperature and the cyclic loading. Pixels at such locations are also susceptible to violating adiabaticity because of high stress gradients. The presented approach disregards unreliable measurements at and near the edge and provides reliable stresses at these locations based on interior TSA data as well as relevant elasticity equations. Moreover, raw TSA results give only a combination of stresses while strength criteria and fatigue analysis typically requires the knowledge of the individual components of stress—which are also provided by the present hybrid technique.

21.2 Predicting Stresses Near Edges

The idea behind improving TSA data at and near edges involves combining the measured thermoelastic information (all except for those near the edges) with an Airy stress function and imposing the traction-free conditions on the edge of the hole. In addition to replacing such disregarded unreliable TSA data at and near an edge, this combination of measured quality interior TSA data with an analytical stress function provides the individual components of stresses.

21.2.1 Airy Stress Function

Under conditions of plane stress, isotropic linear elastic behavior, and based on symmetry about the x- and y-axes, a relevant Airy stress function satisfying the biharmonic equation, $\nabla^4\phi = 0$, can be written as [4, 5]

$$\phi = a_0 + b_0 \ln r + c_0 r^2 + \sum_{n=2,4,\ldots}^{N} \left\{ \left(a_n r^n + b_n \mathrm{r}^{\mathrm{n}+2} + c_n r^{-n} + d_n \mathrm{r}^{-(\mathrm{n}-2)} \right) \cos n\theta \right\} \tag{21.2}$$

where radius r is measured from the center of the irregularly-shaped hole, angle θ is measured counterclockwise from the horizontal x-axis, Fig. 21.1. N is the terminating index value of the series and can be any positive even integer. The individual components of stress in polar coordinates are given in terms of the coefficients as follows:

$$\sigma_r = \frac{b_0}{r^2} + 2c_0 - \sum_{n=2,4,\ldots}^{N} \begin{pmatrix} a_n\, n\,(n-1)\, r^{n-2} \\ +b_n\,(n+1)(n-2)r^n \\ +c_n\, n\,(n+1)\, r^{-(n+2)} \\ +\ d_n\,(n-1)(n+2)r^{-n} \end{pmatrix} cos\,(n\theta) \tag{21.3}$$

$$\sigma_\theta = \frac{-b_0}{r2} + 2c_0 + \sum_{n=2,4,\ldots}^{N} \begin{pmatrix} a_n\, n\,(n-1)\, r^{n-2} \\ +b_n\,(n+1)(n+2)r^n \\ +c_n\, n\,(n+1)\, r^{-(n+2)} \\ +\ d_n\,(n-1)(n-2)\, r^{-n} \end{pmatrix} cos\,(n\theta) \tag{21.4}$$

$$\sigma_{r\theta} = \sum_{n=2,4,\ldots}^{N} \begin{pmatrix} a_n\, n\,(n-1)\, r^{n-2} \\ +b_n\, n\,(n+1)\, r^n \\ -c_n\, n\,(n+1)\, r^{-(n+2)} \\ -\ d_n\, n\,(n-1)\, r^{-n} \end{pmatrix} sin\,(n\theta) \tag{21.5}$$

With the analytical equations of the separate stresses determined in terms of the different Airy coefficients, the TSA's first stress invariant can now be equated as follows:

$$S = \sigma_r + \sigma_\theta = 4\, c_0 + \sum_{n=2,4,\ldots}^{N} \begin{pmatrix} b_n\, 4\,(n+1)r^n \\ -d_n\, 4\,(n-1)r^{-n} \end{pmatrix} \cos\,(n\theta) \tag{21.6}$$

Note that the separate equations for the three individual stresses cannot be obtained by combining the equation above with the recorded TSA data alone since Eq. 21.6 does not involve all of the coefficients present in Eqs. 21.3 through 21.5. Stress separation therefore requires an extra equation that is in terms of at least these other coefficients.

21.2.2 Traction-Free Boundary Conditions

Since a plate with an irregularly-shaped hole is considered here, the normal-tangential coordinate system is used to impose the traction-free conditions at the edge of the cutout, Fig. 21.2. The transformation matrix of Eq. 21.7 is used to convert stresses from polar (r, θ) to normal-tangential (ξ, η) coordinates such that $\sigma_\xi$and $\sigma_{\xi\eta}$ can be imposed to zero on the edge of the hole.

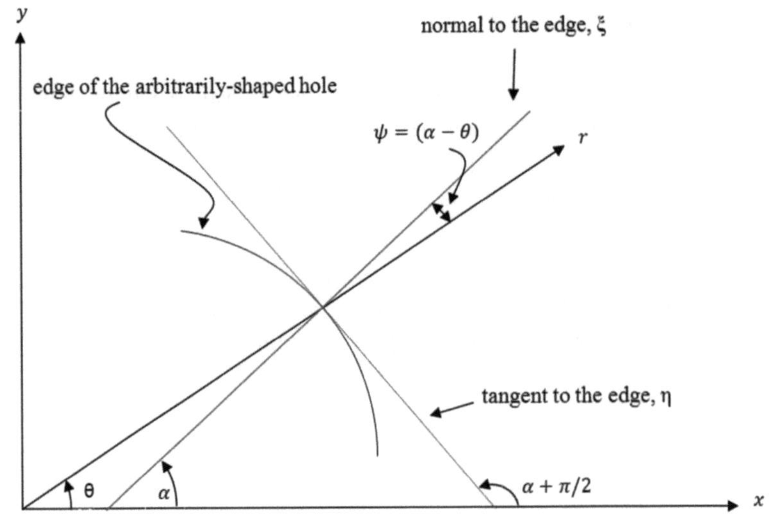

Fig. 21.2 Normal-tangent coordinate system at edge of hole

$$\begin{Bmatrix} \sigma_\xi \\ \sigma_\eta \\ \sigma_{\xi\eta} \end{Bmatrix} = \begin{bmatrix} cos^2\psi & sin^2\psi & 2\,cos\psi\,sin\psi \\ sin^2\psi & cos^2\psi & -2\,sin\psi\,cos\psi \\ -sin\psi\,cos\psi & sin\psi\,cos\psi & (cos^2\psi - sin^2\psi) \end{bmatrix} \begin{Bmatrix} \sigma_r \\ \sigma_\theta \\ \sigma_{r\theta} \end{Bmatrix} \tag{21.7}$$

Angle ψ in the transformation matrix is the angle between the normal ξ—and polar r-directions, Fig. 21.2. In order to determine the angle ψ along the boundary of the hole, 301 discrete points on the edge of the hole were first imported from the CAD model in Fig. 21.1. Then, the local shape defined by a set of three neighboring boundary points was represented by a second order polynomial. Differentiating the polynomial analytically provides the slope, hence the tangential and normal directions and ultimately the angle ψ, at the midpoint. This process was repeated incrementally at overlapping sets of neighboring points along the edge of the hole such that ψ was determined throughout.

21.3 TSA Experiment

The structural member is a finite-width perforated aluminum plate as shown in the CAD model in Fig. 21.1. The plate is 27.94 cm long, 7.62 cm wide, and 0.63 cm thick and at its center is an irregularly-shaped cutout whose exact profile is created by passing a B-spline through three defined points. The specimen was cyclically loaded between 2,669 N and 8,007 N at a frequency of 20 Hz. Figure 21.3 contains a labeled photograph of the test setup as well as the resulting load-induced thermal image. The thermal image was recorded by a liquid nitrogen-cooled infrared DeltaTherm DT1410 Camera (Stress Photonics, Madison, WI; sensor resolution of 256 horizontal × 256 vertical pixels). The TSA image of Fig. 21.3 contains approximately 49,000 pixels of temperature information with a pixel spacing of 1.37 mm. The measured TSA data were averaged throughout the four quadrants of the plate and the analyses performed in one quadrant by making use of symmetry about the x- and y-axes.

21.4 Analysis and Results

Once the TSA image has been recorded and calibrated, a set of linear first stress invariant expressions of the form of Eq. 21.6, as well as the set of linear equations from imposing the local traction-free boundary conditions from Eq. 21.7, can be written in matrix form as follows:

$$[A]_{(m+2h)\ \mathrm{x}\ k}\{c\}_{k\ \mathrm{x}\ 1} = \{d\}_{(m+2h)\ \mathrm{x}\ 1} \tag{21.8}$$

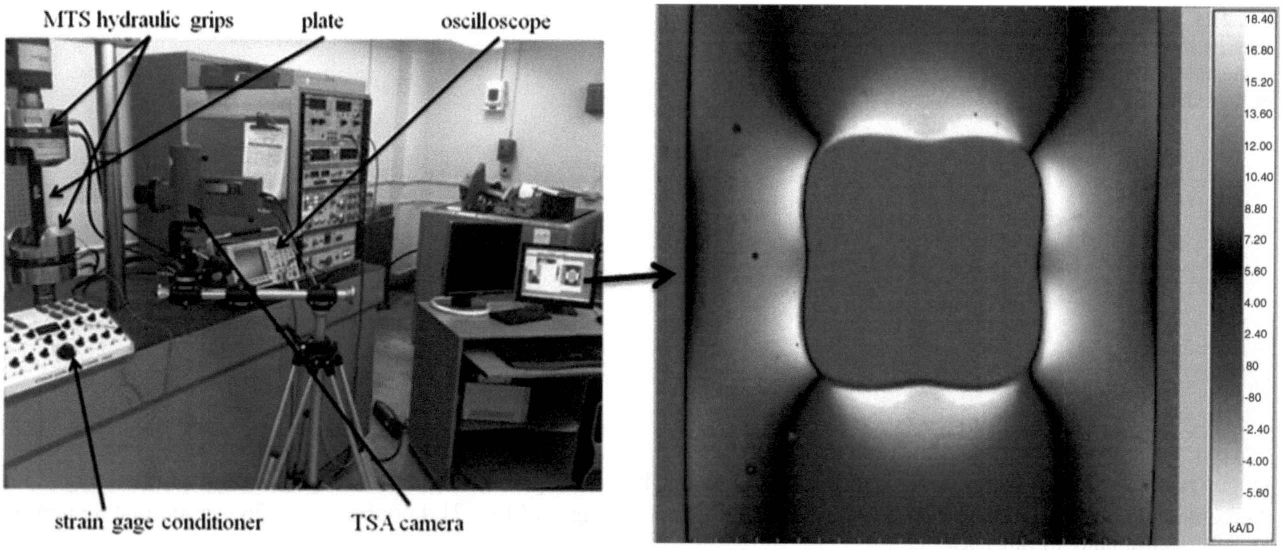

Fig. 21.3 TSA experimental setup and corresponding thermal image

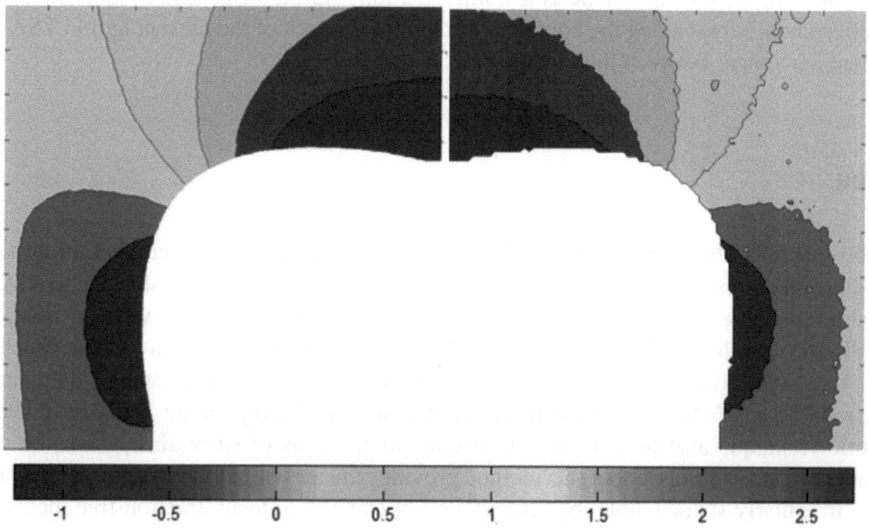

Fig. 21.4 Comparison of reconstructed normalized S (*left*) and that based on the actual measurement (*right*)

where [A] is the Airy matrix composed of $m = 956$ first stress invariant Airy expressions in terms of r and θ associated with the 956 selected source locations of the input values of S as well as $2h = 602$ traction-free boundary conditions of the form of Eq. 21.7, {c} is the set of k unknown Airy coefficients and vector {d} includes the 956 measured values of S as well as 602 zeros from the right hand side of Eq. 21.7. The linear least squares problem associated with Eq. 21.8 was solved using the *backslash* operator in MATLAB. With TSA data typically being unreliable at edges, the input values of S^* selected originated at least two pixels (2.74 mm) away from the edge of the hole. The number of Airy coefficients, k, to retain was assessed by computing the *rms* for a series of different number of Airy coefficients. The *rms* represents the discrepancy between the calculated first stress invariant data {d′} and the measured TSA data. Plotting the *rms* values against multiple values for k yielded a value of 26 to be an appropriate number of Airy coefficients to utilize. Having evaluated all of the Airy coefficients, the first stress invariant and the separate stresses σ_r, σ_θ and $\sigma_{r\theta}$ are can now be made available at locations of the discarded TSA data using Eqs. 21.3 through 21.6.

Figure 21.4 compares the normalized first stress invariant contours, S/σ_o, as reconstructed from using the present hybrid method with those based on the actual measurements. Beyond the improvement of the values of the first stress invariant at the edges of the cutout, data is now provided continuously over the structure and the availability of such information is no

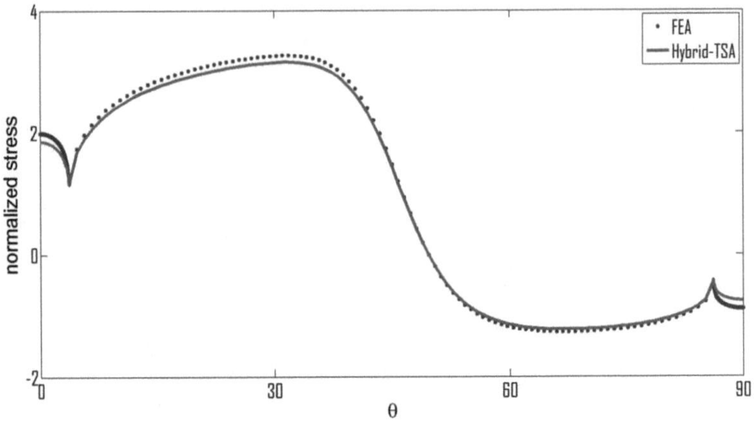

Fig. 21.5 Normalized hoop stress, σ_θ/σ_o, on the boundary of the hole by hybrid-TSA method and FEA

longer restricted to the discrete pixel locations as evident from comparing the left and right contour plots of Fig. 21.4. Agreement between the reconstructed and experimental-based images of Fig. 21.4 confirms $k = 26$ is a reasonable number of Airy coefficients in this case.

A symmetric FEA model of the perforated aluminum specimen was also analyzed using ANSYS for comparative purposes. In particular, FEA is performed here to check the validity of the separate stresses evaluated according to the present hybrid method as the TSA test only provides a combination of stresses, i.e., the first stress invariant. A quarter of the specimen was modeled using 6,704 eight-node isoparametric *Plane-82* elements and 20,807 nodes. Figure 21.5, which compares the normalized hoop stress along the boundary of the hole by each of the present hybrid TSA method and ANSYS demonstrates excellent agreement between the two approaches.

21.5 Conclusion

Experimental methods such as TSA can be employed to determined stresses. Recorded TSA information at and near the edges of geometric discontinuities, often the regions of prime interest, is typically unreliable due to adverse influences of the surrounding ambient temperature and movement associated with the cyclic loading, as well as susceptible to heat transfer associated with local stress gradients. Processing the recorded TSA data away from a hole or notch in loaded isotropic structures with an Airy stress function and imposing the traction-free conditions on the boundary of the geometric discontinuity gives reliable edge stresses. Notwithstanding the recorded temperature signal is related to a combination of stresses, this hybrid technique also provides the individual components of stress throughout the structure. In addition to evaluating accurate edge stress information, this method provides the individual stresses throughout a perforated structure here containing an irregularly-shaped hole by discretely imposing the local traction-free boundary conditions and without knowing the external loading. As well as satisfying load equilibrium, the TSA results agree with those from FEM and strain gages.

References

1. Greene RJ, Patterson EA, Rowlands RE (2008) Chapter 26: Thermoelastic stress analysis. In: Sharpe WM (ed) Handbook of experimental solid mechanics. Springer, New York
2. Stanley P, Chan WK (1987) Assessment and development of the thermoelastic technique for engineering application: four years of progress in stress analysis by thermoelastic techniques. In: Proceedings of SPIE 731, London, pp 17–25
3. Dulieu-Barton JM, Stanley P (1998) Development and applications of thermoelastic stress analysis. J Strain Anal 33:93–104
4. Soutas-Little RW (1973) Elasticity. Dover, Mineola, NY
5. Samad WA (2013) Hybrid full-field stress analysis of structures containing nonconventional cutout geometries. PhD Thesis, University of Wisconsin-Madison

Chapter 22
Measurement of Stress Network in Granular Materials from Infrared Measurement

Pawarut Jongchansitto, Xavier Balandraud, Michel Grédiac, and Ittichai Preechawuttipong

Abstract Infrared thermography (IR) was used in this work that aims to experimentally evidence the stress network in granular media composed of two materials featuring different stiffness, without cohesion and under confined compression. Cylinders of polyoxymethylene (POM) and high-density polyethylene (HDPE) were used to build 2D composite granular systems. Cylinders were placed parallely and mixed together in a square metallic frame. The experiments were performed using a uniaxial testing machine. The granular media were first compacted in order to reach static equilibrium configurations. A cyclic compressive load was then applied. IR camera was employed to measure the temperature changes due to thermoelastic coupling on the cylinder network cross-sections. Temperature variations were then processed to obtain the maps of the amplitude of the sum of the principal stresses during the cycles. Three configurations were tested by changing the ratio between the POM and HDPE diameters and the ratio between the numbers of POM and HDPE cylinders. The experimental technique enables us to identify the stress network within the granular media. The experimental results are compared with numerical results obtained with a molecular dynamics software.

Keywords Granular material • Infrared thermography • Thermoelastic stress analysis • Stress network • Confined compression

22.1 Introduction

Granular materials are composed of particles whose macroscopic mechanical behavior is governed by their interactions. To understand the behavior of such materials at the continuum (macroscopic) scale, an analysis of the force networks must be performed at the grain (microscopic) scale. Photoelasticimetry was first used in 1957 for particles. They had the property to develop birefringence under stress [1]. It was then used in some other studies [2, 3]. Digital image correlation from two-dimensional optical measurements was also employed [4, 5], while X-ray technique was used for three-dimensional measurements [6, 7]. Strain components were also measured using particle image velocimetry [8]. To the best knowledge of the authors, it seems that infrared (IR) thermography was used for granular materials only to analyze heat transfer [9].

IR thermography is a full-field measurement technique which can be used for measuring the temperature changes occurring on the surface of specimens subjected to a mechanical load. These temperature changes result from

P. Jongchansitto
Department of Mechanical Engineering, Faculty of Engineering, Chiang Mai University, 239 Huay Kaew Road, Muang District, Chiang Mai 50200, Thailand

Clermont Université, Université Blaise Pascal, Institut Français de Mécanique Avancée, Institut Pascal UMR 6602, BP 10448, 63000 Clermont-Ferrand, France

X. Balandraud (✉) • M. Grédiac
Clermont Université, Université Blaise Pascal, Institut Français de Mécanique Avancée, Institut Pascal UMR 6602, BP 10448, 63000 Clermont-Ferrand, France
e-mail: xavier.balandraud@ifma.fr

I. Preechawuttipong
Department of Mechanical Engineering, Faculty of Engineering, Chiang Mai University, 239 Huay Kaew Road, Muang District, Chiang Mai 50200, Thailand

N. Sottos et al. (eds.), *Experimental and Applied Mechanics, Volume 6: Proceedings of the 2014 Annual Conference on Experimental and Applied Mechanics*, Conference Proceedings of the Society for Experimental Mechanics Series, DOI 10.1007/978-3-319-06989-0_22,

thermomechanical couplings or to mechanical dissipation due to fatigue for instance. Thermoelastic coupling predicts that temperature changes if the elastic strain changes. Under cyclic loading at sufficiently high frequency, the temperature change in an isotropic material is directly proportional to the sum of principal stresses of the stress tensor. This is the basis of the so-called thermoelastic stress analysis [10], which has been used in various studies devoted to the characterization of materials such as composites [11], polymer coatings [12], sandwich materials and joints [13, 14] for instance, as well as structural components [15]. The present study deals with full-field temperature measurements in 2D composite granular media, i.e. composed of cylinders made of different materials. The objective is to deduce the stress paths from the temperature variations under cyclic loading. Three non-cohesive configurations were prepared. They are different in terms of diameters used for the cylinders and numbers of cylinders for each type of material. After compaction, an IR camera captured the temperature variations due to the thermoelastic coupling during confined compressive cyclic loads. The stress levels in the cylinders were then deduced, analyzed and compared with numerical results obtained with a molecular dynamics software.

22.2 Methodology

Three non-cohesive granular media were prepared using cylinders as particles. Two types of constitutive material were used for the cylinders: polyoxymethylene (POM) and high-density polyethylene (HDPE). The stiffness of POM is four times higher than the stiffness of HDPE: 2.8 GPa and 0.7 GPa, respectively. Different diameters were also used. Cylinders were placed in parallel in a metallic rectangular frame (see Fig. 22.1a). The three tested configurations are given in Table 22.1.

Mechanical tests were performed at ambient temperature with a uniaxial Schenck ±250 kN testing machine. The loading was force-controlled. A cyclic compression loading was imposed with a frequency of 3 Hz and a force amplitude of 50 kN. Note that before applying the cyclic loading, each granular material was compacted in order to reach a static equilibrium configuration. For specimen A which is monodisperse, a periodic stacking was obtained. A Cedip Jade III-MWIR camera was used to capture the temperature fields on the front face of the cylinder network during the cyclic loading (see Fig. 22.1b). A Fourier analysis was performed at each pixel to extract the amplitude ΔT of the temperature oscillation resulting from the thermoelastic coupling. Within the framework of thermoelastic stress analysis, the relationship between temperature amplitude ΔT and amplitude of the sum of the principal stresses $\Delta\sigma$ is given by:

$$\Delta T = \alpha T_0 / \rho C \times \Delta\sigma \tag{22.1}$$

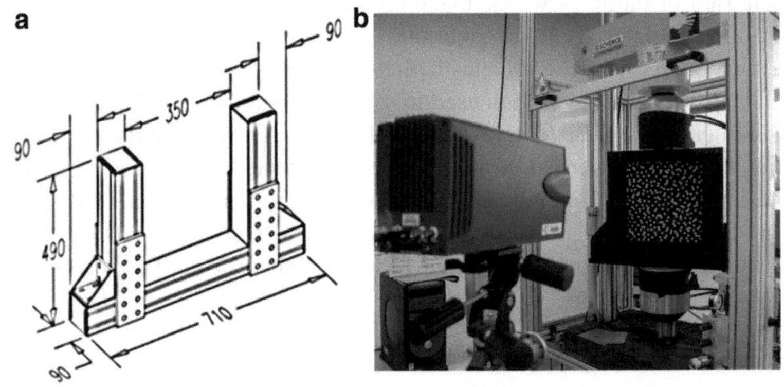

Fig. 22.1 Experimental setup. (**a**) Dimensions of the frame. Two reinforcement plates are not represented in the drawing. They were added at the top the structure to ensure the rigidity of frame. These plates prevented the bending of the two lateral parts of the frame under loading. (**b**) Picture of the experiment. One of the reinforcement plates is visible in the picture

Table 22.1 Three tested specimens

Specimen	Ratio of diameters D_{POM}/D_{HDPE}	Number of cylinders POM:HDPE	Ratio of cylinder numbers POM:HDPE	Total number of cylinders
A	1	435:807	1:1.86	1,242
B	1.6	239:608	1:2.54	847
C	3	86:466	1:5.42	552

where α is the coefficient of thermal expansion, T_0 the mean temperature in Kelvin, ρ the density and C the specific heat. The cylinder cross-sections were not painted in black in order to be able to distinguish POM and HDPE from optical images. As a consequence, the ratio $\alpha T_0/\rho C$ to be used for converting measured temperature in stress was preliminarily identified from uniaxial tensile tests on rectangular sheets for both materials. Noted that as the three tested specimens are non-cohesive (no glue was used to build the granular materials), no tensile contact forces are expected. As a consequence, the value of $\Delta\sigma$ is expected to be proportional to the sum of the amplitudes of the (compression) contact forces applied to each cylinder.

22.3 Experimental Results

Figure 22.2a presents the stress distribution in the monodisperse granular medium (specimen A). A mean value is given in each cylinder. Each cylinder has six neighbors, except those which are along the frame. It can be observed that high values are observed in the cylinders located along the boundary. If these boundary cylinders are removed from the analysis, less than 50 % of the cylinders have a stress level which is higher than the mean value. This property was obtained for the three tested specimens. It is in agreement with simulations performed by molecular dynamics (see below). It can be noted that this property is well known from numerical results on contact forces. It seems to be also achieved in terms of compression stresses in our non-cohesive granular materials. Stress patterns can be identified in the figure: they consist of inclined straight lines for specimen A. These typical patterns result from the fact that specimen A is monodisperse, i.e. composed of cylinders having the same diameter. It can be noted that the stress field is not periodic. This is due to the use of two constitutive materials (POM and HDPE) which were randomly placed in the frame. More complex stress patterns were obtained for the other two specimens. In particular, higher stresses were obtained in specimen C. This increase in stress is related to the percentage of void in the granular material (the higher the percentage of voids in the granular material, the higher the stress intensity in the cylinders for a given macroscopic load), but also to the specific statistical distribution of the stresses in the granular material.

22.4 Simulations

Simulations were performed using a molecular dynamics software [16, 17]. The measurement of the contact stiffnesses between two cylinders was preliminary done from compression tests on two cylinders in contact along a generatrix. The calculation is based on several hypotheses. The particle is considered as a rigid body together with non-conforming surface:

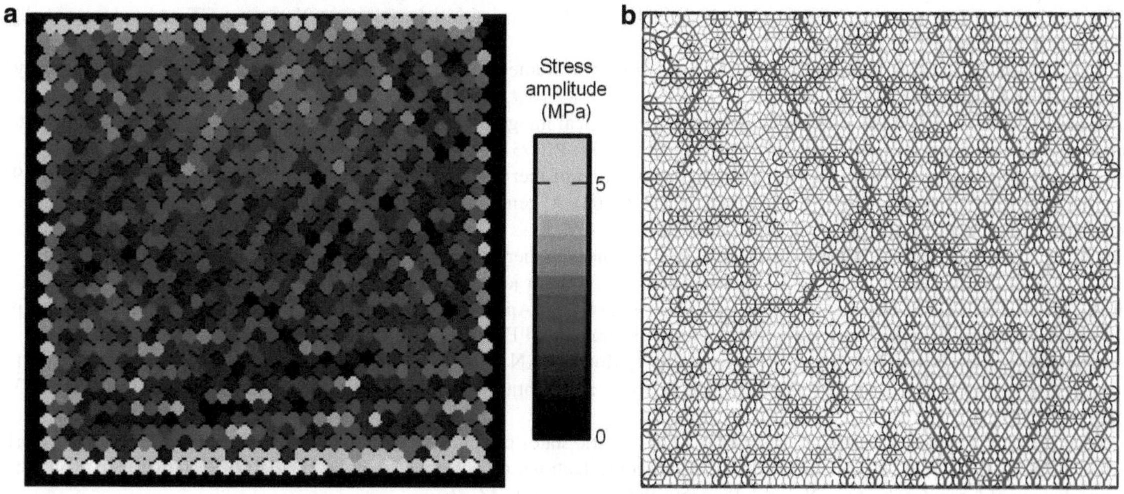

Fig. 22.2 (**a**) Stress distribution for specimen A subjected to confined compression; the color scale was yielded in order to remove the cylinders along the four boundaries from the analysis. (**b**) Force contact network obtained from molecular dynamics simulation for specimen A; the thickness of the red lines is proportional to the force amplitude

the deformation of the particles is very small and perfectly elastic. The contact force is only a function of the virtual overlap depth between particles. A regularized Coulomb's law of friction is considered: the friction force can be expressed as a mono-valued function of the relative tangential displacement. From a numerical point of view, an explicit algorithm is used; it is based on the predictor-corrector scheme with Gear's set of corrector coefficients for the integration of Newton's equations of motion. In practice, the calculation is done following an iterative scheme. Figure 22.2b presents the contact force network obtained for specimen A. The thickness of the red lines is proportional to the force amplitude. It can be observed that the main force chains consist of inclined straight lines as is was observed in the experiment. The maximum stresses which were experimentally observed in the experiment along the four boundaries are not obtained in the simulation. However, simulations and experiments are in global agreement.

22.5 Conclusion

The present study aimed at experimentally evidencing the stress networks in 2D granular materials under confined compression using infrared thermography. Three specimens were prepared with cylinders made of two different constitutive materials and different diameters. A thermoelastic stress analysis was performed. The method enabled us to reveal the stress network under loading. The experimental technique which is used here is new for granular materials to the best knowledge of the authors, and is of course complementary to other full-field measurement methods such as photoelasticimetry, digital image correlation, etc. The technique is also complementary to simulations, as it was initiated in the present study.

Acknowledgements The authors gratefully acknowledge the Thailand Research Fund through the Royal Golden Jubilee Ph.D. Program (Grant No. PHD/0159/2552) and French Embassy in Thailand for their support during this research.

References

1. Dantu P (1957) Contribution à l'étude mécanique et géométrique des milieux pulvérulents. Proceedings of the 4th International Conference on Soil Mechanics and Foundation Engineering, tome 1, Butterworth, London, 144–148
2. Majmudar TS, Behringer RP (2005) Contact force measurements and stress-induced anisotropy in granular materials. Nature 435:1079–1082
3. Dijkstra J, Broere W (2010) New method of full-field stress analysis and measurement using photoelasticity. Geotech Test J 33:469–481
4. Hall SA, Wood DM, Ibraim E, Viggiani G (2010) Localised deformation patterning in 2D granular materials revealed by digital image correlation. Granul Matter 12:1–14
5. Richefeu V, Combe G, Viggiani G (2012) An experimental assessment of displacement fluctuations in a 2D granular material subjected to shear. Geotechnique Letters 2:113–118
6. Wolf H, König D, Triantafyllidis T (2003) Experimental investigation of shear band patterns in granular material. J Struct Geol 25:1229–1240
7. Hall SA et al (2010) Discrete and continuum analysis of localised deformation in sand using X-ray mu CT and volumetric digital image correlation. Geotechnique 60:315–322
8. Słomiński C, Niedostatkiewicz M, Jacek T (2006) Deformation measurements in granular bodies using a particle image velocimetry technique. Arch Hydro-Eng Environ Mech 53:71–94
9. Sepulveda F, Fudym O (2011) Infrared tracking from morphological image processing tools—application to heat transfer characterization in granular media. Heat Transfer Eng 32(9):787–799, article number: PII 931188544
10. Dulieu-Barton JM, Stanley P (1998) Development and application of thermoelastic stress analysis. J Strain Anal Eng Des 33(2):93–104
11. Emery TR, Dulieu-Barton JM, Earl JS, Cunningham PR (2008) A generalised approach to the calibration of orthotropic materials for thermoelastic stress analysis. Compos Sci Technol 68(3–4):743–752
12. Barone S, Patterson EA (1998) Polymer coating as a strain witness in thermoleasticity. J Strain Anal Eng Des 33(3):223–232
13. Boyd SW, Dulieu-Barton JM, Rumsey L (2006) Stress analysis of finger joints in pultruted GRP materials. Int J Adhes Adhes 26(7):498–510
14. Moutrille MP, Balandraud X, Grédiac M, Derrien K, Baptiste D (2008) Applying thermoelasticity to study stress relief and crack propagation in aluminium specimens patched with composite material. J Strain Anal Eng Des 43(6):423–434
15. Fruehmann RK, Dulieu-Barton JM, Quinn S, Peton-Walter J, Mousty PAN (2012) The application of thermoelastic stress analysis to full-scale aerospace structures, Conference on Modern Practice in Stress and Vibration Analysis (MPSVA 2012). J Physics Conf Series 382, Article Number: 012058
16. Preechawuttipong I, Peyroux R, Radjai F, Rangsri W (2007) Static states of cohesive granular media. J Mech Sci Technol 21(12):1957–1963
17. Preechawuttipong I, Peyroux R, Radjai F (2001) Microscopic features of cohesive granular media, 4th International Conference on the Micromechanics of Granular, Kishino Y. (ed.), Powders and Grains 2001, 43–46

Chapter 23
Influence of Relative Humidity on the Thermomechanical Behavior of PA6.6

Adil Benaarbia, André Chrysochoos, and Gilles Robert

Abstract An experimental protocol was developed to achieve complete energy balances associated with low cycle fatigue (LCF) of a polyamide 6.6 matrix (PA6.6). The protocol involves quantitative infrared techniques (IRT), and digital speckle image correlation (DIC). IRT data were used with a local heat diffusion equation to estimate strain-induced heat sources, namely dissipation and coupling sources, while DIC enabled strain and stress assessments. Both techniques were then successfully combined to quantify deformation, dissipated and stored energies and then to estimate the Taylor-Quinney ratio that is widely used in plasticity.

In this paper, the effects of loading frequency and relative humidity were investigated. It was shown that an increase of relative humidity resulted in a decrease in the mean stored energy rate per cycle, while the stored energy ratio was much smaller at low than at high loading frequency. In addition, it was found that this ratio could be negative at the last fatigue stage, just before macroscopic crack inception. These energy properties will act safeguards for the future development of a thermomechanical model of PA6.6 matrix behavior.

Keywords Infrared thermography • PA6.6 matrix • Relative humidity • Stored energy rate per cycle • Digital image correlation

23.1 Introduction

Semi-crystalline polymers represent an important class of materials which are now widespread in various industrial areas. These engineering materials are of particular interest because of their remarkable advantages, notably regarding their low density, high deformability, toughness and long lifespan. Considering the recent growing interest in these engineering materials for automotive structures, it is essential to enhance the overall knowledge of their thermomechanical constitutive behavior during repetitive cyclic loadings. The thermomechanical behavior of these thermo-hygro-sensitive materials can be completely characterized and conceivable when all factors and mechanisms affecting their lifespan are identified. Several deformation mechanisms involved during the fatigue process in such materials have been thoroughly studied in several experimental investigations. However, some physical concepts tightly related to these deformation mechanisms are yet to be completely understood and require full clarification. One promising approach for assessing these concepts is based on energy considerations. A combined description of mechanical and energy phenomena occurring during the deformation process may shed greater light on the behavior of these polymeric materials.

A. Benaarbia (✉)
LMGC, CNRS, Université Montpellier 2, Place Eugène Bataillon, 34095 Montpellier, France

Solvay Engineering Plastics, Technyl Innovation Center-Simulation et Validation des Applications, Avenue Ramboz – BP64, 69192 Saint Fons, France
e-mail: abenaarb@univ-montp2.fr

A. Chrysochoos
LMGC, CNRS, Université Montpellier 2, Place Eugène Bataillon, 34095 Montpellier, France

G. Robert
Solvay Engineering Plastics, Technyl Innovation Center-Simulation et Validation des Applications, Avenue Ramboz – BP64, 69192 Saint Fons, France

N. Sottos et al. (eds.), *Experimental and Applied Mechanics, Volume 6: Proceedings of the 2014 Annual Conference on Experimental and Applied Mechanics*, Conference Proceedings of the Society for Experimental Mechanics Series,
DOI 10.1007/978-3-319-06989-0_23,

Many experimental surveys based on energy concepts are needed to improve constitutive material models, thus facilitating the use and application of such engineering materials. More precisely, estimations on stored and dissipated energies may provide crucial information on microstructural changes and the nature of their irreversibility. The most significant developments in the computation and interpretation of stored and dissipated energies are closely related to calorimetric procedures. The earliest published studies on this subject were focused on metallic materials. *Farren and Taylor (1925)* [1] and *Taylor and Quinney (1934)* [2] were the first to build an apparatus to measure dissipated energy during the deformation of metallic materials subjected to quasi-static monotonous tensile tests. *Williams (1967)* [3] and *Leach (1970)* [4] reviewed calorimetric methods in detail and assessed calorimetry procedures and equipment applicable for heat energy measurement. Most investigations have focused on the Taylor-Quinney ratio, which expresses the fraction of the anelastic deformation energy rate irreversibly converted into heat, i.e. dissipated. In addition, these early works were limited by the fact that they were mostly carried out with standard and traditional measurement devices that could not reliably obtain high precision estimates of this ratio. However, with the introduction of high resolution infrared scanners in the early 1980s, extensive experimental studies rectified these deficiencies and assessed the Taylor-Quinney coefficient with greater accuracy (see e.g. [5–10]).

In the present paper, we document some findings pertaining to the thermomechanical behavior of a PA6.6 matrix having different hygrothermal states (relative humidity: dry-as-molded, RH50 and RH80). Specifically, we used infrared and CCD cameras to simultaneously record, during cyclic loading, fields corresponding to temperature variations and displacement over the sample gauge part. We focused particularly on the energy balance form associated with the mechanical hysteresis loop. In particular, we obtained energy balances at two different loading frequencies, thus providing the deformation, dissipated and stored energy time patterns. We finally derived time courses of the Taylor-Quinney ratio in order to quantify the relative importance of both mean dissipated and stored energy rates over each complete loading cycle.

In the following, we briefly outline the theoretical framework used to interpret the experiments. We then describe the experimental setup and the tests performed. Finally, the main energy results are shown and discussed with respect to the mechanical hysteretic response.

23.2 Governing Equations

23.2.1 Energy Balance

Energy storage depicts the internal energy changes induced by anelastic transformations. An essential feature of this stored energy is that it remains in the deformed material, at least temporarily, until possible release during restoration processes. Generally, the stored energy amounts are obtained as the difference between the deformation energy spent during the material transformation and the dissipated energy. The total deformation energy $\mathcal{W}_{\mathrm{def}}$ is usually determined from force-displacement data (and/or from stress–strain data for local analyses). This deformation energy can be split into the elastic $\mathcal{W}_{\mathrm{e}}$ and anelastic $\mathcal{W}_{\mathrm{an}}$ energies, where $\mathcal{W}_{\mathrm{an}}$ is both stored and dissipated. In this context, the stored energy rate $\mathcal{W}_{\mathrm{s}}'$ reads [5]:

$$\mathcal{W}_{\mathrm{s}}' = \mathcal{W}_{\mathrm{def}}' - \mathcal{W}_{\mathrm{e}}' - \mathcal{D}_1 \tag{23.1}$$

According to the energy rate balance equation Eq. 23.1, a correct estimate of $\mathcal{W}_{\mathrm{def}}', \mathcal{D}_1, \mathcal{W}_{\mathrm{e}}'$ leads to indirect assessment of $\mathcal{W}_{\mathrm{s}}'$. Since the mechanical and heat energies are measured by two different techniques, systematic errors in one of the quantities do not cancel systematic errors in the measurement of stored energy. In the following, only the assessment over each complete cycle of the above volume energy rates will be considered. Thus, the elastic component per cycle remains close to zero and it can be considered that the mechanical and anelastic energy rates per unit volume per cycle are approximately the same $\left(\mathcal{W}_{\mathrm{def}}'\right)^{(n)} \approx \left(\mathcal{W}_{\mathrm{an}}'\right)^{(n)}$ (*n* representing the considered cycle). The key issue here not only concerns computing the stored energy rate per unit volume and per cycle but also evaluating its contribution in the global energy balance. We thus define the variable stored energy rate ratio which is expressed by $\beta_{\mathrm{s}}^{(\mathrm{n})} = \left(\mathcal{W}_{\mathrm{s}}'\right)^{(\mathrm{n})} / \left(\mathcal{W}_{\mathrm{def}}'\right)^{(\mathrm{n})}$. The time courses of this ratio will be shown and analyzed in the following. Note that $\beta_{\mathrm{s}}^{(\mathrm{n})}$ can be related to $\beta_{\mathrm{diff}}^{(\mathrm{n})}$, introduced by Rittel in [8], to describe the dissipated part of the deformation energy rate: $\beta_{\mathrm{diff}}^{(\mathrm{n})} = 1 - \beta_{\mathrm{s}}^{(\mathrm{n})}$.

23.2.2 Thermodynamic Foundations of the Heat Equation

The Thermodynamics of Irreversible Processes framework [11] is hereafter used to introduce the energy terms. Under the hypothesis of small deformations, the thermal, mechanical and microstructural states of the material are described by the following observable state variables: the absolute temperature $T = \alpha_o$, the small strain tensor $\boldsymbol{\varepsilon} = \boldsymbol{\alpha}_1$ and a set of internal variables $\{\alpha_k\}_{k=1,\ldots,m}$. The chosen thermodynamic potential is the specific Helmholtz free energy $\psi = \psi(T, \boldsymbol{\varepsilon}, \alpha_k)$. Using the scalar product of the following "flux" vector $(\dot{\boldsymbol{\varepsilon}}, \dot{\alpha}_k, \mathbf{q})$ and its conjugate vector of thermodynamic forces $(\boldsymbol{\sigma}-\rho\psi_{,\boldsymbol{\varepsilon}}, \rho\psi_{,\alpha_k}, -T^{-1}\mathbf{grad}T)$, the classical Clausius-Duhem inequality leads to [11]:

$$\mathcal{D} = \underbrace{(\boldsymbol{\sigma} - \rho\psi_{,\boldsymbol{\varepsilon}}) : \dot{\boldsymbol{\varepsilon}} - \rho\psi_{,\alpha_k}\dot{\alpha}_k}_{\mathcal{D}_1} \underbrace{-T^{-1}\mathbf{grad}T.\mathbf{q}}_{\mathcal{D}_2} \geq 0 \tag{23.2}$$

where $\boldsymbol{\sigma}$, $\mathbf{q}$ and ρ are the total stress tensor, the heat flux vector and the material density, respectively, while $\dot{\mathbf{x}}$ stands for the time derivative of the state variable $\mathbf{x}$, and the operator : denotes the tensorial contraction over two indices.

We assume generally that the intrinsic $\mathcal{D}_1$ and thermal $\mathcal{D}_2$ dissipation are separately non-negative, i.e. $\mathcal{D}_1 \geq 0$ and $\mathcal{D}_2 \geq 0$. Next, we define the specific heat as $C_{\boldsymbol{\varepsilon},\alpha} = -T\psi_{,TT}$ at constant $\boldsymbol{\varepsilon}$ and α_k. By combining both principles of thermodynamics, the standard local heat diffusion equation can be directly obtained as follows:

$$\rho C_{\varepsilon,\boldsymbol{\alpha}}\dot{T} + \mathrm{div}\left(\underbrace{-\boldsymbol{k}\mathbf{grad}T}_{\mathbf{q}}\right) = \underbrace{(\boldsymbol{\sigma} - \rho\psi_{,\varepsilon}) : \dot{\boldsymbol{\varepsilon}} - \rho\psi_{,\alpha_k}\dot{\alpha}_k}_{\mathcal{D}_1} + \underbrace{\rho T\psi_{,T,\varepsilon} : \dot{\boldsymbol{\varepsilon}} + \rho T\psi_{,T,\alpha_k}\dot{\alpha}_k}_{w'_c} + r_e \tag{23.3}$$

where $\boldsymbol{k}$ denotes the heat conduction tensor and r_e the heat supply term characterizing the volume heat exchange with the surroundings.

The terms on the left hand side represent a partial differential operator applied to the temperature. The thermal inertia $\rho C_{\boldsymbol{\varepsilon},\boldsymbol{\alpha}}\dot{T}$ reflects the stored or released heat rate. The term - div($\boldsymbol{k}$ **grad**T) stands for the volume heat loss by conduction. The terms on the right hand side pool the different heat sources. In turn, we find the intrinsic mechanical dissipation $\mathcal{D}_1$, which reflects the irreversible transformation into heat of the deformation energy, the thermomechanical coupling sources $\mathcal{W}'_c$, which are the calorific signatures of possible interactions between temperature and the other state mechanical or microstructural variables.

Under some reasonable hypotheses, the previous heat balance equation Eq. 23.3 can be reduced to three different simplified formulations, i.e. two, one and/or zero-dimensional diffusion equations. Readers interested in these formulations are referred to [12, 13]. The choice of a particular model strongly depends on the type of experimental investigation under consideration. The passage from a 2D to a 0D model is possible since heat source fields remain homogeneous throughout the test. By contrast with the 2D and/or 1D diffusion models, the 0D heat equation is less sensitive to measurement noise and less time consuming in terms of computation duration. This approach is well adapted to so-called homogeneous tests when a rapid experimental estimate of heat sources is required. This global approach is generally based on the hypothesis that for uniaxial tests the heat source distributions are uniform throughout the test duration within the specimen gauge area. In the following, θ denotes the spatial average of temperature variations over the gauge part (x, y and z coordinates). The local heat diffusion equation Eq. 23.3 can be simplified to the following differential equation:

$$\rho C_{\boldsymbol{\varepsilon},\boldsymbol{\alpha}}\left(\dot{\theta} + \frac{\theta}{\tau_{\mathrm{th}}^{0\mathrm{d}}}\right) = \mathcal{D}_1 + \mathcal{W}'_{\mathrm{ther}} \tag{23.4}$$

where $\tau_{\mathrm{th}}^{0\mathrm{d}}$ represents a time constant characterizing the linear heat losses perpendicular to the specimen surface, while the $\mathcal{W}'_{\mathrm{ther}}$ term denotes the thermoelastic coupling source.

23.3 Experimental Procedure

The experimental setup used for conducting the thermomechanical tests is shown in Fig. 23.1. Samples were loaded using an MTS 810 hydraulic testing machine equipped with a load cell of ±25 kN. Temperature variations on one side and displacements on the other side of the sample were measured and captured simultaneously using both a high resolution infrared focal plane array camera and a high resolution visible CCD camera, respectively.

23.3.1 IRT Techniques

Infrared thermography is a convenient non-contact imaging technique that uses an infrared matrix element to visualize and estimate temperature fields of material surfaces. The principle of the technique is based on use of the intensity and wavelength of radiation emitted by the loaded material to measure its surface temperature according to *Planck's law*. In this survey, the infrared thermal imaging system used for thermographic measurement was a *CEDIP Jade* camera, cooled with a *Stirling cooler*. This contained InSb matrix elements which are sensitive to infrared radiation in the 3–5 μm wavelength range. The camera was equipped with a 25 mm lens and positioned on a tripod approximately 0.65 m from the specimen surface. We decided to operate the detector with a 160 × 128 pixel window size, 29 μm pixel size and 400 μs integration time. The IR camera frame rate was set at tenfold the loading frequency to obtain enough sampled data per cycle and to improve the signal-to-noise ratio using local least-squares fitting.

During the experiments, several factors complicated measurement of the surface temperature of the investigated materials. Several experimental precautions were thus taken to overcome these difficulties. For instance, the heat flux density transferred to the infrared sensor included noise components due to the reflected incident energy from the surroundings. The IR measurements were therefore carried out indoors, and the experimental tools were screened and isolated far from the radiation sources of the surroundings. Besides, the camera was turned on for 4 h before starting the experiments in order to reach thermal equilibrium. The temperature was quantitatively measured after improving the surface emissivity. The tested specimens were coated with matt black paint, and we assumed that the specimens were as perfect as possible black bodies with an emissivity value close to 1. Even though these precautions were taken, the measured data were not necessarily quantitatively reliable and the true temperature value could not be derived directly. An appropriate calibration protocol was thus used to determine the true temperature value. This allowed conversion of the thermal radiation digitized by the infrared camera into temperature using pixel calibration functions [14]. This pixel protocol was based on a polynomial fitting of the digital levels $\mathcal{DL}_{(m,n)}$ delivered by each individual detector element (m, n) using a black body

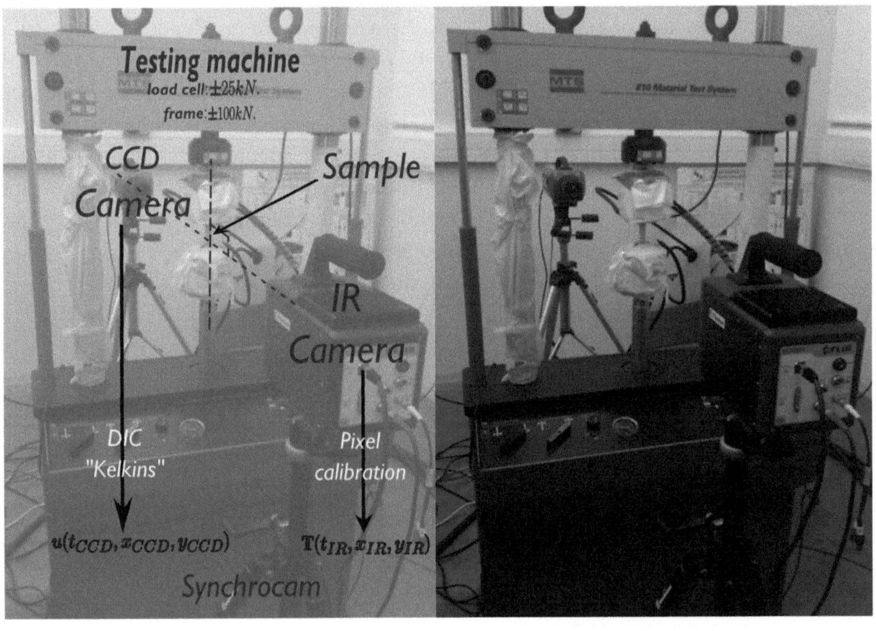

Fig. 23.1 Experimental setup showing both cameras used for recording the in-plane displacement and temperature fields

with a uniform high emissivity coating, maintained at cold and hot temperatures $T_{(m,n)}$. The measurement accuracy could thus be assessed for each pixel individually in the resulting image.

$$T_{(m,n)} = \sum_{k=0}^{d_f} \varkappa_{(m,n)_k} \mathcal{DL}_{(m,n)} \tag{23.5}$$

where d_f denotes the degree of the polynomial fitting function and $\varkappa_{(m,n)_k}$ are coefficients of the calibration law for the element (m, n). These coefficients were derived from least squares fitting approximations.

Two bad pixel detection criteria were used. The first consisted of the elimination of pixels whose response predicted by the polynomial fitting $T_{(m,n)}$ was far from the ordered temperature of the used black body T_{bb}. Whereas the thermal response of the second eliminated pixels was too far from the calculated average temperature throughout the detector T_{av}. In agreement with the foregoing, both criteria were written as:

$$\left|T_{(m,n)} - T_{bb}\right| \geq \delta T_1, \quad \left|T_{(m,n)} - T_{av}\right| \geq \delta T_2 \tag{23.6}$$

where δT_1 and δT_2 were respectively the threshold values of the first and second criteria.

The room temperature was taken into account by recording the temperature data of an unloaded specimen (*dummy*) kept nearby the test specimen. The temperature images were then acquired using *Altair* software and stored on a computer hard drive. A local least-square fitting method using an optimized approximation function set was developed to filter the IR images from the noise [15].

23.3.2 DIC Techniques

Digital image correlation techniques (DIC) are now well-known non-contact full-field measurement tools which offer the possibility of determining displacement and strain fields at the surface of loaded specimens. It consists of capturing consecutive images of a random surface pattern recorded using a charge coupled device (CCD) camera. To apply this method, the random pattern (local optical signature of a material surface element) was achieved by spraying the specimen surface with a painted speckle pattern (pulverization of black and white paints). These patterns were chosen in order to obtain well illuminated visible images with high contrast. The change in surface characteristics of the specimen was evaluated by matching different sub-cells of two images captured before and after loading the specimen. As it is almost impossible to find the matched zone using few pixels, a subset with multiple pixel points was defined to perform the matching process. Adequate DIC software was required to obtain information on the positional changes in the material surface element under investigation. In our case, we used the tailored *Kelkins* software package developed by LMGC. Displacement fields were thus determined with suitable algorithms after digitizing images captured by the visible camera [16]. Since the experimental data were inherently noisy, appropriate smoothing methods were needed to reduce the effect of noise on the strain estimate. As for the thermal data, we used projections of data on a local analytical function set in a least-squares sense: (1) to approximate the experimental displacement values in the selected regions, (2) to reduce the noise present in these displacements, and (3) to derive the measured strain and strain rate.

23.3.3 Synchronization System

To monitor the energy behavior of studied materials, it is crucial to assess the applied load, temperature and strain at the same time. Therefore, it is of paramount importance to synchronize the IR and visible cameras with the load cell sensor of the hydraulic testing machine (MTS). A specific tailored electronic device named *synchrocam* was designed to simultaneously capture the three data types. Both imaging systems were then triggered from a digital pulse to capture data concurrently. This ensured that all data captured by the imaging systems and *MTS* were initiated at the same time. Fortunately, in the current investigation both systems were initiated at the same time with identical frame rates, which facilitated the image processing.

In addition, the geometric transformation between the frames of reference of the two cameras was determined using a calibrated target. The correspondence between the spatial coordinates of the thermal image (x_{IR}, y_{IR}) and those of the visible image (x_{CCD}, y_{CCD}) was established by trigging the motion of the calibrated target using the following equations:

$$\begin{pmatrix} x_{CCD} \\ y_{CCD} \end{pmatrix} = \underbrace{\begin{pmatrix} \gamma_x & \eta_x \\ \gamma_y & \eta_y \end{pmatrix}}_{\mathcal{R}} \begin{pmatrix} x_{IR} \\ y_{IR} \end{pmatrix} + \underbrace{\begin{pmatrix} \lambda_x \\ \lambda_y \end{pmatrix}}_{\mathcal{D}} \quad (23.7)$$

where $\mathcal{R}$ is the rotation matrix and $\mathcal{D}$ a displacement vector. This spatial correspondence was carried out before each new test.

23.4 Experimental Results

The experimental study involved observation and analysis of the thermomechanical and energy properties of the PA6.6 matrix involved during cyclic fatigue tests. The investigations were essentially divided into two parts. At two different loading frequencies, the first part examined the contribution of the stored and dissipated energies in the global energy balance associated with dry specimens of polyamide 6.6. The second part was focused on highlighting the effect of relative humidity on PA6.6 specimens having different hygrothermal states (relative humidity: dry-as-molded, RH50 and RH80). The cyclic tensile-tensile loads were carried out under a stress ratio of 0.1. The loading frequency ranged from 1 to 10 Hz, while the number of cycles was 10^4. The cycle number was high enough to raise the kinematic and thermal properties of the investigated materials.

23.4.1 Thermal Response

For the sake of simplicity, the mean temperature variations averaged over the sample gauge part are first considered. Typical plots associated with these thermal variations as a function of time (image number) and at loading rates of 1 and 10 Hz and for the same stress range $\Delta\sigma$ are shown in Fig. 23.2 . For confidentiality reasons, the stress range will not be given.

Conventionally, temperature variations involve low drifts with an oscillation in opposite phase with the monochromatic loading signal. We associated two independent heat sources with both of these thermosignal characteristics. The oscillation is induced by the thermoelastic source coming from the material thermodilatability. The progressive heating of the specimen is associated with the dissipated energy. It is often characterized by a gradual global increase in temperature variations (blue dashed line in Fig. 23.2). A decrease in mean temperature is associated with a decrease in the dissipation intensities, with the heat exchange coefficients between the specimen and the surroundings assumed to be constant throughout the test. Comparing the trend of thermal curves in dry-as-moulded speciments subjected to low (Fig. 23.2, plot (a)) and high loading rates (Fig. 23.2, plot (b)), it was shown that the temperature variation values were similar throughout the test. For the same

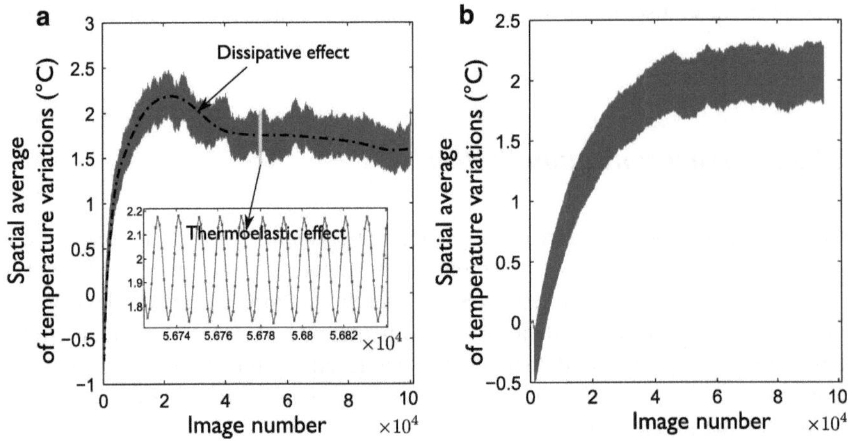

Fig. 23.2 Spatial averages of temperature variations showing both dissipation and coupling sources for dry PA6.6 matrix specimens subjected to (**a**) low loading frequency of 1 Hz, (**b**) high loading frequency of 10 Hz, while the stress ratio was $R_\sigma = 0.1$

stress range, the thermoelastic amplitude was the same for both low and high loading frequencies with induced temperature variations of 0.4 °C. However, the self heating induced by the dissipative effect was about 2 °C. From a calorimetric viewpoint, it was found that the intrinsic dissipation intensities remained very low compared to the thermoelastic source ranges.

23.4.2 *Energy Rate Balance*

23.4.2.1 Loading Frequency Effect

The energy balances were established for 1 and 10 Hz loading frequencies. All experimental data were converted and used to compute the previous energies used in Eq. 23.4. The dark curve corresponds to the mean deformation energy rate per cycle, the green curve shows values of the stored energy rate per cycle and the red curve indicates the values of the mean intrinsic dissipation per cycle. All energies are plotted as a function of the cycling number, and the time courses associated with their variable ratios are given in Figs. 23.3, plots (b, d) and 23.4. Note that the heat sources were divided by ρC and thus expressed in °C/s in order to define, for each energy, an equivalent heating rate, thus facilitating comparisons between the dissipative and stored energy rates.

From all the plots, it can be observed that the overall shape of the energy rate balance closely depends on the loading frequency. For a dry matrix at a high loading rate ($f_L = 10Hz$), the results indicate a substantial part of the mean stored energy rate per cycle with a ratio near 83 % (Fig. 23.3, plot (b)). This part dominates the mean intrinsic dissipation per cycle whose

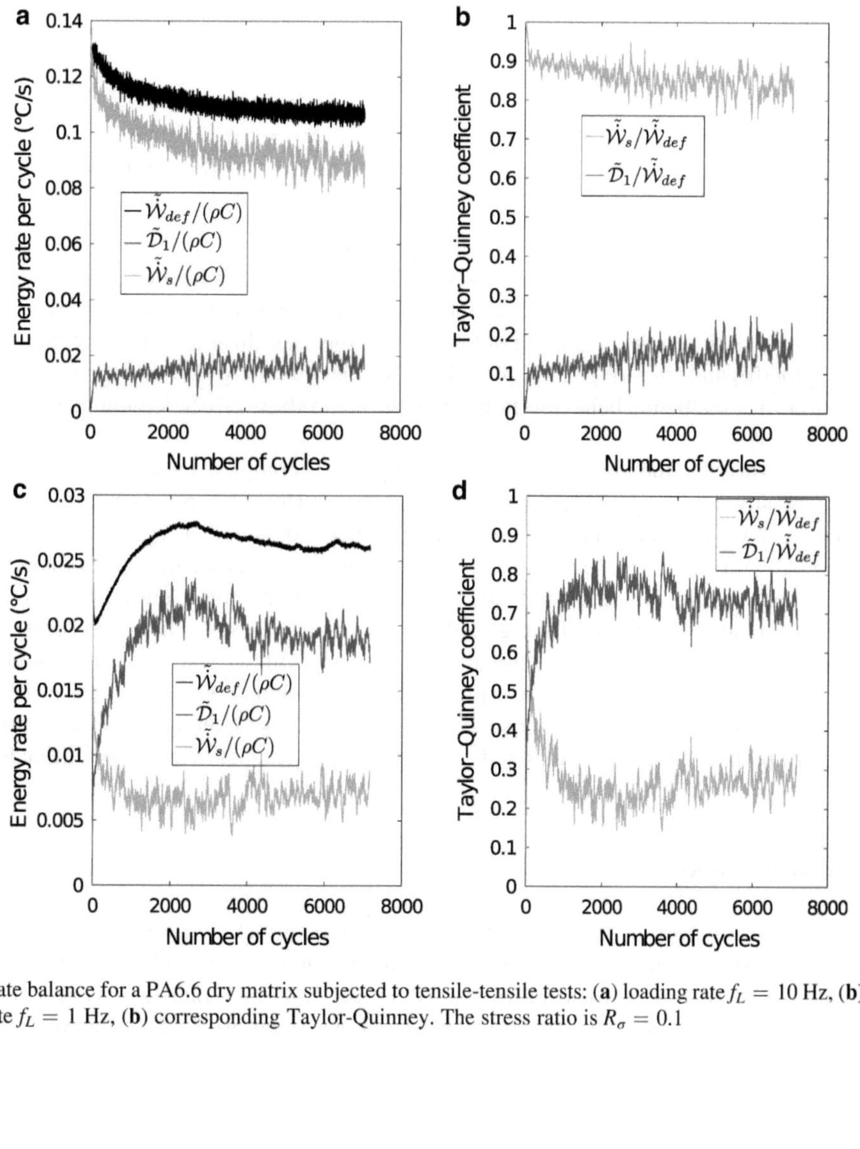

Fig. 23.3 The energy rate balance for a PA6.6 dry matrix subjected to tensile-tensile tests: (**a**) loading rate $f_L = 10$ Hz, (**b**) corresponding Taylor-Quinney, (**c**) loading rate $f_L = 1$ Hz, (**b**) corresponding Taylor-Quinney. The stress ratio is $R_\sigma = 0.1$

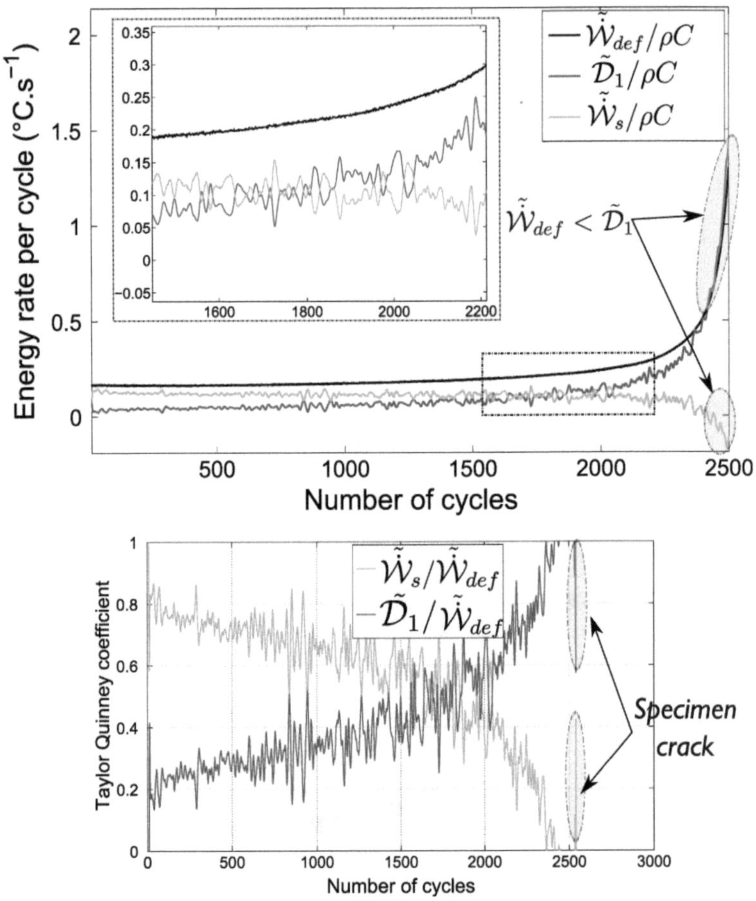

Fig. 23.4 The energy rate balance for a PA6.6 dry matrix from the beginning of cyclic loading until the last few cycles before the macroscopic crack. The stress ratio is $R_\sigma = 0.1$

ratio seems to level off at around 17 %. However, in the case of low loading frequencies ($f_L = 1\ Hz$), the trend of these variable ratios changes completely (Fig. 23.3). The mean intrinsic dissipation per cycle is now predominant compared to the stored energy rate. It starts increasing at the beginning of loading and then stabilizes after the first thousands of cycles, with a ratio of 73 % (Fig. 23.3, plot (b)) and 65 % (Fig. 23.3, plot (d)). Note that this energy stabilization was never accompanied by a stabilized cyclic shakedown, a ratcheting behavior induced by the mean stress effect, leading to a regular drift in the cyclic hysteresis loop.

Our studies so far have shown that the mean mechanical energy over a complete cycle exceeded the heat generated. The strained sample could generally not dissipate more heat than it received. However, in some cases the heat was found to be generated at a higher rate than the mechanical energy, indicating a release of stored energy (see e.g. [8, 17, 18] for polymers and [19] for metals). Fig. 23.4 illustrates a case of a dry PA6.6 specimen that released more heat than the mechanical energy at the last stages of its fatigue life. Thermal and kinetic images were recorded until final rupture of the specimen (2,531 cycles), allowing us to study the trend of the global energy balance close to the macroscopic crack. In this figure, the ratio of the stored energy rate over a complete cycle falls continuously with the test time. The stored energy rate is high at the beginning of loading and then decreases as the test proceeds until reaching negative values at the end of loading. This highlights a rapid increase in intrinsic dissipation, thus indicating an amount of internal energy converted into heat. Even at low frequencies, the stored proportion remains high and could thus represent a slow but regular material degradation.

23.4.2.2 Relative Humidity Effect

In what follows, we set the loading rate at 10 Hz and we highlighted the effect of relative humidity upon the kinetics and contributions of stored and dissipated energy rates. For a dry-as-molded matrix, the results indicated a substantial part of the mean stored energy rate per cycle with a stored ratio near 70 % (see Fig. 23.5, plot (a)). This part dominated the mean intrinsic dissipation per cycle whose ratio seemed to be stabilized at around 30 %. This finding is compatible with that

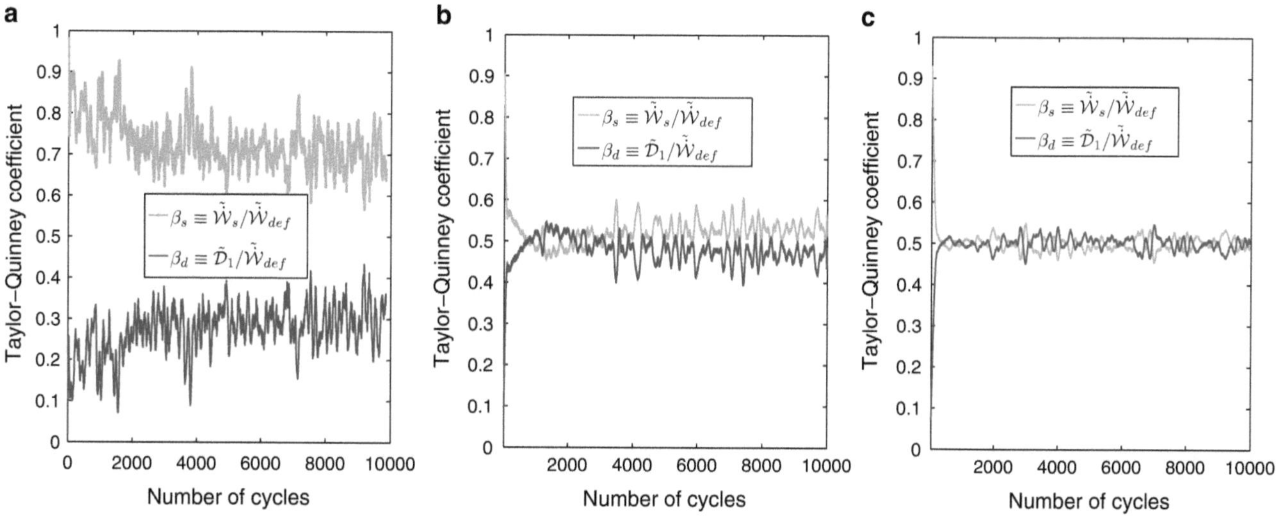

Fig. 23.5 Taylor-Quinney coefficients for the PA6.6 matrix subjected to a loading frequency of 10 Hz. The plots show results coming from (**a**) dry-as molded PA6.6 matrix, (**b**) specimen preconditioned at RH50 and (**c**) specimen preconditioned at RH80. The stress ratio $R_\sigma = 0.1$

obtained in Fig. 23.3, plot (a). Whereas, the results indicated in Fig. 23.5, plots (b, c) showed that ratios decreased at the beginning of loading and stabilized continuously after a thousand of cycles. The comparison in terms of Taylor-Quinney ratios for dry-as-molded and conditioned specimens indicated that for a certain relative humidity level, the stored energy contribution was not modified by the changing water content, as demonstrated by experiments on both RH50 and RH80. For RH50 specimens, the mean intrinsic dissipation ratio increased at the beginning of loading and then stabilized after the first few cycles, with a value of almost 50 % (see Fig. 23.5, plot (b)). This ratio had approximately the same value for specimens conditioned at RH80 (see Fig. 23.5, plot (c)). The deformation energy rate appeared to be equally converted to both dissipated and stored energy rates.

23.5 Concluding Comments

Some aspects related to thermal and energy responses associated with PA6.6 matrix specimens subjected to tensile-tensile tests were investigated here. The first aspect was the influence of the loading frequency on the energy balance form. The investigations of energy stored during deformation indicated that the stored ratio was significantly lower at low loading rates (e.g. 25 %) than those obtained at high loading rates (e.g. 80 %). In addition, it was shown that this ratio could be negative at the last fatigue stages, heralding the macroscopic crack. This negative value was interpreted as a transformation of stored into dissipated energy.

Taylor-Quinney ratios were also determined for three different hygrothermal states (RH0, RH50 and RH80). This generated some insight into the water plasticizing effect through values of the stored energy rate. For dry-as-molded and conditioned specimens, the Taylor-Quinney ratios indicated a limited influence of the relative humidity which seemed to saturate for RH greater than RH50. In other words, the stored energy amounts were no longer modified by the water content change, as shown in both experiments of conditioned specimens (RH50 and RH80), for which we obtained the same stored energy ratio. Further experiments are required to obtain more information on the humidity effects, especially in fiber-reinforced polymeric materials (DURAFIP project) in which interactions between the fiber orientation and thermo-hygro-mechanical matrix behavior have to be explored. Experiments are currently under way to gain further insight into the role of loading frequency and humidity on the cyclic behavior of these industrial composites.

Acknowledgements The authors gratefully acknowledge Solvay Engineering Plastics for supporting this work and for providing material data and specimens. This work benefited from the financial support of the French Minister for Research (ANRT) and was performed in the framework of the European DURAFIP project

References

1. Farren WS, Taylor GI (1925) The heat developed during plastic extension of metals. Proc R Soc A 107:422–451
2. Taylor GI, Quinney H (1934) The latent energy remaining in a metal after cold working. Proc R Soc A 14(3):307–326
3. Williams RO (1967) In: Herman H (ed) Experimental methods of materials research, vol 1. Interscience, New York, NY
4. J. S. Ll. Leach (1970) Physicochemical measurements in metals research, vol 4. R.A. Rapp (ed) Interscience, New York, NY
5. Chrysochoos A (1985) Energy balance for elastic plastic deformation at finite strain (in French). J Theor Appl Mech 5:589–614
6. Chrysochoos A, Maisonneuve O, Martin G, Caumon H, Chezeaux JC (1989) Plastic and dissipated work and stored energy. Nucl Eng Des 114:323–333
7. Mason J, Rosakis A, Ravichandran G (1994) On the strain and strain rate dependence of the fraction of plastic work converted to heat: : an experimental study using high speed infrared detectors and the Kolsky bar. Mech Mater 17:135–145
8. Rittel D (1999) On the conversion of plastic work to heat during high strain rate deformation of glassy polymers. Mech Mater 31:131–139
9. Rosakis P, Rosakis AJ, Ravichandran G, Hodowany J (2000) A thermodynamic internal variable model for the partition of plastic work into heat and stored energy in metals. J Mech Phys Solids 48:581–607
10. Oliferuk W, Maj M, Raniecki B (2004) Experimental analysis of energy storage rate components during tensile deformation of polycrystals. Mater Sci Eng A 374:77–81
11. Halphen B, Nguyen QS (1975) On the generalized standards materials (in French). Journal de Mécanique 14(I):39–63
12. Chrysochoos A, Louche H (2000) An infrared image processing to analyse the calorific effects accompanying strain localisation. Int J Eng Sci 38:1759–1788
13. Berthel B, Chrysochoos A, Wattrisse B, Galtier A (2008) Infrared image processing for the calorimetric analysis of fatigue phenomena. Exp Mech 48:79–90
14. Honorat V, Moreau S, Muracciole JM, Wattrisse B, Chrysochoos A (2005) Calorimetric analysis of polymer behaviour using a pixel calibration of an IRFPA camera. Qirt J 2:153–171
15. Boulanger T, Chrysochoos A, Mabru C, Galtier A (2004) Calorimetric analysis of dissipative and thermoelastic effects associated with the fatigue behavior of steels. Int J Fatigue 26:221–229
16. Wattrisse B, Chrysochoos A, Muracciole J-M, Némoz-Gaillard M (2001) Analysis of strain localization during tensile tests by digital image correlation. Exp Mech 41:29–39
17. Dillon OWJ (1966) The heat generated during the torsional oscillations of copper tubes. Int J Solids Struct 2:181–204
18. Benaarbia A, Chrysochoos A, Robert G (2014) Kinetics of stored and dissipated energies associated with cyclic loadings of dry polyamide 6.6 specimens. Polym Test 34:155–167
19. Oliferuk W, Korbel A, Bochniak W (2001) Energy balance and macroscopic strain localization during plastic deformation of polycrystalline metals. Mater Sci Eng A 319:250–253

Chapter 24
Temperature Field in FSW Process: Experimental Measurement and Numerical Simulation

C. Casavola, A. Cazzato, V. Moramarco, and C. Pappalettere

Abstract Friction Stir Welding (FSW) is a relatively new welding process, which was developed at The Welding Institute (TWI), United Kingdom, in 1991. FSW is a solid-state joining process, i.e. no melting occurs. The welding process is promoted by the rotation and translation of an axis-symmetric non-consumable tool along the weld centreline. Thus the FSW process is performed at much lower temperatures than the conventional fusion welding. Nevertheless the control of the temperature field is fundamental to guarantee a high quality joint. In the present work the temperature field during the welding process was measured using an infrared camera. The test was conducted on 6 mm thick 5754 H111 aluminium alloy plates, in bead on plate configuration, with constant tool rotation rate and feed rate. Furthermore a finite element model was implemented and validated on experimental measurement data to evaluate the temperature field also into the plate.

Keywords Friction stir welding • Welded joint • Temperature field • Finite element method • Infrared thermography

24.1 Introduction

The friction stir welding (FSW) is a solid state welding process invented in 1991 by The Welding Institute [1]. In this technique, the material is not led to fusion but the joint is the combined result of the plasticization due to the heat generated by both friction and stirring action of the pin on the material, as consequence of the rotation and the displacement of the tool.

First application of the FSW was with light alloy such as aluminum, magnesium, copper and nickel [2–7]. Nowadays this technique, with some changes [8], is under study and in some cases it is experimentally used with various types of metals such as titanium [9, 10], carbon steel [11, 12], metal matrix composites (MMC) [13] and polymeric materials [14]. As FSW is expected to have several advantages than other traditional welding technique that induce a complete fusion of metal (i.e. it does not require protective atmospheres or filler materials) [15, 16], it seems to be suitable for realizing various types of joints [2] in several industrial fields like shipbuilding, offshore platforms, aerospace or railway or automotive industry [17]. Anyway, some aspects related to mechanical performance and residual stress level should be deeply investigated [8, 18, 19].

Temperature distributions and thermal histories have an important role in FSW. Some studies [20–22] have shown that the control of temperature during the FSW process is essential to obtain a good welded joint and also to control the residual stress distribution, the microstructure and the strength of weld. Xu et al. [20], using thermocouples, found that the temperature distribution during the tool inserting is only related to the rotational speed and that the temperature increases with increasing the rotational speed or decreasing the transverse speed during the stable phase of welding. They also found that the tensile strength and yield strength of joints increase with increasing the rotational speed, while the percentage of elongation decreases. Maeda et al. [21] have applied several thermocouples on both top and bottom surface of the work plates carried out that the near FSW tool zone generally shows asymmetric temperature distribution between the advancing side and the retreating side. Also Hwang et al. [22] conducted a study on temperature distribution using thermocouples and find out that the temperatures on the advancing side are slightly higher than those on the retreating side. Although experimental measures are fundamental to understand the thermal phenomenon during the FSW process, they present some limitations (e.g. internal temperature field measure). For this reason, the implementation of numerical models, that can

C. Casavola (✉) • A. Cazzato • V. Moramarco • C. Pappalettere
Dip. di Meccanica Matematica e Management (DMMM), Politecnico di Bari, Bari, Italy
e-mail: caterina.casavola@poliba.it

N. Sottos et al. (eds.), *Experimental and Applied Mechanics, Volume 6: Proceedings of the 2014 Annual Conference on Experimental and Applied Mechanics*, Conference Proceedings of the Society for Experimental Mechanics Series,
DOI 10.1007/978-3-319-06989-0_24,

predict the temperature distributions, thermal histories and stress distributions, could support the design of the tool and could estimate the correct weld parameters in order to improve mechanical properties of the welded joints. Several papers, in the past years, have addressed the numerical models in different way. Soundararajan et al. [23] have developed a thermo-mechanical numerical model validated by thermocouples and Schmidt et al. [24] have defined an analytical model for heat generation in FSW, based on different assumptions of the contact condition between the rotating tool surface and the weld piece. Song et al. [25] have carried out a three-dimensional numerical model in a moving coordinate system validated by thermography. In the above mentioned models [23, 24], the use of thermocouples is a limiting way to proceed because this allows to measure in single points losing the entire view of temperature distribution and, in special manner, the temperature in the front zone of the tool. Song et al. [25] used the thermography technique but have not simulated the real tool movement on the specimen but they realized a moving coordinate system to reduce the difficulty of modelling the moving tool.

In view of understanding the mechanical response of friction stir welded joints on 5754 H111 aluminum alloy, in particular how temperature gradient generated during welding process could affect the residual stress field and microstructural modification of the specimen, in this paper an infrared camera will be used to capture the temperature field during the welding process. Furthermore a three-dimensional finite elements thermal model will be implemented to simulate the temperature field also into the plate thickness. The FE model will be then validated on experimental measurement data.

24.2 Material and Methods

In order to study the thermal field generated during FSW process, 5754 H111 aluminum alloy plates with 6 mm thickness has been studied. Tables 24.1 and 24.2 summarize the chemical composition, thermo-physical properties and mechanical characteristics of the material at room temperature (20 °C). This material has good characteristics of weldability, cold workability, corrosion resistance and machinability making it suitable for nuclear, chemical and automotive structural applications.

In order to study the thermal gradient inside the material, bead on plate specimens were considered. Specimen geometry was 200 mm length, 100 mm width and 6 mm thick. FSW process was produced by a 4 kW machine with a threaded tool. Process parameters are summarized in Table 24.3.

Table 24.1 Chemical composition Al5754

Al [%]	Si [%]	Fe [%]	Mn [%]	Mg [%]
96.1–95.5	0.4	0.4	0.5	2.6–3.2

Table 24.2 Physic and mechanical characteristics

Parameter	Value
Density [kg/m^3]	2,650
Melting temperature [°C]	600
Thermal conductivity [W/mK]	138
Thermal expansion coefficient [1/K]	24×10^{-6}
Young's modulus [GPa]	68
σ_y [MPa]	100
σ_u [MPa]	215
HV	55

Table 24.3 Process parameter of FSW

Traverse speed [cm/min]	Dwell time [s]	Rotational speed [rpm]
20	3	500

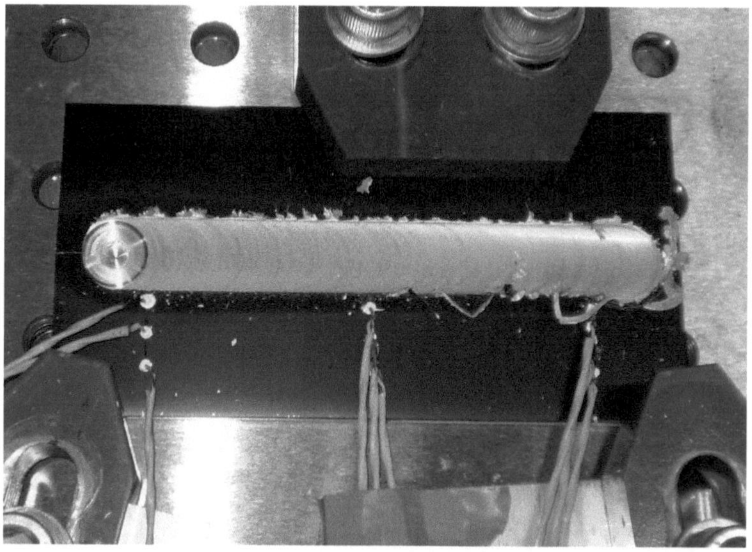

Fig. 24.1 Bead on plate welding and clamping system

Fig. 24.2 Experimental setup

24.2.1 Temperature Measurement

The thermal history and the temperature field have been acquired during welding by the infrared camera NEC H2640 (Range 0–2,000 °C; Resolution 0.06 °C; Accuracy ±2 °C or ±2 %). In Fig. 24.1, the specimen painted with matte black acrylic spray paint is shown. The coating reduces the reflection of the aluminium and increases the emissivity of the surface ($\varepsilon = 0.95$ after painting). Figure 24.2 shows the experimental setup. The infrared camera has been placed in front of welding machine, normal to the weld direction with an angle of 30° between the optical axe of the camera and the normal direction at the aluminium sheet.

24.2.2 Numerical Model

To describe the FSW thermal history, a three-dimensional transient thermal analysis has been developed using the finite elements commercial code ANSYS 14.5. Due to the symmetry of the problem, just a half plate model has been implemented, in order to decrease the elements number and reduce computational time. The model has been meshed using 6000 SOLID90 elements [26]. In order to describe more accurately the thermal behavior near the tool area, to take into account the higher gradient of temperature, a thicker mesh has been used near the plane of symmetry (Fig. 24.3).

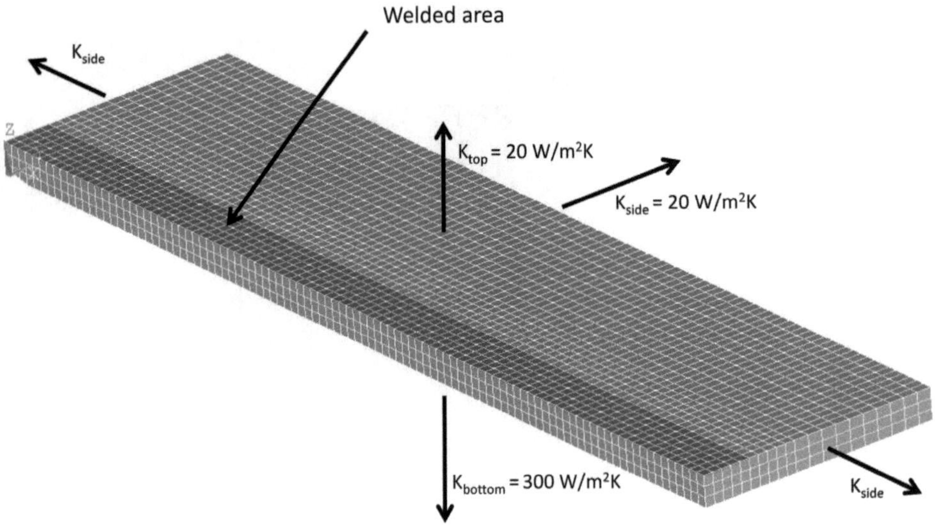

Fig. 24.3 FSW model mesh and convection coefficient

Table 24.4 Temperature dependent material properties of AA5754 [28]

Temperature [°C]	Density [kg/m³]	Specific heat [J/kgK]	Thermal conductivity [W/mK]
20		891.6	138
100		950	147.2
200	2,650	988.4	152.7
300		1015.2	162.7
400		1045.6	152.7
500		1106.6	158.75

In the present numerical model only the thermal effect of the tool has been considered (i.e. thermo-mechanical effects will not be implemented). The natural convection between aluminum and air on the top surface and on the lateral side of the specimen is equal to 20 W/m^2 °C. Furthermore, the conduction between the bottom surface of the specimen and the backing plate has been simulated with a convection coefficient of 300 W/m^2 °C. The numerical value of this one has been selected in the way to match as well as possible the maximum temperature of the numerical model with experimental data. Because the SOLID90 element does not accept at the same time thermal loads and convection, it has not been possible to insert the convection heat transfer on the welded area. The radiation heat loss has been neglected because temperature hardly exceeds 500 °C in the FSW process [27]. The AA5754 thermo-physical material properties (e.g. thermal conductivity and specific heat) have been implemented as a function of temperature (Table 24.4).

The welding process has been simulated by a moving thermal source along the welded zone. According to Schmidt et al. [24], the analytical expression used for simulate FSW tool heat generation is:

$$Q = \frac{2}{3}\mu\omega\frac{F_n}{R^2_{shoulder}}\left(\left(R^3_{shoulder} - R^3_{probe}\right)(1 + \tan\alpha) + R^3_{probe} + 3R^2_{probe}H_{probe}\right) \qquad (24.1)$$

Equation 24.1 takes in account the specific tool geometry: $R_{shoulder}$ and R_{probe} are the tool shoulder and probe radius while α and H_{probe} are the tool shoulder cone angle and probe height (Fig. 24.4). The other parameters are related to FSW process: μ and F_n are respectively the friction coefficient and the normal plunge force while ω is the rotational speed.

According with Schmidt et al. [24], the friction coefficient μ was set equal to 0.3. The tool geometrical parameters are listed in Table 24.5.

The normal force F_n has been set equal to 20,000 N in accordance with that measured during the experimental tests in the steady state conditions. Finally, the penetration and the tool pullout phase have been described in the numerical model as a linear increment of the FSW heat source in the time. The penetration speed has been set in 0.637 mm/s in the model on the base of experimental observation and the corresponding penetration time is 6.7 s. The numerical description of this behavior finds evidence in the experimental data.

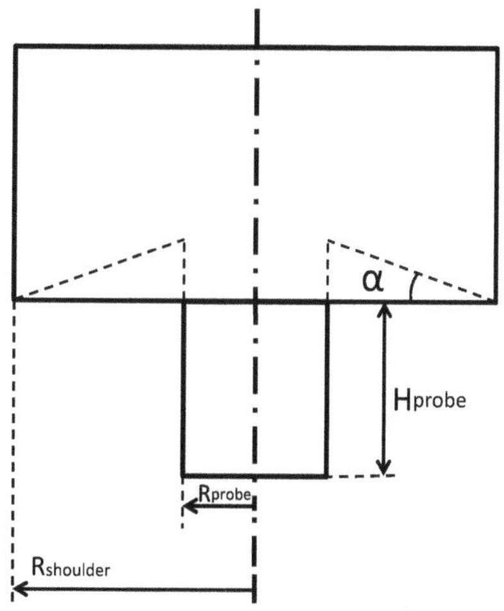

Fig. 24.4 Tool scheme

Table 24.5 Tool geometrical parameters

Parameter	Value
$R_{shoulder}$ [mm]	10.75
R_{probe} [mm]	3
α [°]	0
H_{probe} [mm]	5.8

24.3 Results

In order to verify and validate the previous described model, a FSW bead on plate specimen have been carried out. The temperature of the specimen have been measured by an infrared thermo-camera and compared with the numerical data of the model. In Fig. 24.5 has been reported a scheme of the points that have been analysed.

Three different steps have been studied and reported in the later graphs. In each step the tool has been aligned at the measured points and the temperature has been acquired in both numerical and experimental data. The start moment corresponds to the welding initial step when the FSW tool has just started the weld and it is about 30 mm from the right edge of the plate. The middle step has been selected in order to be in the half of the specimen. Finally, the end step is located at 30 mm from the left of the plate. For each step the temperature of four different points has been acquired, the firsts three points have been spaced of 10 mm from each other, the last one has been positioned at the edge of the specimen. It has not been possible to collect experimental data nearer the center line for the presence of the FSW tool and the processing burrs that modifies the emissivity of the surface and thus the temperature.

The peak temperature reached at a specific position during welding is a very important issue that determines the material microstructure and also the mechanical properties of the joints. Therefore, the graphs of the maximum temperatures versus distance from the weld line have been plotted in Figs. 24.6, 24.7 and 24.8 describing the start, middle and end step. In these have been plotted both numerical and experimental data to compare and validate the model. Generally, these show that the temperature decreases moving away from the welding line and that the numerical model fits the experimental data with low errors. In fact, in the start step (Fig. 24.6) the maximum error between experimental and numerical data is 5.9 %, in the middle step it is 6.2 % (Fig. 24.7) and in the end step, Fig. 24.8, the error is 6.3 %. The temperatures in the start step are generally lower than that the other two phases (that are almost the same) and this implies that, after the start step, a steady heat transfer could be reached in the FSW process. The temperature gradient is maximum in the area near the tool while, moving away from this, it decrease. For example, for the middle step, the temperature gradient near the tool is 9.1 °C/mm while, at the edge of the plate, it is 2.35 °C/mm.

In order to accurately validated the numerical model, in Fig. 24.9 it has been plotted the temperature vs. time of the experimental data and the numerical model. This represents the temperatures of a central point distant 16 mm from the center (point A in Fig. 24.5). The comparison shows generally a good agreement between the numerical results and the experimental data with some discrepancies in the cooling phase. This could be due to the use of heat convection instead of heat conduction to simulate the interface between the specimen and backing plate and also the assumption of a constant temperature of the "simulated" backing plate. Actually, the temperature of the worktable rises during welding and this decreases the cooling rate of the weld. The maximum error between experimental and numerical data is 37.1 % at the end of the cooling phase.

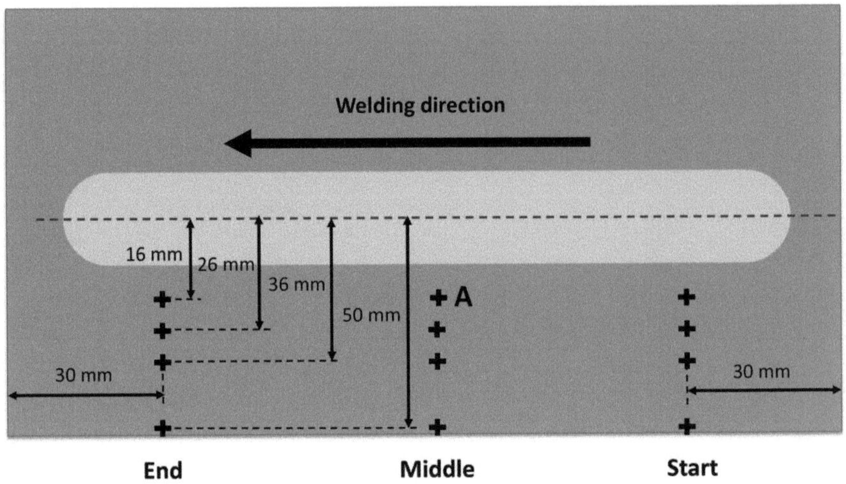

Fig. 24.5 Location scheme of experimental measured points

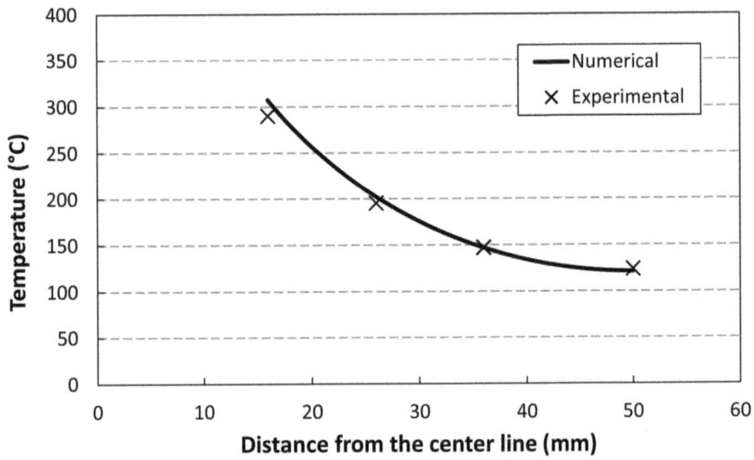

Fig. 24.6 Temperature trend in the start phase

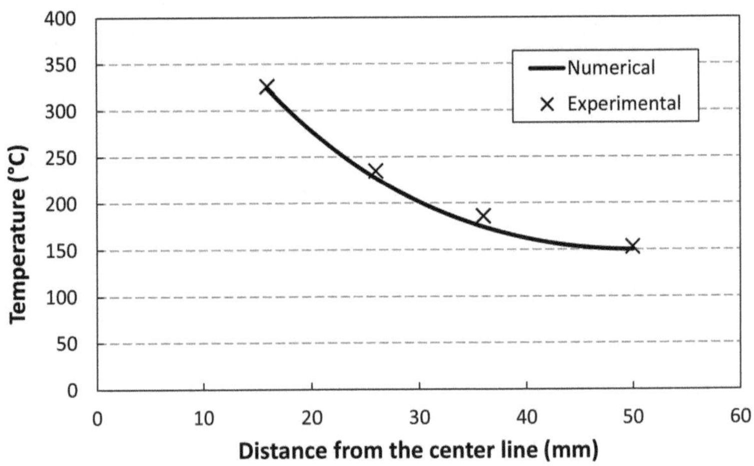

Fig. 24.7 Temperature trend in the middle phase

The permanence time at high temperatures and the cooling rate at a given point during welding is an important issue that conditions the mechanical properties, material microstructure formation and induces distortion in the ended welded structure [27]. In order to determine the permanence time at high temperature and because the typical aluminum temperature of recrystallization is between 200 °C and 300 °C, a temperature of 250 °C has been chosen for discussion. This time, in the

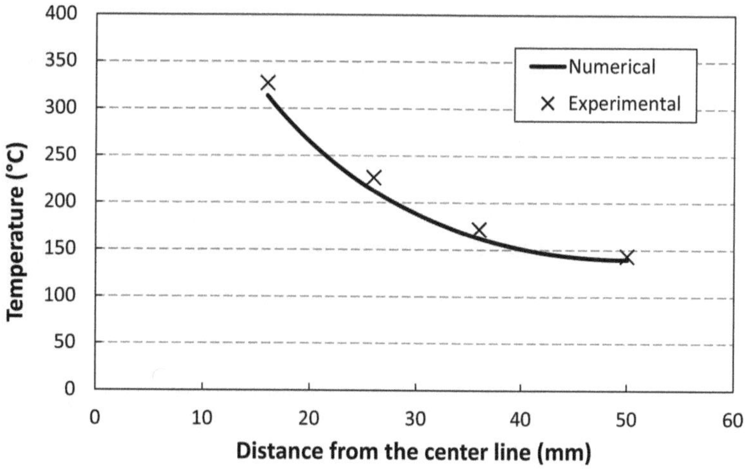

Fig. 24.8 Temperature trend in the end phase

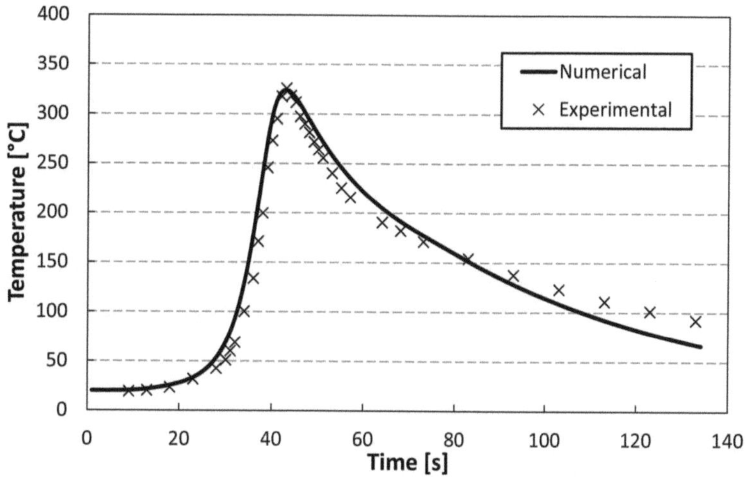

Fig. 24.9 Temperature vs. time of a middle point distant 16 mm from the center line

analyzed case, is about 12 s. This value increase for lower speed of welding and then there is more time for the specimen to conduct the heat to the surrounding materials. This situation could lead to a larger heat affected zone and weld nugget.

The maximum temperature reached in the point A is 325.8 °C. This value is the same for both numerical and experimental data as result of the method used to determine the heat flow between specimen and backing plate. Finally, the model is validated and its results are good considering all the parameters, such as coefficient of convection, friction coefficient, variability of normal force, that are difficult to take into account.

Figure 24.10 shows the comparison between the numerical isothermals at the top surface of the specimen and the image of the surface temperature distribution taken by the infrared camera. The white area in this figure describes the zone of the plate where the temperature exceeds 250 °C. Furthermore, in the thermal images of Fig. 24.10, the temperature of the back welded area of the FSW tool is not reliable because there is a variation of the emissivity after the transit of FSW tool. This is due to the paint removal and change of roughness. To describe and represent the data in a better way and because a semi plate model has been made, the numerical results have been mirrored to the joint line in the next figure. Overall, there is a good agreement between the numerical and experimental data on the whole plate has highlighted in Fig. 24.10. Only in the area near the clamping system, not implemented in the FEM model, the surface temperature is slight different.

Figure 24.11 shows the calculated temperature contours (cross section) at the three steps analyzed in this paper. The figure illustrates the temperature distribution in the specimen during the tool pin crossing. The grey zone indicates the area which over 550 °C. The highest temperature has been observed in the center of the weld region where the rotation of the shoulder and probe contributes the highest heat flux. Also, the heat dissipation between the welded plate and the backing plate is higher than the dissipation between the top surface and air. This phenomenon causes the temperature contour in the weld nugget area to follow a "V" shape. At 20 mm from the weld centerline, the temperatures become uniform through thickness and the isothermal lines become almost vertical moving away from the center line, according with that observed by

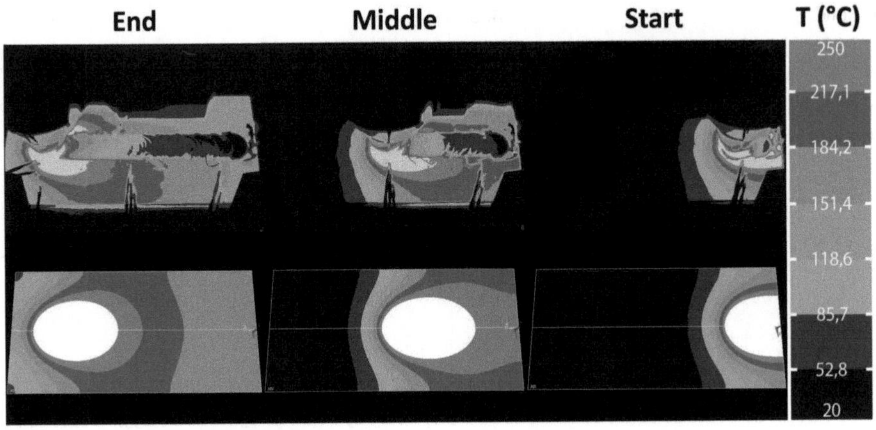

Fig. 24.10 Graphical comparison between numerical data and the experimental ones

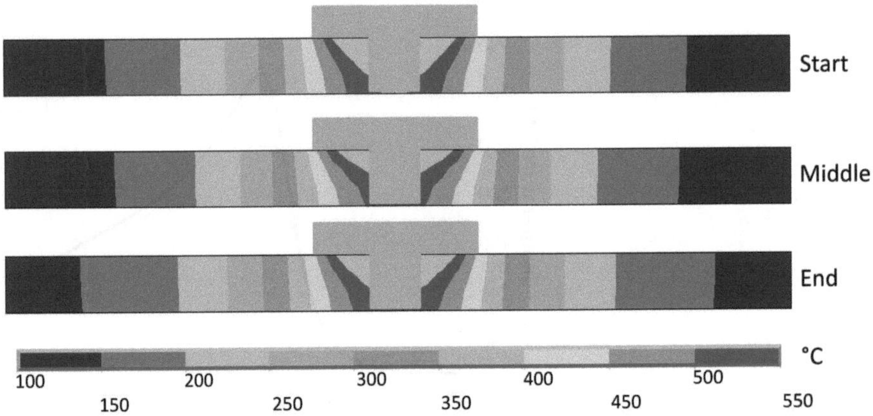

Fig. 24.11 Comparison between model cross section of start, middle and end phase

Song et al. [25]. Figures 24.10 and 24.11 show that detailed information can be provided by FEM model results also in areas (e.g. under tool shoulder, temperature fields in the plate) where the experimental data cannot provide information. This is one of the most important advantage of numerical modeling together with the possibility of predict thermo-mechanical behavior of the joints that can be used for predict the correct parameters of the weld process.

24.4 Conclusion

In the present work the temperature field during the friction stir welding process has been measured using an infrared camera. A 6 mm thick 5754 H111 aluminium alloy plate, in bead on plate configuration, with constant tool rotation rate and feed rate has been used to carried out the test. Furthermore a three-dimensional thermal finite elements model has been implemented and validated on experimental measurement data to evaluate the temperature field and thermal history into the plate. The following conclusions can be drawn:

- The model has been validated and it can accurately describe the heat-transfer process in FSW. The difficulty of determining many parameters (e.g. coefficient of convection, friction coefficient, variability of normal force) has been exceeded. This should allow extending the study to different process parameters, in view also of understanding the residual stress generation.
- The temperature gradient is higher in the area near the tool while decreases progressively from the center to the borders of the plate.

- The final cooling phase behavior of the welding process does not perfectly fit the numerical model due to the assumption of (i) using heat convection instead of heat conduction and also of (ii) a constant temperature for the "simulated" backing plate. Anyway, this should not affect the mechanism of residual stress generation that is supposed to be ruled by the temperature gradient during the initial phase of cooling.
- The heat dissipation between the welded plate and the backing plate is higher than the dissipation between the top surface and air. This phenomenon causes the temperature contour in the weld nugget area to follow a "V" shape.

One of the most important advantage of this numerical model is the possibility of predict thermal behavior of the joints and to provide information on the whole plate where experimental data cannot give results.

References

1. Thomas WM, Nicholas ED, Needham JC, Murch MG, TempleSmith P, Dawes CJ (1991) The Welding Institute, TWI, International Patent Application No. PCT/GB92/02203 and GB Patent Application No. 9125978.8
2. Mishraa RS, Ma ZY (2005) Friction stir welding and processing. Mater Sci Eng R 50:1–78
3. Padmanaban G, Balasubramanian V (2009) Selection of FSW tool pin profile, shoulder diameter and material for joining AZ31B magnesium alloy—An experimental approach. Mater Des 30(7):2647–2656
4. Lee WB, Jung SB (2004) The joint properties of copper by friction stir welding. Mater Lett 58(6):1041–1046
5. Sun YF, Fujii H (2010) Investigation of the welding parameter dependent microstructure and mechanical properties of friction stir welded pure copper. Mater Sci Eng A 527(26):6879–6886
6. Shen JJ, Liu HJ, Cui F (2010) Effect of welding speed on microstructure and mechanical properties of friction stir welded copper. Mater Des 31 (8):3937–3942
7. Fernández JR, Ramirez AJ (2013) Dissimilar friction stir welding of steel to Ni-based alloy 625 - Butt and lap joints. Proceedings of the international offshore and polar engineering conference 2013, pp 207–210
8. Campanelli SL, Casalino G, Casavola C, Moramarco V (2013) Analysis and comparison of friction stir welding and laser assisted friction stir welding of aluminum alloy. Materials 6(12):5923–5941
9. Farias A, Batalha GF, Prados EF, Magnabosco R, Delijaicov S (2013) Tool wear evaluations in friction stir processing of commercial titanium Ti–6Al–4 V. Wear 302(1–2):1327–1333
10. Buffa G, Ducato A, Fratini L (2013) FEM based prediction of phase transformations during Friction Stir Welding of Ti6Al4V titanium alloy. Mater Sci Eng A 581:56–65
11. Esmailzadeh M, Shamanian M, Kermanpur A, Saeid T (2013) Microstructure and mechanical properties of friction stir welded lean duplex stainless steel. Mater Sci Eng A 561:486–491
12. Khodir SA, Morisada Y, Ueji R, Fujii H (2012) Microstructures and mechanical properties evolution during friction stir welding of SK4 high carbon steel alloy. Mater Sci Eng A 558:572–578
13. Cioffi F, Fernández R, Gesto D, Rey P, Verdera D, González-Doncel G (2013) Friction stir welding of thick plates of aluminum alloy matrix composite with a high volume fraction of ceramic reinforcement. Compos Part A: Appl S 54:117–123
14. Bagheri A, Azdast T, Doniavi A (2013) An experimental study on mechanical properties of friction stir welded ABS sheets. Mater Des 43:402–409
15. Casavola C, Lamberti L, Pappalettere C, Tattoli F (2010) A comprehensive numerical stress - Strain analysis of laser beam butt-welded titanium compared with austenitic steel joints. J Strain Anal Eng Des 45(7):535–554
16. Casavola C, Campanelli SL, Pappalettere C (2008) Experimental analysis of residual stresses in the selective laser melting process. Society for experimental mechanics - 11th international congress and exhibition on experimental and applied mechanics 2008, vol 3, pp 1479–1486
17. Gibson BT, Lammlein DH, Prater TJ, Longhurst WR, Cox CD, Ballun MC, Dharmaraj KJ, Cook GE, Strauss AM (2014) Friction stir welding: Process, automation, and control. J Manufact Processes 16(1):56–73
18. Peel M, Steuwer A, Preuss M, Withers PJ (2003) Microstructure, mechanical properties and residual stresses as a function of welding speed in aluminium AA5083 friction stir welds. Acta Mater 51(16):4791–4801
19. Ma Yu E, Xia ZC, Jiang RR, Li WY (2013) Effect of welding parameters on mechanical and fatigue properties of friction stir welded 2198 T8 aluminum–lithium alloy joints. Eng Fract Mech 114:1–11
20. Xu W, Liu J, Luan G, Dong C (2009) Temperature evolution, microstructure and mechanical properties of friction stir welded thick 2219-O aluminum alloy joints. Mater Des 30(6):1886–1893
21. Maeda M, Liu H, Fujii H, Shibayanagi T (2005) Temperature field in the vicinity of FSW-tool during friction stir welding of aluminium alloys. Welding in the World 49(3–4):69–75
22. Hwang YM, Kang ZW, Chiou YC, Hsu HH (2008) Experimental study on temperature distributions within the workpiece during friction stir welding of aluminum alloys. Int J Mach Tool Manuf 48(7–8):778–787
23. Soundararajan V, Zekovic S, Kovacevic R (2005) Thermo-mechanical model with adaptive boundary conditions for friction stir welding of Al 6061. Int J Mach Tool Manuf 45(14):1577–1587
24. Schmidt H, Hattel J, Wert J (2004) An analytical model for the heat generation in friction stir welding. Model Simul Mater Sci Eng 12(1):143–157

25. Song M, Kovacevic R (2003) Thermal modeling of friction stir welding in a moving coordinate system and its validation. Int J Mach Tool Manuf 43(6):605–615
26. The ANSYS® 14.5 Users' Manual, Swanson Analysis System Inc. (2012)
27. Chao YJ, Qi X, Tang W (2003) Heat transfer in friction stir welding - experimental and numerical studies. J Manufact Sci Eng, Trans ASME 125(1):138–145
28. Long X (2005) Finite element analysis of residual stress generation during spot welding and its effect on fatigue behavior of spot welded joint, PhD Thesis, University of Missouri, Columbia, USA

Chapter 25
Dynamics of Strain Localization Associated with Lüders Deformation: An Insight

Srinivasan Nagarajan, Raghu Narayanaswamy, and Venkatraman Balasubramaniam

Abstract This work aims at studying the dynamics of strain localization associated with Lüders instability in mild steel under monotonic, uniaxial tensile testing using full field measurement techniques such as infrared thermal imaging and digital image correlation. Based on the spatiotemporal evolutions of temperature and strain an enhanced understanding on the Lüders deformation behavior is achieved by addressing band nucleation and propagation, band growth mechanism, inhomogeneity in stress, strain and strain rate distributions within the band and the thermomechanical coupling associated. The experimental results are also compared and explained with some of the proposed models and concepts on Lüders instability.

Keywords Lüders band • Thermomechanical coupling • Strain rate field • Stress-strain distribution • Band growth mechanism

25.1 Introduction

Plastic instabilities commonly observed in materials such as carbon steels and Al-Mg alloys cause surface roughness and also reduce their ductility during metal forming operations and service conditions. Though instabilities in materials have been studied since long time, due to the associated detrimental effects on the materials performance and lack of complete understanding on its behavior it is of high concern to researchers even today. Lüders deformation is one such plastic instability that occurs due to the pinning of dislocations by solute atoms and their unpinning under the influence of stress characterized by discontinuous yield points (upper and lower) in the load-elongation curve. The unpinning of dislocations or generation of new dislocations corresponds to the upper yield point and their movement in a collective and self-organized manner corresponds to lower yield point manifesting on the surface of the material as Lüders band propagation. Lüders deformation has been studied in many materials addressing the effects of strain rate, grain size, testing temperature and specimen geometry on the nature of bands, critical strain for band propagation, delay time, Lüders stress, Lüders strain, band width, orientation and the propagation velocity of the band front using various techniques such as photography, microscopy, strain gauges, acoustic emission and Barkhausen noise etc [1–5]. Developments in full field measurement techniques such as infrared thermal imaging and digital image correlation offer the possibility of quantitative measurement and visualization of strain localizations associated with plastic instabilities. In this work, we focus on the application of these techniques in studying the band nucleation and growth mechanism, thermal-strain distribution across the band front, strain rate variation within the band and validation of the proposed models and concepts in these areas, where less works have been reported [6–8].

S. Nagarajan (✉) • R. Narayanaswamy • V. Balasubramaniam
Quality Assurance Division, Indira Gandhi Centre for Atomic Research, Kalpakkam 603 102, India
e-mail: srini0497@gmail.com

N. Sottos et al. (eds.), *Experimental and Applied Mechanics, Volume 6: Proceedings of the 2014 Annual Conference on Experimental and Applied Mechanics*, Conference Proceedings of the Society for Experimental Mechanics Series, DOI 10.1007/978-3-319-06989-0_25,

25.2 Material and Experimental Study

The experimental setup involved infrared camera (CEDIP Silver 420) with 25 mK temperature sensitivity and 320 × 256 pixels resolution, digital camera (Marlin- F131) for image correlation with 1,380 × 1,035 pixels resolution and a 100 kN displacement controlled, electromechancial, screw driven tensile testing machine as shown in Fig. 25.1. Uniaxial tensile deformation of the specimen with dimension 216 × 25 × 1.6 mm was performed monotonically at room temperature (298 K). The specimen used in this study was prepared from IS 2062 grade-E250 B mild steel sheet which is a widely used structural material in many industries, particularly as pipes, storage tanks, plates, angles and hydraulic press structures withstanding heavy loads by the nature of use etc. The deformation of the specimen was recorded at 10 Hz frame rate using both infrared and digital camera placed on either side of the flat surface of the specimen. The specimen surface facing the infrared camera was coated with black paint to achieve uniform emissivity and that facing digital camera was coated with random dot pattern to achieve varying gray levels. For the focal distances set in this study, spatial resolution achieved with infrared camera was 584 μm and that with the digital CMOS camera was 153 μm. Strain fields were obtained by performing image correlation through associated 2D-image correlation processing software. The subset size used for image correlation was 21 × 21 pixels. Strain rate fields are computed by performing time derivatives of strain fields.

25.3 Results and Discussion

The load-time behavior of the specimen deformed at 3.3×10^{-4}/s strain rate is shown in Fig. 25.2 from which the load plateau involving strain localizations associated with Lüders instability is identified to be in the interval of 49.8–110 s.

The spatiotemporal evolutions of temperature and strain over the entire gauge length during nucleation and growth of Lüders bands are shown in Fig. 25.3. At 47 s thermal field reveals the decrease in temperature over the entire gauge length due to thermoelastic effect. However, at the lower end of the gauge length, temperature localization is observed manifesting the nucleation of Lüders band. The corresponding strain field is also found to reveal the band nucleation through strain localization on the same location. With further increase in load, at 49.8 s corresponding to upper yield point nuclei covers the complete cross section and develops as a fully grown band as observed in the thermal and strain fields. The band formation behavior revealed by the thermal and strain fields is found to be in line with band formation mechanism proposed by Friedel [9], according to which band nucleation accompanied by microscopic slip lines occurs in a zone of stress concentration. In the present experiment nucleation is observed in the lower end of the gauge length due to stress concentration induced in the lower fillet region of the specimen by gripping. It is to be noticed that zone of band nucleation is observed in both thermal

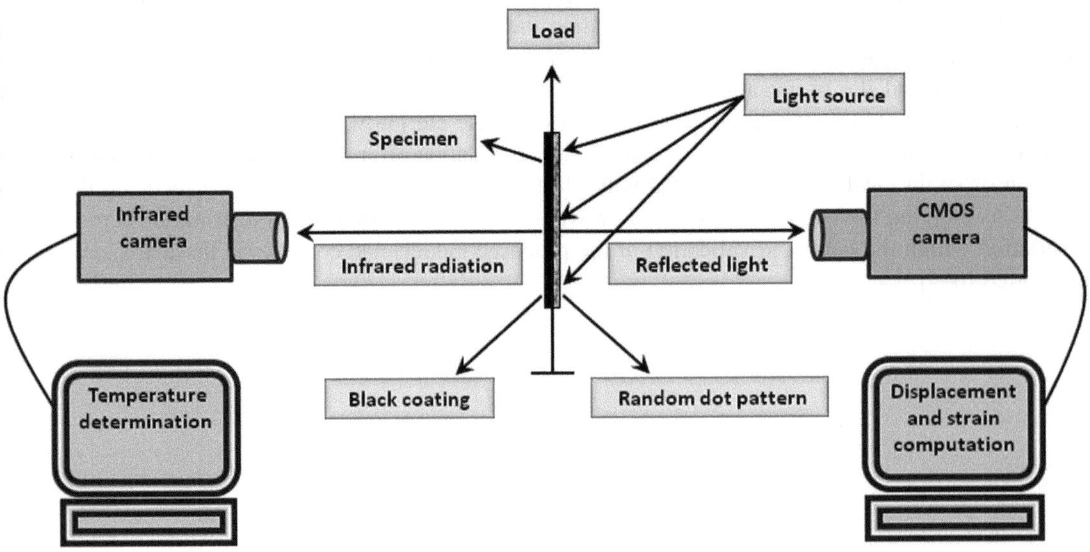

Fig. 25.1 Schematic of experimental setup

Fig. 25.2 Load-time curve

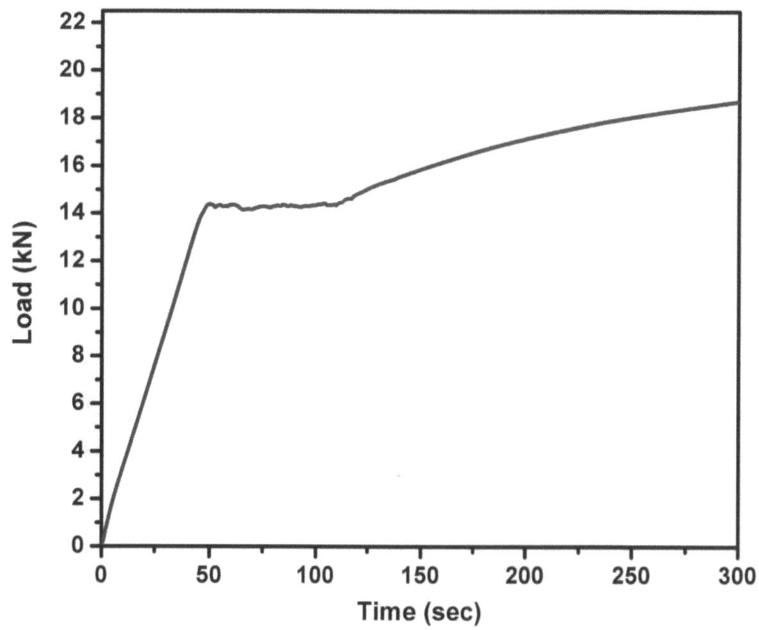

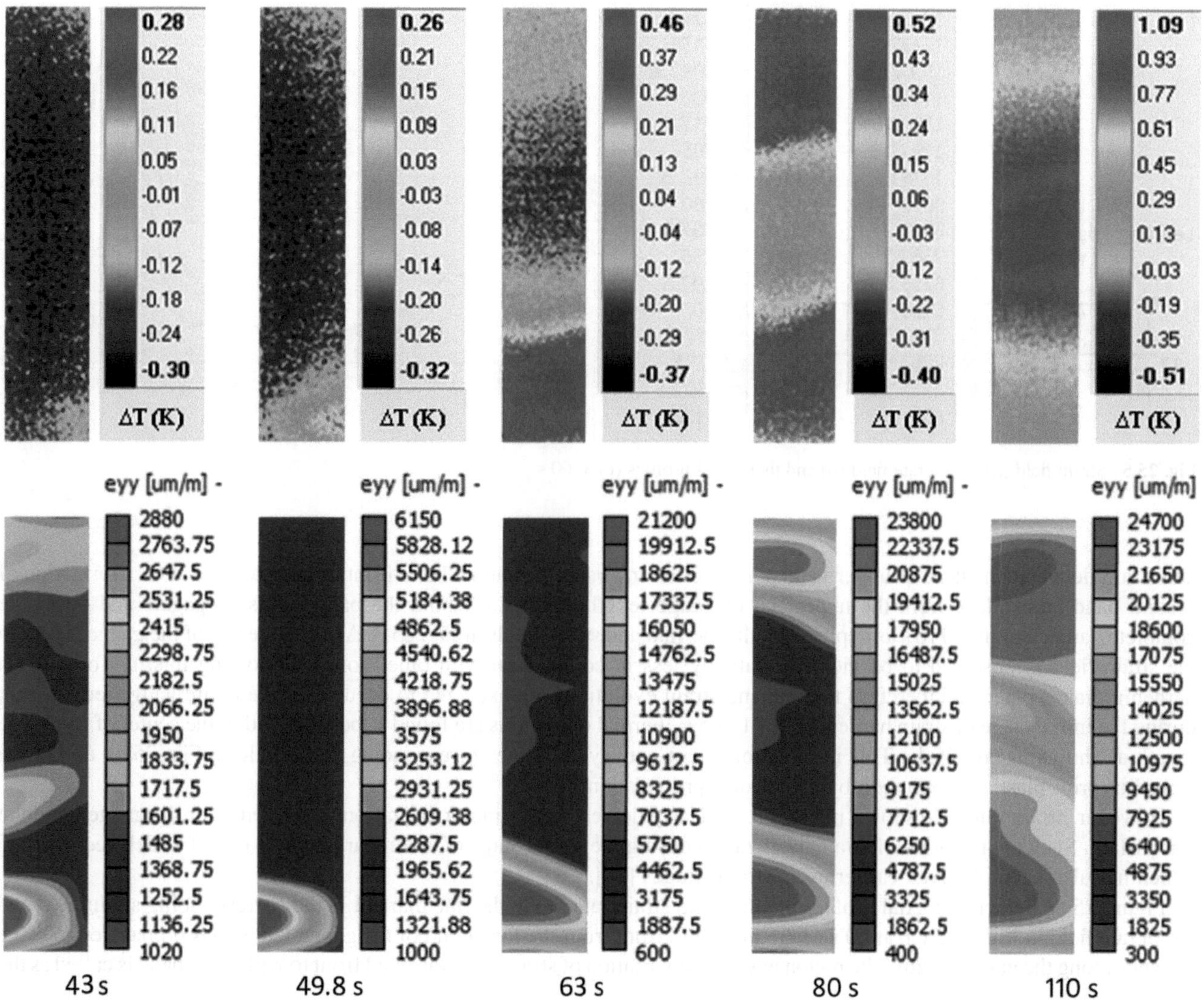

Fig. 25.3 Thermal and strain fields associated with bands nucleation and growth

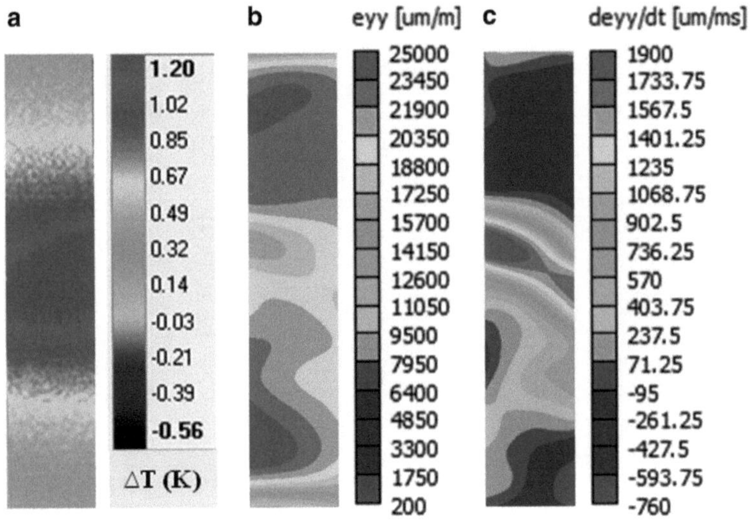

Fig. 25.4 Thermal field (**a**), strain field (**b**) and strain rate field (**c**) at 114.5 s

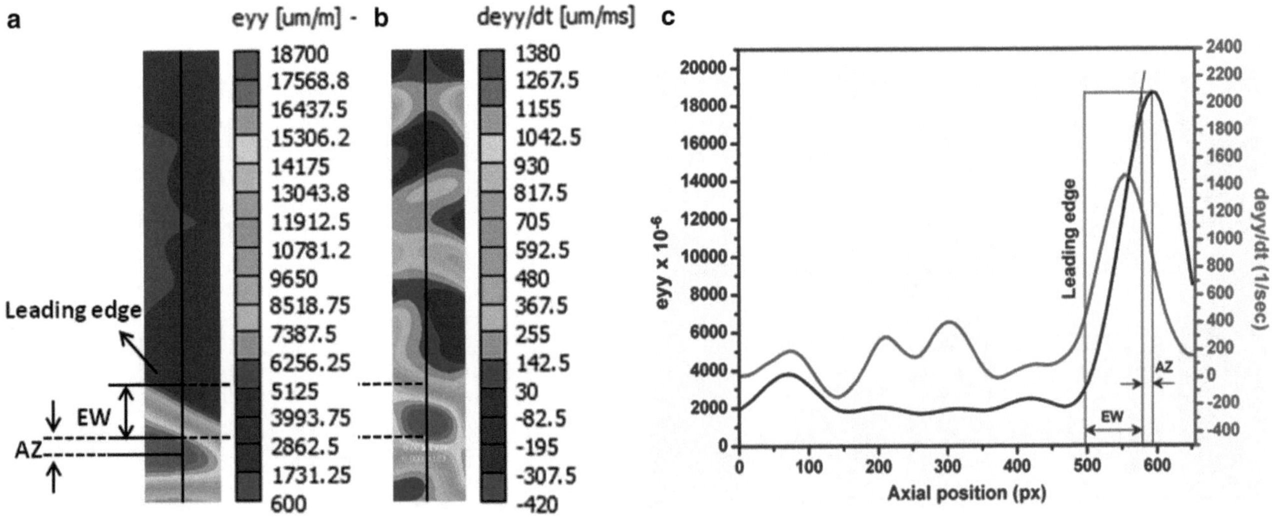

Fig. 25.5 Strain field (**a**), strain rate field (**b**) and their axial profiles (**c**) at 60 s

and strain fields much ahead of upper yield point. Following band formation, band front propagates and at 63 s formation of second band takes place from the upper end of gauge length after which both the band fronts propagate and cover the complete gauge length at 110 s as depicted by the thermal and strain fields in Fig. 25.3. At 110 s, center of the gauge length in the strain field is observed to be yielded relatively less in comparison with other zones. Following this, uniform strain hardening takes place and therefore thermal and strain evolutions are expected to occur over the entire gauge length. But, during the initial stage of strain hardening, till 114.5 s, thermal evolutions are found to be localized in the center of the gauge length as shown in Fig. 25.4a. This information is not conveyed by the corresponding strain field in Fig. 25.4b due to the record of prior plastic deformation by band fronts propagation.

However, strain rate field shown in Fig. 25.4c clearly depicts such strain localization in the center of the gauge length at this instant. Such continuation of strain localization during the initial stage of strain hardening is called as delayed yielding which has also been observed earlier using strain gauges [10].

Figure 25.5c depicts the axial profiles of strain and strain rate along the line marked in the corresponding strain field and strain rate field in Fig. 25.5a, b at 60 s where first band front from the lower end of the gauge length has propagated to some distance along the gauge length. The region with linear variation of strain from the band front to within the band is called as the

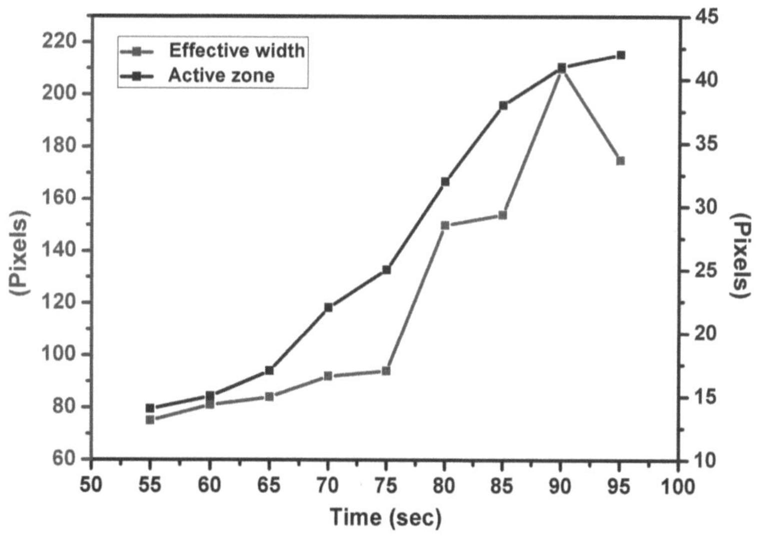

Fig. 25.6 Effective width and active zone variation during propagation of first band front

effective width (EW) of Lüders band and the following region where strain saturates to a maximum is called as active zone (AZ) of Lüders band. The effective width and active zone are identified in the axial strain profile based on their definitions as shown in Fig. 25.5c. The peak strain rate is observed in the effective width which indicates that deformation is localized in the effective width. This can also be visualized from the corresponding strain rate field shown in Fig. 25.5b. In the active zone strain rate varies significantly contradicting the assumption of constancy of strain rate in this zone by Sleeswyk and Verel [11]. According to Van Rooyen [2], the Lüders band growth mechanism involves two stages, namely 1. Yielding of the material in the elastic region close to the band front, resulting in growth of effective width, 2. strain hardening of material within the effective width, resulting in the growth of the active zone. It has been proposed that these two stages are continuous and occur simultaneously characterized with constant effective width and increasing active zone with constant magnitude during the band growth in the load plateau. Figure 25.6 represents the variation in the effective width and irregular growth of active zone measured during the propagation of first band front at various instants while covering the lower half of the gauge length. This observation confirms that the region which is yielded in the effective width is not completely strain hardened immediately following the movement of the band front. In other words two stages involved in band growth mechanism do not compensate each other to the same magnitude conveying the irregularity in their occurrence.

Van Rooyen [2] has proposed a model featuring the inhomogeneity in stress-strain distribution across the Lüders band front as shown in Fig. 25.7. According to the model, B is the band front, C is the zone which is yielded by band front propagation, but not strain hardened to the extent of full Lüders strain as seen in Fig. 25.7a. Since the zone C has mobile dislocations, it has been proposed that the load carrying capacity of this zone is very less and most of the applied load is carried by the elastic region in the immediate vicinity of band front B, where stress ranges from lower yield stress to true upper yield stress (σ_1 to σ_2). In other words, flow stress of zone C is less than that of the elastic zone close to the band front. Stress increases from C to D because of strain hardening after which it saturates and E represents the plastic end. The thermal-strain distribution across the gauge length at 96.6 s where two band fronts tend to approach the center of the gauge length is shown in Fig. 25.8. From the strain field image in Fig. 25.8c the band front B of the second band from the upper end of gauge length is identified and its axial position is marked on the thermal and strain profiles. The point E represents the plastic end and is marked as the rear end of the second band. The local zones C and D are identified in the thermal and strain profiles based on their definitions proposed by the model and its correlations to the thermal and strain evolutions. According to the model, stress remains constant in the elastic region and on approaching the band front increases drastically to true upper yield stress σ_2. At the center of the gauge length where no band front has crossed, temperature is remains minimum and on approaching the second band front which has initiated from the upper end of gauge length temperature increases and reaches 0.34 K as seen in Fig. 25.8b. Because of the diffusion effect, temperature rise has been observed much before the band front. Strain has been proposed to increase only from the zone C. But from the strain profile measured using DIC, slight variation in strain is observed much ahead of the band front as seen in Fig. 25.8d. This is because of the stress variation from lower to upper yield stress (σ_1 to σ_2) in the elastic zone close to the band front. From the band front to within the band (plastic region) stress initially decreases and reaches the minimum C followed by its increase and saturation at D corresponding to lower yield stress (σ_2). From the axial thermal profile in Fig. 25.8b, it is observed that

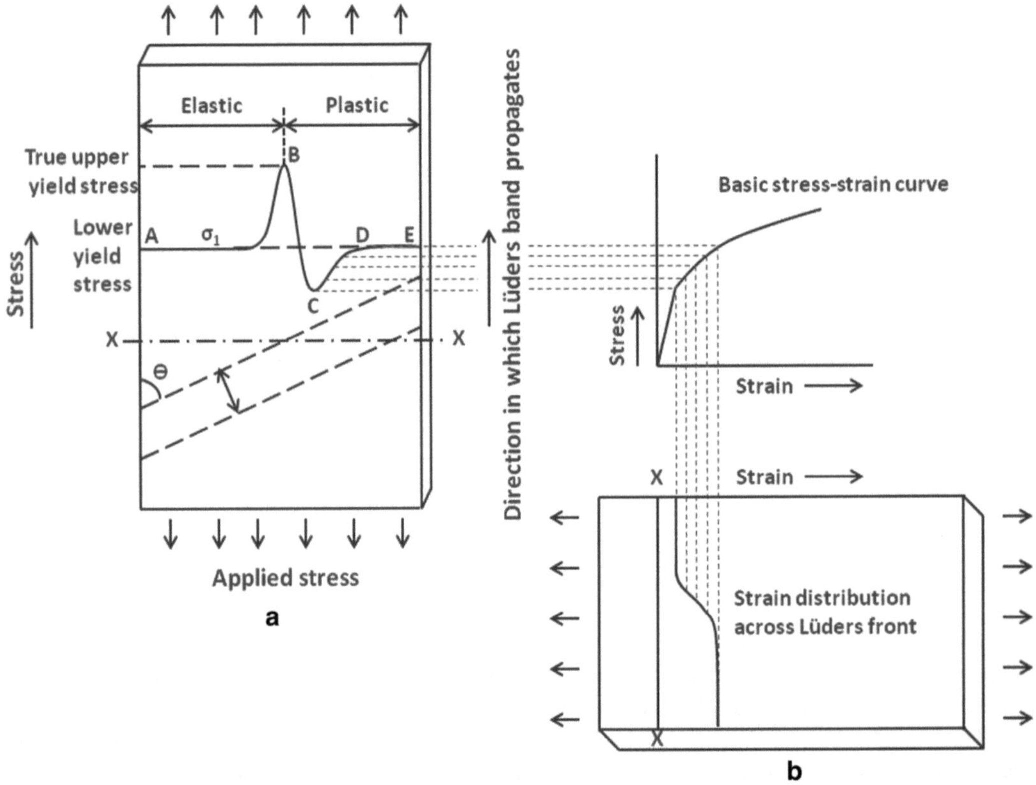

Fig. 25.7 Stress-strain distribution across the band front, Van Rooyenn [2]

temperature increases from the band front B, reaches the peak C followed by a decrease to D and E. The peak temperature observed at C corresponds to the zone of mobile dislocations which has less load carrying capacity and hence most of the applied mechanical energy is dissipated as heat in this zone.

The zone C identified in the thermal profile is used in locating its position in the corresponding strain profile. Similarly, from the peak strain D in the strain profile its corresponding location in the thermal profile is identified. C to D involves strain hardening and therefore it is expected to show considerable thermal evolutions. But due to thermal diffusion from zone C involving peak temperature, such thermal evolutions are masked and a decreasing trend is observed from C to D. However, from C to D strain increases. From D to E corresponding to the region of initiation of the band, the model proposes constant value of stress and strain as shown in Fig 25.7a, b. But temperature is found to decrease from D to E because of diffusion loses associated with prior deformation. Strain decreases after reaching the peak D contradicting the model because of the symmetry in distribution of strain on either side of the nucleation zone as observed in the strain field image in Fig. 25.8c. It should also be noticed that in the present case nucleation of the band is not from the exact upper end of gauge length as assumed by the model. Above observations based on thermal and strain distribution measured using IRT and DIC confirms the inhomogeneity in stress-strain distribution across the Lüders band front and reveals additional information such as symmetrical strain distribution on either side of nucleation zone and the strain evolutions much before the band front.

25.4 Summary and Conclusion

To summarize, IRT and DIC have been applied simultaneously to study the dynamics of strain localization during Lüders instability in mild steel. Spatiotemporal evolutions of temperature and strain accompanying Lüders deformation clearly revealed the band formation and growth mechanism, inhomogeneity in stress-strain distribution across the band front and the continuation of strain localization following band front propagation during the initial stage of uniform strain hardening.

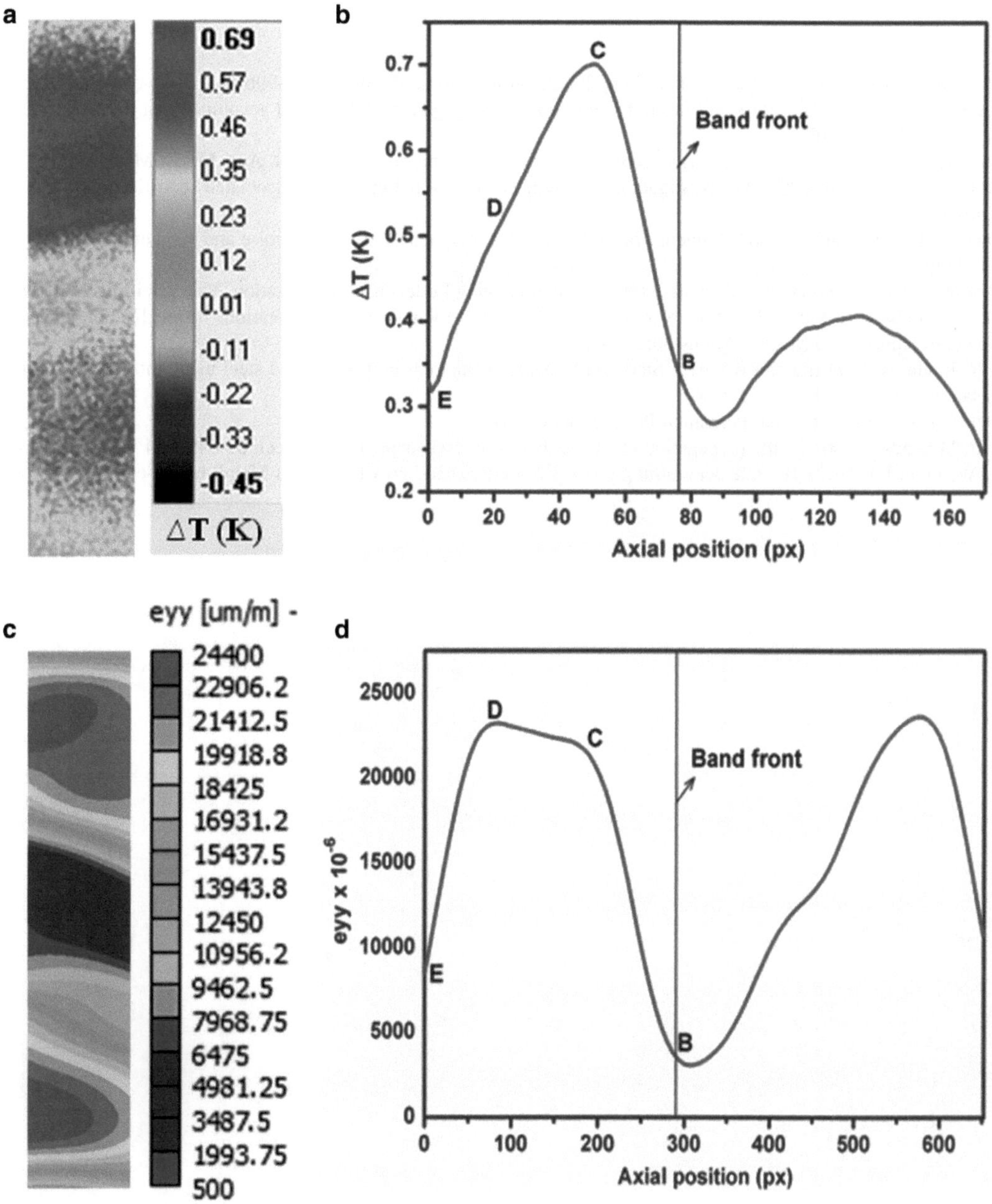

Fig. 25.8 Thermal-strain distribution across the band front

The zone of band formation has been observed in the thermal and strain fields much before the macroscopic yield point. Thermal and strain evolutions during band nucleation and formation are found to match with the band formation model proposed by Friedel [9] thereby serving as an experimental validation of the model. The phenomenon of delayed yielding following band front propagation during initial stage of strain hardening in a zone which is relatively less yielded has also been successfully characterized. Effective width of Lüders band has been found to vary in the load plateau and active zone is found to increase with irregular magnitude. This observation leads us to conclude that the two stages of band growth mechanisms involving yielding and strain hardening do not compensate each other immediately and hence are irregular. Strain rate has been observed to peak in the effective width and vary in the active zone contradicting the assumption of Sleeswyk et al. [10]. Thermal-strain distribution measured across the band front revealed the inhomogeneity in the stress-strain distribution proposed by Van Rooyen [2]. Interesting features such as symmetrical strain distribution on either side of nucleation zone and strain evolutions much before the band front which was not addressed in the model has also been observed.

References

1. Moon DW, Vreeland T (1969) The initiation of yielding in silicon-iron. Acta Metall 17:989–996
2. Van Rooyen GT (1968/1969) The stress and strain distribution in a propagating Lüders front accompanying the yield-point phenomenon in iron. Mater Sci Eng 3: 105–116
3. Ananthan VS, Hall EO (1991) Macroscopic aspects of Lüders band deformation in mild steel. Acta Metall Mater 39:3153–3160
4. Armentrout DL, Carpenter SH (1997) An investigation on Lüders band deformation and the associated acoustic emission in Al-4.5 Mg alloys. J Acoust Emission 15:43–52
5. Dhar A, Clapham L, Atherton DL (2002) Influence of Lüders band on magnetic Barkhausen noise and magnetic flux leakage signals. J Mater Sci 37:2441–2446
6. Herve L, Andre C (2001) Thermal and dissipative effects accompanying Lüders band propagation. Mater Sci Eng A 307:15–22
7. Watrrise B, Chrysochoos A, Muracciole J-M, Nemoz-Gaillard M (2001) Kinematic manifestations of localization phenomena in steels by digital image correlation. Eur J Mech A Solids 20:189–211
8. Srinivasan N, Raghu N, Venkatraman B (2012) Study on Lüders deformation in welded mild steel using infrared thermography and digital image correlation. Adv Mater Res 585:82–86
9. Friedel J (1967) Dislocations, 1st edn. Pergamon Press, Elmsford, NY
10. Kyriakides S, Miller JE (2000) On the propagation of Lüders bands in steel strips. J Appl Mech 67:645–654
11. Sleeswyk AW, Verel DJ (1972) The rate determining process in Lüders band growth. Scripta Metall 6:677–679

Chapter 26
Raman Spectroscopy-Enhanced IIT: In Situ Analysis of Mechanically Stressed Polycrystalline Si Thin Films

Yvonne B. Gerbig, Chris A. Michaels, and Robert F. Cook

Abstract Exposed to mechanical stress, semiconductor materials may phase transform, resulting in changes of crystallographic structure and material properties, rather than deform by plastic flow. As a consequence, prediction of the state and distribution of strain in semiconductors has become crucial for the evaluation of performance and reliability of structures made of these materials. Indentation-induced phase transformation processes were studied by in situ Raman imaging of the deformed contact region of silicon, employing a Raman spectroscopy-enhanced instrumented indentation technique (IIT). This is, to our knowledge, the first sequence of Raman images documenting the evolution of the strain fields and combined changes in the phase distributions of a material under contact load.

Keywords Indentation • In situ • Raman imaging • Phase transformation • Silicon

26.1 Introduction

Silicon (Si) in its pristine diamond cubic (dc) phase can be transformed to different crystallographic structures through the application of mechanical stress [1]. Given its importance to electronic devices and microelectromechanical systems and given that modifications of the Si crystallographic structure are connected to its performance related properties, phase transformations have been intensively studied. At hydrostatic pressures of 10–16 GPa, Si undergoes a nonmetallic-metallic transition, as the dc structure of Si-I transforms to the denser body centered tetragonal (bct) β-tin structure of the Si-II phase [2–4]. Further compression leads to a sequence of transitions to other crystallographic phases according to several theoretical studies and quasi-hydrostatic diamond anvil cell (DAC) experiments [1]. Shear stress is critical in the initiation of phase transitions [5, 6] and has been reported to lower the threshold pressure for the phase transformation in monocrystalline Si [7]. Furthermore, it has been speculated that shear stress may promote the formation of phases not observable under hydrostatic conditions [8, 9]. Simulation studies suggest the crystal lattice of dc Si transforms into various bct structures with sixfold (β-tin) and fivefold coordination (bct5) under the presence of shear stress during loading [10–14].

Instrumented indentation technique (IIT) has been widely used to investigate the phase transformation of Si under the presence of shear stress. However, in conventional indentation experiments on Si, the structure and phase of the material is probed only after the completion of the indentation test (*ex situ*), which leaves the exact path of transformation to conjecture, as the presence of intermediate phases cannot be directly observed. In order to analyze the indentation-induced phase transformation in situ, auxiliary characterization techniques must be employed in indentation studies. Raman spectroscopy has become a versatile tool, enabling determination of the kinetics and physics involved in the mechanical deformation of materials at the crystallographic and molecular level (*e.g.*, strain build-up in crystal lattices, phase transformations, and

Y.B. Gerbig (✉)
Material Measurements Laboratory, National Institute of Standards and Technology (NIST), 100 Bureau Drive, Gaithersburg, MD 20899, USA

Department of Mechanical Engineering, University of Maryland, College Park, MD 20472, USA
e-mail: yvonne.gerbig@nist.gov

C.A. Michaels • R.F. Cook
Material Measurements Laboratory, National Institute of Standards and Technology (NIST), 100 Bureau Drive, Gaithersburg, MD 20899, USA

N. Sottos et al. (eds.), *Experimental and Applied Mechanics, Volume 6: Proceedings of the 2014 Annual Conference on Experimental and Applied Mechanics*, Conference Proceedings of the Society for Experimental Mechanics Series, DOI 10.1007/978-3-319-06989-0_26,

changes in crystallinity). The utility of IIT and Raman spectroscopy has led researchers at NIST to explore the combination of these methods for in situ measurements, Raman spectroscopy-enhanced IIT, and to develop an indentation device that is coupled with a Raman microscope to conduct spectroscopic and optical analysis (probing and mapping) of mechanically strained regions of transparent samples under contact loading [15].

In this study, Raman spectroscopy-enhanced IIT in a laser scanning Raman imaging configuration is employed to study in situ the generation and evolution of the phase transformation of Si under contact load. The experimental findings are compared to relevant simulations and contact models. This is, to our knowledge, the first sequence of Raman images of a material held under contact load and undergoing phase transformation induced by mechanical deformation.

26.2 Experimental Details

26.2.1 Experimental Setup

The in situ Raman mapping experiments were performed using an instrumented indentation device developed at NIST [15] that is coupled with a laser scanning Raman microscope to conduct in situ spectroscopic analyses of mechanically deformed regions of optically transparent materials under contact loading.

The force transducer of the indentation device allows adjustment of experimental parameters, such as indentation loads and (un)loading rates. An incorporated displacement sensor allows for collection of force-displacement data comparable to conventional IIT devices. The indentation device, along with a specimen holder featuring an aperture in the center, are mounted on the X-Y translation stage of an inverted optical microscope that is configured for Raman microscopy, allowing optical access to the mechanically deformed regions of transparent samples.

The Raman microscope has been described in considerable detail elsewhere [15], only a brief discussion of the general set up of recent modifications will follow. Laser scanning capability has been added that enables the acquisition of Raman images of the specimen while held under constant load. These images are generated by raster scanning the beam across the sample region of interest while acquiring Raman spectra at each pixel. Image maps can then be constructed on the basis of parameters of particular spectral features (*e.g.*, peak width or amplitude). The optical arrangement used is conventional for laser scanning microscopes in which two galvanometer scan mirrors (X, Y) are co-located at a plane conjugate to the microscope objective pupil in a conventional 4f imaging system. It is worth noting that this approach, the principal advantage of which is the rapid scan speed, is unusual in Raman microscopy where slow spectral acquisition times are the typical bottleneck. In this case however, the advantage of scanning the laser beam lies in the avoidance of moving the sample and indenter as would be required in a sample scanning configuration. Such movement would undoubtedly compromise the measurement stability and the quality of the force and displacement measurements. The conversion between the angular motion of the galvanometer and the beam displacement in the focal plane is calibrated using a Ronchi grating. A Labview [16] program is used to coordinate the motion and hold periods for the galvanometers and the spectral acquisition by the CCD (charge-coupled device) camera.

26.2.2 Test Specimen

The in situ mapping experiments were performed on a Si on sapphire (SoS) specimen (manufacturer: Valley Design Corp., Vera Cruz, CA) [16], which is comprised of a Si(100) film epitaxially grown on a sapphire (r-plane) substrate with both sides polished. The thicknesses of film and substrate were 600 nm ± 60 nm[1] and 530 μm ± 50 μm,[1] respectively. The $\langle 001 \rangle$ direction of the Si film was rotated away from the substrate normal by 5° ± 1[2]. The sample (20 mm × 20 mm) was cut from a SOS wafer along with a slight misalignment of 3.5° ± 1°[2] with respect to the directions $\langle 110 \rangle$ and $\langle 1\bar{1}0 \rangle$ of the film, as shown by electron backscattering diffraction measurements.

---

[1] Uncertainties represent manufacturer specifications.

[2] Uncertainties represent intrinsic measurement errors and mounting errors.

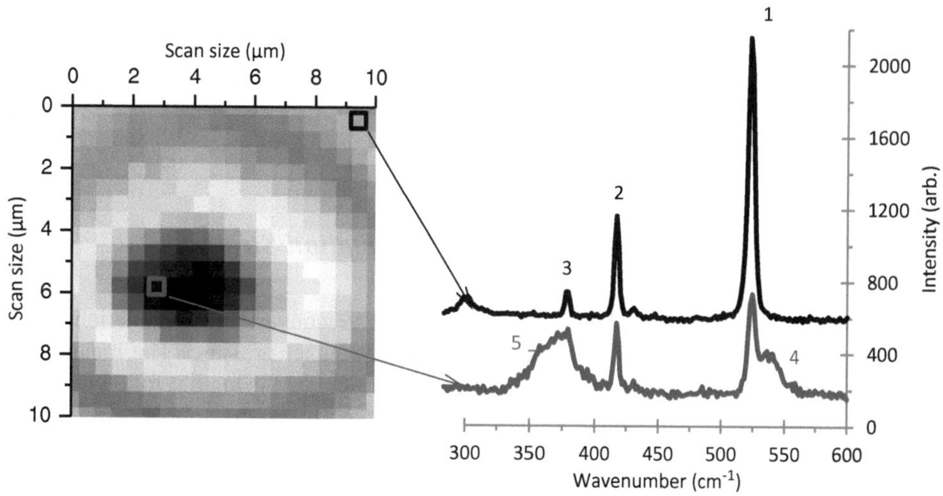

Fig. 26.1 Grey scale image showing the distribution of the intensity of peak 1. The *dark colored circular area marks* the contact region between indenter tip and SoS sample. Raman spectra taken at the indicated spots show the difference in the Raman modes observed outside the contact region and inside the contact region. The spectra are offset along the intensity axis for better illustration

Table 26.1 Bounding conditions applied in fitting model of the Raman spectra

Fitting parameter	Peak 1	Peak 2	Peak 3	Peak 4	Peak 5	Peak 6	Peak 7
Frequency (cm^{-1})	510–530	290–310	415–420	375–380	525–550	340–380	250–300
FWHM	1–50	1–50	1–20	1–20	1–50	1–50	1–300

26.2.3 Mapping Experiment

A series of Raman images from the contact region between indenter tip and SoS sample were collected for increasing contact loads in stepped indentation experiments. In these tests, the indenter probe was brought into contact with the sample and a load of 40 mN applied and held constant for the time required to collect the Raman image of the region of interest (ROI). The applied indentation force was then increased (loading rate 6 mN/s) to 80 mN and a second Raman image of the same ROI was acquired. The indentation loads were further increased twice more (to 120 mN and 200 mN, loading rate 6 mN/s) and another Raman image was recorded each time. The indenter probe used in the experiments was a conospherical tip made of diamond. The effective radius of tip curvature as a function of indentation depth was determined prior to the mapping experiments based on to the procedure described in [17].

To collect a Raman image, an ROI of 10 μm × 10 μm was scanned. Twenty individual spectra were collected at evenly spaced locations along the 10 μm scan range in each direction, generating a 20 pixels × 20 pixels Raman image. The collection time for an individual Raman spectrum was set to 15 s. Although these images may take several hours to acquire, tests of displacement and force stability have been performed that indicate that the system is sufficiently robust to perform such measurements. The estimated depth of field for the collection of Raman photons was about 10 μm. The illumination power of the 785-nm light source was set to 5 mW at the sample surface.

26.2.4 Analysis of Raman Spectra

The individual Raman spectra (examples are shown in Fig. 26.1) were fitted employing a multiple-peak model with Pearson VII functional form for the observed peaks and an iterative optimization routine available in the Peak Analyzer function of the commercial software OriginPro 9 [16] to determine position of the center frequency, height and FWHM (full width at half maximum) of the various peaks. To facilitate the fitting procedure, the bounding conditions summarized in Table 26.1 were applied to the multiple-peak model.

Table 26.2 Values for average contact pressure p_i and indentation strain ε_i calculated from the indentation data at indentation forces F_i, at which Raman images were collected

	Image 1	Image 2	Image 3	Image 4
F_i (mN)	44	83	122	201
p_i (GPa)	5.8	6.3	6.4	6.4
ε_i	0.25	0.38	0.51	0.72

26.2.5 Determination of the Contact Pressure and Indentation Strain

To facilitate the analysis of the experimental data, it is necessary to convert the indentation force (specific to the experimental set-up) to the more "generalized" parameter of average contact pressure. The mean contact pressure p_i at the indentation force F_i (at which a Raman image was recorded) was calculated using the following equation:

$$p_i = F_i / \left(\pi \cdot a_{c,i}{}^2\right) \tag{26.1}$$

where $a_{c,i}$ is the radius of the circle of contact at the force F_i. The contact radius $a_{c,i}$ was determined at a specific indentation force according to the approach employed by Weppelmann et al. [18].

$$a_{c,i} = \sqrt{R_i^2 - \left(R_i - h_{p,i}\right)^2} \tag{26.2}$$

where R_i is the effective radius of indenter tip curvature and $h_{p,i}$ is the plastic deformation at the indentation force F_i. The plastic deformation $h_{p,i}$ at the indentation force F_i was calculated with the following equation:

$$h_{p,i} = h_i - \frac{\delta_i}{2} \tag{26.3}$$

where h_i is the total displacement and δ_i is the elastic deformation at the indentation force F_i. The elastic deformation δ_i at the indentation force F_i was calculated with the following equation:

$$\delta_i = h_i - h_{r,i} \tag{26.4}$$

where h_i is the total displacement and $h_{r,i}$ is the residual displacement at the indentation force F_i. The values for F_i, h_i and $h_{r,i}$ were determined from the force displacement data recorded during the indentation test.

The indentation strain ε_i was estimated using the following equation:

$$\varepsilon_i = h_i / t \tag{26.5}$$

where h_i is the indentation depth at the indentation force F_i and t is the film thickness.

As can be seen from the calculated values for p_i and ε_i given in Table 26.2, the strain increases significantly with greater indentation forces, whereas the average contact pressure changes only slightly.

26.3 Results and Discussion

Figure 26.1 shows two Raman spectra, one of which is taken outside the contact zone between indenter probe and sample and the other inside the contact zone under the first contact load. The spectrum taken outside the contact zone is representative of the pristine SoS featuring three peaks. Peak 1 observed at 522 cm^{-1} is associated with first-order phonon of the dc phase of Si [19, 20]. The peaks located at 378 cm^{-1} and 418 cm^{-1} (peaks 2 and 3) can be assigned to the E_g and A_{1g} modes of the sapphire substrate, respectively [21]. In the center of the contact region, significant changes in the Raman spectrum of the SoS sample were observed: Peak 4 appeared in the spectrum at 535 cm^{-1} and formed a shoulder on the right-hand side of the dc peak. Peak 5 emerged at around 370 cm^{-1} forming a shoulder on the left-hand side of the sapphire E_g mode. As neither of those peaks seems to be associated with the dc Si structure or sapphire, their presence is a manifestation of phase transformation processes in the contact zone at the first contact load (p_i: 5.8 GPa, ε_i: 0.25). The contact pressure of

5.8 GPa is much less than the values (10–16 GPa) for the primary phase transformation of single crystal Si reported in quasi-hydrostatic DAC tests [2–4]. However, it confirms previous suggestions that a combination of isotropic compressive and shear strains (as in case of indentation) induces phase transformation at much smaller stresses than those required for compression alone (as in DAC test) in semiconductors [2, 7]. Also, our observation is in good agreement with a recently published model for the stress-induced phase transformation of micro-crystalline Si films, which predicts a primary phase transformation event at a contact stress of 6.4 GPa and contact strain of about 0.2 [22].

In a previous study [23], the shift of the Raman peaks 4 and 5 as a function of contact pressure were investigated and compared to data available in the literature. In this comparison, the wavenumber shift for peak 5 aligned reasonably well with data for the transverse optical (TO) mode of β-tin reported for quasi-hydrostatic compression [24–27]. However, in indentation simulations (compressive stress + shear strain), the formation of another phase, bct 5 was observed [11–14]. Two Raman-active modes with E_g and A_{1g} symmetries were predicted for the bct5 phase with wavenumbers of 520 cm^{-1} and 340 cm^{-1} [8, 28]. The calculated E_g frequency for the bct5 phase is in the wavenumber range of peak 5, suggesting the formation of this phase in the contact zone. The wavenumbers for the A_{1g} mode of bct5 and the TO mode of β-tin are similar, which makes an unambiguous assignment of peak 5 to either of these phases extremely difficult. It was assumed that both phases were present in the contact zone, as recent simulations predict the concurrent transformation of the dc phase to β-tin and bct5 structures during indentation loading [10–14]. To determine unequivocally the formation of β-tin in the contact region, the occurrence of the longitudinal optical mode for β-tin has to be verified, which is located at around 120 cm^{-1} [24, 27] and currently outside the detection range of the experimental set up.

Images assembled from the hyperspectral Raman data cube are summarized in Fig. 26.2 for peak 1 (amplitude and wavenumber) and peaks 4 and 5 assigned to the newly formed, high-pressure phases (wavenumber only) for the different contact loads. Peak 2 and 3 originating from the substrate were considered as not being relevant for this study. As can be seen in Fig. 26.2, the center of the contact region was not exactly aligned with the center of the scan area, so that, unfortunately, the affected region was not entirely mapped, especially at greater indentations forces (strains). However, important conclusions and insights can be drawn from these partial maps.

The low intensity of peak 1 indicates where the phase transformation from the dc phase to high-pressure phases has taken place, as it correlates with the occurrence of peaks 4 and 5 in the wavenumber maps.[3] The wavenumber maps also show a circular, isotropic distribution of the high-pressure phases (at least in lateral direction) for the indented polycrystalline Si, which is different from the pronounced anisotropic phase distribution found in residual indentations on monocrystalline Si [6]. The phase (bct5) associated with peak 4 covers a larger portion of the contact area than the phases associated with peak 5 at the first contact load, which is presumably due to the fact that the bct5 phase is formed earlier than the β-tin phase [23, 29]. The structure associated with peak 4 continues to occupy a larger portion of the deformed area than the Si phases associated with peak 5 even as the indentation experiment progresses, making it the predominant of the high pressure phases.

With increasing indentation strain (the average contact pressure changes only marginally), the region of reduced peak-1 intensity and the areas where peaks 4 and 5 were detected, grow larger, indicating that an increasing portion of the dc structure in the contact area was transformed to high pressure phases. This seems to give experimental evidence for previous predictions regarding the shear strain being the critical factor in initiating phase transformation rather than the compressive stress [5, 7].

As shown in Fig. 26.2, the wavenumbers of the analyzed Raman modes shifted (significantly) within the deformed area. The wavenumber of the dc Raman peak is shifted to larger wavenumber within the center of the contact zone, indicating the exposure of this phase to compressive stress [30]. However, with greater contact loads, the dc phase seems to relax (less compressively stressed) as the relative Raman shift of this phase decreases in the contact zone. At the same time, the relative Raman shifts for peaks 4 and 5 increases, as the further indentation-induced stresses and strains are compensated by the deformation of the expanding, newly formed phases rather than the diminishing dc structure in the contact zone. As result, the region of the highest Raman shifts of the high-pressure phases enlarges in the center of the contact zone. For peaks 4 and 5, the Raman modes harden from the edge towards the center of the contact area.

[3] The areas colored in black in the frequency plots for peaks 4 and 5 indicate that no signal for the particular peak was collected at these locations.

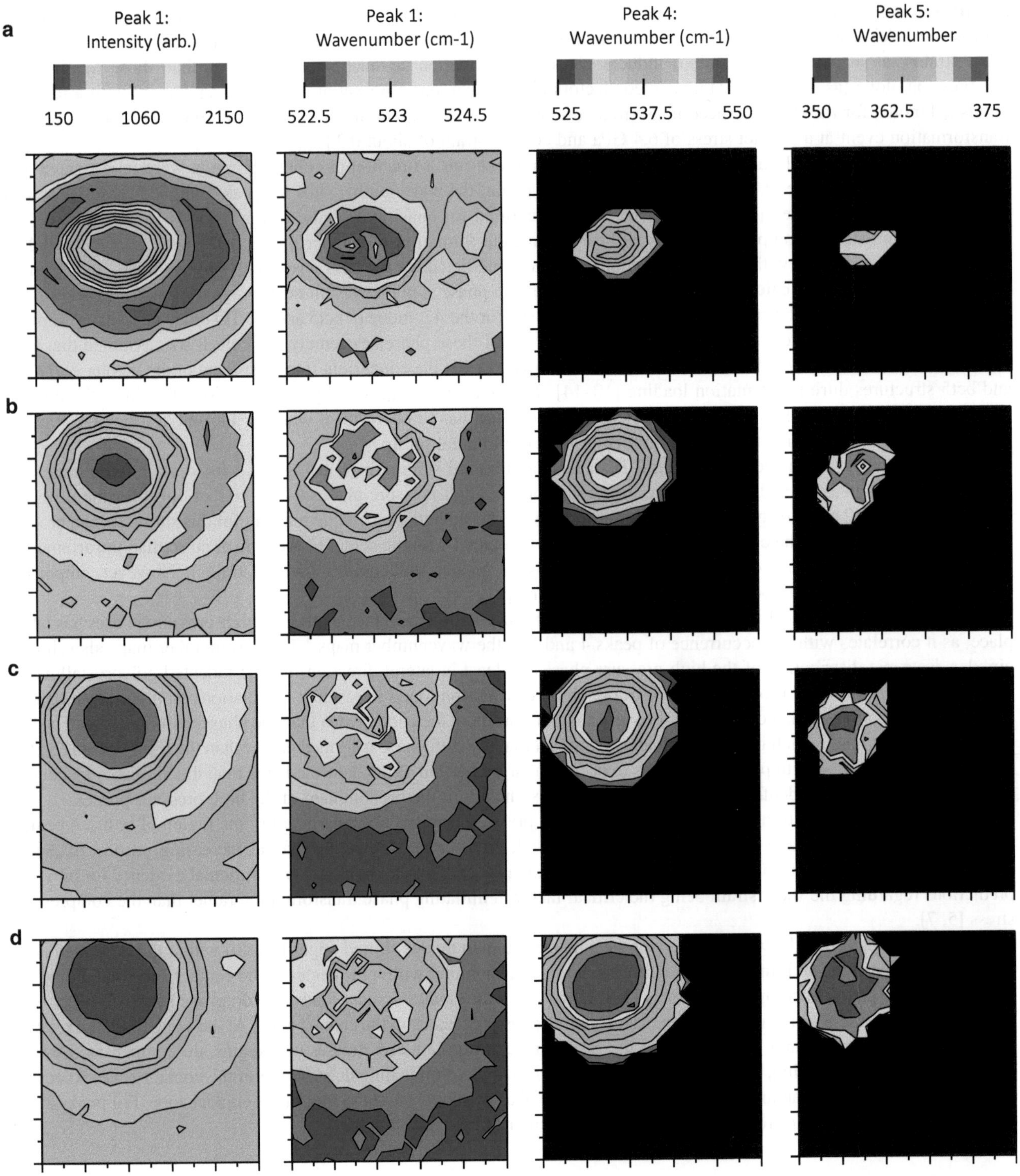

Fig. 26.2 Raman maps imaging the distribution of intensity (peak 1 only) and wavenumber of particular Raman bands as a function of contact load (expressed in indentation force F_i, contact pressure p_i and indentation strain ε_i) (**a**) F_i: 44 mN, p_i: 5.8 GPa. ε_i: 0.25, (**b**) F_i: 83 mN, p_i: 6.3 GPa, ε_i: 0.38, (**c**) F_i: 122 mN, p_i: 6.4 GPa, ε_i: 0.51, (**d**) F_i: 180 mN, p_i: 6.4 GPa, ε_i: 0.72. The scan size is 10 μm × 10 μm. The *black colored areas* in the maps for peak 4 and 5 indicate the absence of the particular Raman band in the collected spectrum

26.4 Conclusion

Indentation-induced phase transformation processes were studied by in situ Raman imaging of the deformed contact region of Si, employing a Raman spectroscopy-enhanced instrumented indentation technique (IIT). The evolution of the stress field and the distribution of the high-pressure phases generated in the transformation processes were qualitatively analyzed as a function of the contact conditions. The reported in situ experiments provide insights to the transformation processes in Si during loading, confirming and providing the experimental evidence of some of the previous assumptions made on this subject. In this context, the developed experimental setup coupling indentation with in situ Raman microscopy has shown its potential in advancing the understanding of the deformation mechanisms.

References

1. Mujica A, Rubio A, Munoz A, Needs RJ (2003) High-pressure phases of group-IV, III-V, and II-VI compounds. Rev Mod Phys 75:863–913
2. Gupta MC, Ruoff AL (1980) Static compression of silicon in the [100] and in the [111] directions. J Appl Phys 51(2):1072–1075
3. Jamieson JC (1963) Crystal structures at high pressures of metallic modifications of silicon and germanium. Science 139(3556):762–764
4. Hu JZ, Merkle LD, Menoni CS, Spain IL (1986) Crystal data for high-pressure phases of silicon. Phys Rev B 34(7):4679–4684
5. Cheong WCD, Zhang LC (2003) Stress criterion for the β-tin transformation in silicon under indentation and uniaxial compression. Key Eng Mater 233–236:603–608
6. Gerbig YB, Stranick SJ, Cook RF (2001) Direct observation of phase transformation anisotropy in indented silicon studied by confocal Raman spectroscopy. Phys Rev B 83(20):205209
7. Gilman J (1993) Shear-induced metallization. Philos Mag 67(2):207–214
8. Boyer LL, Kaxiras E, Feldman JL, Broughton JQ, Mehl MJ (1991) New low-energy crystal structure for silicon. Phys Rev Lett 67(6):715–718
9. Mylvaganam K, Zhang LC, Eyben P, Mody J, Vandervorst W (2009) Evolution of metastable phases in silicon during nanoindentation: mechanism analysis and experimental verification. Nanotechnology 20:305705
10. Eyben P, Clemente F, Vanstreels K, Purtois G, Clarysse T, Duriau E, Hantschel T, Sankaran K, Mody J, Vandervorst W, Mylvaganam K, Zhang L (2010) Analysis and modeling of the high vacuum scanning spreading resistance microscopy contact on silicon. J Vac Sci Technol B 28(2):401–406
11. Kim DE, Oh SI (2008) Deformation pathway to high-pressure phases of silicon during nanoindentation. J Appl Phys 104:013502
12. Lin Y-H, Jian S-R, Lai Y-S, Yang P-F (2008) Molecular dynamics simulations of nanoindentation-induced mechanical deformation and phase transformation in monocrystalline silicon. Nanoscale Res Lett 3:71–75
13. Sanz-Navarro CF, Kenny SD, Smith R (2004) Atomistic simulations of structural transformations of silicon surfaces under nanoindentation. Nanotechnology 15:692–697
14. Smith GS, Tadmor EB, Bernstein N, Kaxiras E (2001) Multiscale simulations of silicon nanoindentation. Acta Mater 49:4089–4101
15. Gerbig YB, Michaels CA, Forster AM, Hettenhouser JW, Byrd WE, Morris DJ, Cook RF (2012) Indentation device for in situ Raman spectroscopic and optical studies. Rev Sci Instrum 83:125106
16. Any mention of commercial products within this paper is for information only; it does not imply recommendation or endorsement by NIST
17. Bushby AJ, Jennet NM (2001) Determining the area function of spherical indenters for nanoindentation. Mater Res Soc Symp Proc 649: Q7.17.1–Q7.17.6
18. Weppelmann ER, Field JS, Swain MV (1993) Observation, analysis, and simulation of the hysteresis of silicon using ultra-micro-indentation with spherical indenters. J Mater Res 8(4):830–840
19. Aggarwal RL, Farrar LW, Saikin SK, Aspuru-Guzik A, Stopa M, Polla DL (2011) Measurement of the absolute Raman cross section of the optical phonon in silicon. Solid State Commun 151:553–556
20. Windl W, Pavone P, Karch K, Schütt O, Strauch D, Giannozzi P, Baroni S (1993) Second-order Raman spectra of diamond from ab initio phonon calculations. Phys Rev B 48(5):3164–3170
21. Watson GH Jr, Daniels WB, Wang CS (1981) Measurement of Raman intensities and pressure dependence of phonon frequencies in sapphire. J Appl Phys 52:956–958
22. Han C-F, Lin J-F (2010) The model developed for stress-induced structural phase transformations in micro-crystalline silicon films. Nano Micro Lett 2(2):68–73
23. Gerbig YB, Michaels CA, Forster AM, Cook RF (2012) In situ observation of the indentation-induced phase transformation of silicon thin films. Phys Rev B 85(10):104102
24. Olijnyk H (1992) Raman scattering in metallic Si and Ge up to 50 GPa. Phys Rev Lett 68(14):2232–2234
25. Chang KJ, Cohen ML (1985) Solid-solid phase transitions and soft phonon models in highly condensed Si. Phys Rev B 31(12):7819–7826
26. Lewis SP, Cohen ML (1993) Theoretical study of Raman modes in high-pressure phases of Si, Ge, and Sn. Phys Rev B 48(6):3646–3653
27. Gaál-Nagy K, Schmitt M, Pavone P, Strauch D (2001) Ab initio study of the high-pressure phase transition from the cubic-diamond to the β-tin structure of Si. Comput Mater Sci 22:49–51
28. Kaxiras E, Boyer LL (1994) Energetics of large lattice strains: application to silicon. Phys Rev B 50(3):1535–1540
29. Hale LM, Zhou X, Zimmerman JA, Moody NR, Ballarini R, Gerberich WW (2011) Phase transformations, dislocations and hardening behavior in uniaxially compressed silicon nanospheres. Comput Mater Sci 50:1651–1660
30. De Wolf I (1993) Micro-Raman spectroscopy to study local mechanical stress in silicon integrated circuits. Semicond Sci Technol 11 (2):139–154

MIX
Papier aus verantwortungsvollen Quellen
Paper from responsible sources
FSC® C105338
www.fsc.org

If you have any concerns about our products,
you can contact us on
ProductSafety@springernature.com

In case Publisher is established outside the EU,
the EU authorized representative is:
Springer Nature Customer Service Center GmbH
Europaplatz 3, 69115 Heidelberg, Germany

Printed by Libri Plureos GmbH
in Hamburg, Germany